ABOUT ISLAND PRESS

Island Press is the only nonprofit organization in the United States whose principal purpose is the publication of books on environmental issues and natural resource management. We provide solutions-oriented information to professionals, public officials, business and community leaders, and concerned citizens who are shaping responses to environmental problems.

In 2004, Island Press celebrates its twentieth anniversary as the leading provider of timely and practical books that take a multidisciplinary approach to critical environmental concerns. Our growing list of titles reflects our commitment to bringing the best of an expanding body of literature to the environmental community throughout North America and the world.

Support for Island Press is provided by The Nathan Cummings Foundation, Geraldine R. Dodge Foundation, Doris Duke Charitable Foundation, Educational Foundation of America, The Charles Engelhard Foundation, The Ford Foundation, The George Gund Foundation, The Vira I. Heinz Endowment, The William and Flora Hewlett Foundation, Henry Luce Foundation, The John D. and Catherine T. MacArthur Foundation, The Andrew W. Mellon Foundation, The Moriah Fund, The Curtis and Edith Munson Foundation, The New-Land Foundation, Oak Foundation, The Overbrook Foundation, The David and Lucile Packard Foundation, The Pew Charitable Trusts, The Rockefeller Foundation, The Winslow Foundation, and other generous donors.

The opinions expressed in this book are those of the author(s) and do not necessarily reflect the views of these foundations.

ABOUT THE SOCIETY FOR ECOLOGICAL
RESTORATION INTERNATIONAL

The Society for Ecological Restoration International is an international nonprofit organization comprised of members who are actively engaged in ecologically sensitive repair and management of ecosystems through an unusually broad array of experience, knowledge sets, and cultural perspectives.

The mission of SER International is to promote ecological restoration as a means of sustaining the diversity of life on Earth and reestablishing an ecologically healthy relationship between nature and culture.

SER International, 285 W. 18th Street, #1, Tucson, AZ 85701. Tel. (520) 622-5485, Fax (520) 622-5491, E-mail info@ser.org, www.ser.org.

ASSEMBLY RULES AND RESTORATION ECOLOGY

SOCIETY FOR ECOLOGICAL RESTORATION INTERNATIONAL

The Science and Practice of Ecological Restoration
James Aronson, EDITOR
Donald A. Falk, ASSOCIATE EDITOR

*Wildlife Restoration: Techniques for Habitat Analysis and
Animal Monitoring,* by Michael L. Morrison

Ecological Restoration of Southwestern Ponderosa Pine Forests,
edited by Peter Friederici and Ecological Restoration
Institute at Northern Arizona University

Ex Situ Plant Conservation: Supporting Species Survival in the Wild,
edited by Edward O. Guerrant Jr., Kayri Havens, and Mike Maunder

Great Basin Riparian Areas: Ecology, Management, and Restoration,
edited by Jeanne C. Chambers and Jerry R. Miller

*Assembly Rules and Restoration Ecology: Bridging the Gap
between Theory and Practice,*
edited by Vicky M. Temperton, Richard J. Hobbs, Tim Nuttle,
and Stefan Halle

Assembly Rules and Restoration Ecology

Bridging the Gap

between Theory and Practice

Edited by

Vicky M. Temperton, Richard J. Hobbs,
Tim Nuttle, and Stefan Halle

SOCIETY FOR ECOLOGICAL RESTORATION INTERNATIONAL

ISLAND PRESS

Washington · Covelo · London

ISLAND PRESS is a trademark of The Center for Resource Economics.

Library of Congress Cataloging-in-Publication data.
    Assembly rules and restoration ecology : bridging the gap between theory
    and practice / edited by Vicky M. Temperton...[et al.].
       p.cm.—(Science and practice of ecological restoration)
    ISBN 1–55963–374–3 (cloth : alk. paper)—ISBN 1–55963–375–1 (pbk. : alk. paper)
    1. Restoration ecology. 2. Biotic communities. I. Temperton, Vicky M. II. Series.
    QH541.15.R45A88   2004
    333.71'53—dc22
    2003024788

British Cataloguing-in-Publication data available.

Printed on recycled, acid-free paper ♺

Design by Teresa Bonner

Manufactured in the United States of America

10 9 8 7 6 5 4 3 2 1

This book is dedicated to all those resilient plants, animals, fungi, and bacteria that almost always manage to reassemble in some form or other, no matter what we humans do to a landscape. It is also dedicated to the people who spend their working lives trying to understand how ecological communities and ecosystems work and how they regenerate after human-induced degradation.

All nature is but art unknown to thee;
All chance, direction which thou canst not see;
All discord, harmony not understood;
All partial evil, universal good.

—Alexander Pope

CONTENTS

This book was inspired by material presented and discussed at an international workshop titled "Assembly Rules in the Regeneration of Ecosystems" held in Dornburg near Jena, Germany, in October 2001. It brought together ecologists and environmental scientists representing a variety of disciplines, locations, methodological approaches, and backgrounds—all linked in one way or another to the central issue of ecological assembly rules and their applicability to restoration ecology and ecological restoration projects. A specific aim was the inclusion of material that covered diverse geographic areas and represented a variety of perspectives. The workshop was held in one of the three Dornburg castles, a picturesque Renaissance chateau overlooking the Saale Valley and often visited by Goethe. The opportunity to meet and converse in such an environment greatly enhanced the creative process. The workshop provided an excellent venue for developing ideas about community assembly in a restoration context.

The Dornburg workshop was inspired by the work of the Jena Graduate Research Group (*Graduiertenkolleg* GRK 266) on "Functioning and Regeneration of Degraded Ecosystems," which has involved the search for conceptual models applicable to many different regenerating ecosystems, some of the fruits of which are presented in this book. We thank the *Deutsche Forschungsgemeinschaft* (DFG; German Research Council) for providing generous and full funding for this group since 1996, including the workshop that led to this book.

The workshop provided a very productive opportunity for an exchange of ideas between younger and more experienced scientists from a variety of backgrounds. The following people participated at the Dornburg workshop: Jens Arle, Fakhri Bazzaz, Lisa Belyea, Hans Bergmann, Anthony Bradshaw,

P.1 Just north of Jena, Germany, in the village of Dornburg, stand the three Dornburg castles, overlooking the Saale Valley. The workshop took place in the Renaissance Castle, on the left (the Rococo Castle was under renovation, right, and the Old Castle is not pictured).

P.2 The Renaissance Castle, built around 1540. At least two important literary events have taken place here: Germany's most famous poet, Johann Wolfgang Goethe, spent the summer of 1828 writing here, and the "Assembly Rules in the Regeneration of Ecosystems" workshop took place here in October 2001.

P.3 Attendees of the "Assembly Rules in the Regeneration of Ecosystems" workshop. From left to right: Peter White, Richard Hobbs, Miroslav Vosátka, Anthony Bradshaw, Vicky Temperton, Jan Rothe, Eva-Barbara Meidl, Anke Jentsch, Christoph Matthaei, Carsten Renker, Martin Zobel, Kerstin Zirr, Marzio Fattorini, Falko Wagner, Gottfried Jetschke, Matthias Held, François Buscot, Stefan Halle, Birgit Klein.

François Buscot, Ingo Ensminger, Marzio Fattorini, Tamara Gigolashvili, Jens Gutsell, Stefan Halle, Matthias Held, Richard Hobbs, Anke Jentsch, Gottfried Jetschke, Birgit Klein, Jill Lancaster, Christoph Matthaei, Eva-Barbara Meidl, Jörg Perner, Carsten Renker, Jan Rothe, Hartmut Sänger, Vicky Temperton, Mona Vetter, Winfried Voigt, Miroslav Vosátka, Falko Wagner, Markus Wagner, Peter White, Kerstin Zirr, and Martin Zobel.

All chapters in this book have been reviewed both by workshop participants and by other scientists not present at the workshop. As the book's editors, we extend our thanks to all the participants for the stimulating discussions (much helped by the exquisite dumplings and red cabbage prepared by the Dornburg Castle cook, Frau Scheffel) that led to the idea for this book; to the reviewers of manuscripts; and to the authors themselves, of course, for their varied and important contributions. Last but not least, we are grateful to Barbara Dean and Barbara Youngblood at Island Press for their excellent collaboration in making this book become material reality.

Chapter 1

Introduction: Why Assembly Rules Are Important to the Field of Restoration Ecology

Vicky M. Temperton, Richard J. Hobbs, Tim Nuttle, Marzio Fattorini, and Stefan Halle

How are ecosystems assembled? How do the species that make up a particular biological community arrive in an area, survive, and interact with other species? Why do only some species succeed in particular places? Why are similar assemblages of species seen in different parts of the landscape? These questions are fundamental elements of the science of ecology, and ecologists have been asking questions like these for centuries. This set of questions has been formulated into a search for what have been termed assembly rules: if only certain species can establish and survive in any given area, and if species tend to occur in recognizable and repeatable combinations or temporal sequences, then maybe we can identify a set of rules governing the assembly of ecosystems and communities.

Another group of people have been asking a different set of questions. How do we repair the damage caused to natural and managed ecosystems by overexploitation, misuse, pollution, mining, and so forth? How can we return a biological assemblage to mine-spoil heaps? How can we return a degraded area to a functioning ecosystem that can serve as habitat or perform useful ecosystem services? These are the questions asked by restoration ecologists, who aim to tackle the problems arising from the increasing human use — and misuse — of the planet's ecosystems.

In this book, we work from the premise that these two sets of questions are not really very different; indeed, there is considerable overlap between them. In fact, ecological restoration involves putting lost parts back into a system, and obviously any assembly rules that may exist must be considered to ensure success. Nevertheless, there have been few attempts so far to explore and develop that overlap.

1

Recent ideas about assembly rules have come primarily from community ecology (for example, Belyea and Lancaster 1999, Weiher and Keddy 1999), but this work has not yet been translated into practical outcomes. This fact is not really surprising, however. If one looks at the history of ecologists' work on constraints to community development, one finds intense debates running from the beginning of the twentieth century up to the present (see Chapter 3), none of which are adequately resolved as yet (Booth and Larson 1999). Two central questions have been at the root of this whole century of debate: How do communities of organisms come to be the way they are, and what are the constraints on membership in a community? Today's ecologists, who are having another shot at these two biggest questions in ecology, can be considered either the perseverant, ever-curious descendants of Darwin, Clements, Gleason, MacArthur, and Wilson, or quixotic adventurers trying to tame a windmill, as the complexity of nature has notoriously evaded simple description. Whatever your opinion, the questions are still there, and we need the answers. These answers are becoming more and more important in a world dominated and transformed by humanity, in which the understanding and repair of damaged ecosystems will be essential to our future survival (Hobbs and Harris 2001). Given that ecological restoration, as defined by the Society for Ecological Restoration International (SERI), is an intentional activity that initiates or accelerates the recovery of an ecosystem, it seems almost essential to consider possible rules or principles that can guide how components should be added to an ecosystem.

The concepts of succession and assembly rules are related, and yet there are important distinctions between the two (Young et al. 2001). One of the main distinctions is focus. Succession follows the dynamics of changes in a community's development, often with particular reference to an endpoint or deviations from this endpoint (climax). Assembly theory and the search for assembly rules focuses more on interactions among organisms within a community and the actual pathways a community can take in response to such interactions. Ecologists seek the mechanisms behind organism assemblages in different situations. This approach is directly relevant to ecological restoration, in which one seeks to guide an ecosystem toward a specific stable state after a disturbance. Lockwood (1997) and Young (2000) suggested that assembly and succession are the core concepts in restoration ecology. Yet very little overt attention has been paid to this suggestion, either in the scientific literature or in practical restoration. In this book, we aim to fill this gap and to provide an introduction, overview, and synthesis of the potential role of assembly rules in restoration ecology.

Themes to be Explored

Ecologists have worked on ecological assembly and assembly rules in a number of different ways, and a comparison of the current approaches is provided in Chapter 3. Certain aspects of the ecological debate about assembly rules have recently attracted increasing attention. One of these issues involves abiotic and biotic filters through which species or organisms trying to enter a community must pass in order to arrive, establish themselves, and survive there (Keddy 1992). How do these filters act and interact, and how do they change over time, especially as a system regenerates? Are they inextricably linked, or do the abiotic environmental conditions of an ecosystem act separately from the biotic interactions? This is one of the central issues addressed in this book (see Part II: "Ecological Filters as a Form of Assembly Rule"). A critique of the filter concept is provided in Chapter 7.

Another main theme of this book is a question of level of abstraction: Are assembly rules constraints imposed by interactions among organisms alone, or does the sum of the interactions between organisms and their environment limit membership to certain species only? In other words, do assembly rules include abiotic effects and biotic effects, or do the abiotic conditions form a backdrop against which biotic interactions form the "real" rules of assembly? (See Chapter 7.) There are two schools of thought among ecologists on this issue (see Chapter 3), and we have deliberately included proponents from both sides of the debate in this book.

Another important issue is that of disturbance. Disturbance produces change, such as alterations in the availability of resources, and provides opportunities to organisms. It is increasingly being seen as a positive (as well as a negative) force that is necessary to many species in ecosystems (Pickett and White 1985, White and Jentsch 2001). Disturbance, of course, is also an integral part of restoration ecology. It has not generally been considered much in the assembly rules debate, just as the dynamics of assembly rules over time have generally been neglected until now (see Chapter 3). Disturbance and assembly are strongly related, of course, since the disturbance regime of an ecosystem exists over a period of time and affects assembly of the community differently at different times. Thus, the role disturbance plays in assembly needs to be an integral part of the search for generality in the assembly of communities and ecosystems. This issue is addressed primarily in Part V, "Disturbance and Assembly."

The idea of thresholds (Hobbs and Norton 1996) between alternative stable states of a system is the final theme woven into the book. If a threshold divides different stable equilibria, then ecosystem restoration aims to allow a system to overcome this threshold when it cannot be overcome without intervention. It is important to know what intensity, frequency, and quality of disturbance might be necessary to achieve the desired goal, and to what extent the ecosystem requires a helping hand in returning to or reaching a desired stable state. Knowing more about how disturbance relates to such system thresholds can help us move forward and improve our restoration of ecosystems.

This book is unique in that it attempts to develop ideas arising from theoretical and empirical community ecology and places them in a practical restoration context. All too often, restoration and conservation management practice lags well behind current developments in theoretical ecology and related fields; hence, this book attempts to bring current ideas to the forefront and interpret them from a restoration perspective. In this regard, this book is a combination of theoretical and empirical research on ecological assembly with an attempt to use the concepts and findings arising from the research reviewed herein to guide future restoration efforts. Thus, rather than being a restorationists' handbook, this book is a summary of the current status of an emerging field and will be of immediate use to practitioners. We hope that it will be used widely to develop the ideas further and apply them in a wider array of ecosystems and situations.

We wrote this book for academic restoration ecologists and restoration practitioners, particularly those employed by government agencies, nongovernmental organizations (NGOs), and others who have supervisory roles in restoration projects. We assume that the reader will already have some understanding of the concepts and practice of restoration ecology, although these will be summarized or referenced as necessary. The book may be attractive as a text for university courses in restoration ecology and professional training courses. It also may be useful as a supplemental text in more basic ecology courses.

How This Book Is Organized

What follows is an overview of the book's structure. After this introductory chapter, which outlines why the subject of ecological assembly is inherently related to the subject of restoration ecology, the book is divided into five parts.

In Part I, "Assembly Rules and the Search for a Conceptual Framework for Restoration Ecology," we examine the theoretical background for ecological assembly rules and how this background might be relevant to ecological restoration. The discipline of restoration ecology aims to provide a scientifically sound basis for the recovery of degraded ecosystems and to produce self-sustaining systems. However, despite some recent attempts to consolidate the scientific theory (Hobbs and Norton 1996, Urbanska et al. 1997; also see Chapter 3), we are still far from achieving a conceptual framework in restoration ecology. Part I includes one provocative contribution on restoration ecology (Chapter 2) and two reviews of the current status of assembly rules research and its relevance to restoration ecology (Chapters 3 and 4). The search for assembly rules in ecosystems has taken on many different forms and has involved a number of quite different approaches. One of these approaches looks at the constraints on species membership in a community in terms of a filtering out of those species unable to establish themselves (under the conditions reigning in that community).

Part II, "Ecological Filters as a Form of Assembly Rule," explores different facets of this approach to assembly and its relevance to restoration ecology. The composition of a biotic community in any particular place arises because of the action of a number of biotic and abiotic filters on the arrival and survival of species. How can an understanding of such filters help in restoration? For instance, can filters be modified to speed up restoration? Part II explores the idea in three chapters that bring together important aspects of filters in relation to ecology and restoration science. Chapter 5 deals with ecological filters as gradients in resistance to restoration. Chapter 6 proposes a dynamic filter model that could be implemented at the beginning of restoration projects to assess the current status of an ecosystem. Chapter 7 is a critique of the ecological filters approach and a plea to consider systems approaches to assembly and restoration as well.

An important aspect of ecological communities is how they are structured and how we as humans perceive this structure. Part III, "Assembly Rules and Community Structure," looks at various methods and approaches in elucidating community structure and their potential relevance to restoration ecology projects. At what level of abstraction should one look for assembly rules in a community? Are species, functional groups, or other levels of organization most fruitful in, for example, plankton communities (Chapter 8)? Do consumer guilds in a regenerating grassland follow specific rules that might provide a potential guide in restoration efforts (Chapter 9)? Chapter 10 explores a group of organisms—arbuscular mycorrhizae—that fails to be simply categorized into species units and yet forms an indispensable contribution to the recovery of ecosystems. Chapter 11 provides a model of how plant community structure responds to disturbance. Chapter 12 examines how stable isotopes may be useful to investigate food web development in regenerating ecosystems.

When restoration practitioners are faced with restoring a severely disturbed or degraded ecosystem, any knowledge of rules or guidelines applicable to ecosystem assembly will be welcomed. A critical question, however, is whether such severe environments are governed by the same rules as govern natural or seminatural environments. Do such environments require specific, concerted actions when guiding reassembly of ecological communities?

Part IV, "Assembly Rules in Severely Disturbed Environments," presents a series of examples from a variety of ecosystems that are in the process of regenerating after considerable human disturbance, such as mining or severe air pollution. What can we learn from these systems that will be relevant when we come to assemble ecosystems in a restoration context? The focus here is on the dynamics of the systems over time and on how different organism groups play essential roles at various steps in the regeneration process. Systems that have been reduced to a handful of species through severe disturbance form an excellent opportunity to investigate invasion and the reassembly of a system.

Restoration efforts are often directed at areas that have been severely disturbed or impacted by mining activity, pollution, or other highly disruptive activities. The physical and chemical properties of these sites often impose severe limitations on the ability of biota to recolonize. This section examines the recolonization process on such sites (Chapters 13 and 15) and how it is affected by manipulations of various sorts (Chapter 14), from which we can gain important insights into both intrinsic colonization processes and restoration methods. Chapter 16 reviews the importance of nutrients and ecosystem function in the assembly and restoration of ecosystems.

Finally, in Part V, "Disturbance and Assembly," disturbance as a driver of change in ecological communities is explored in relation to its role in assembly as well as in ecological restoration. Disturbances such as fires and storms are natural parts of many ecosystems, and ecosystem reassembly following such events is an important process. Increasingly, humans are altering disturbance regimes with respect to frequency, intensity, and type of disturbance. Chapter 17 explores the relationship among disturbance, succession, and community assembly. Chapters 18 and 19 take a look at disturbance in different zones of rivers and how it relates to community assembly and restoration of rivers. Most ecosystems in need of restoration have been disturbed beyond the point where they can reassemble unaided. However, targeted disturbances at definite phases of natural succession may be just the tool needed in restoration management. In the case of rivers, this may involve restoration of disturbance regimes via restoration of natural flooding regimes. Hence, this section examines what we can learn from examining ecosystem response to disturbances, which can be of benefit when we aim to assist the assembly process.

The final synthesis (Chapter 20) draws together the major conclusions from the other parts of the book and critically assesses the likely value of ecosystem assembly rules in ecological restoration. Key components of an approach to incorporating the ideas presented in this book into practical restoration projects are identified, as are future directions for research and development of these ideas.

We hope that this volume will contribute new knowledge and ideas to the issues of assembly rules and restoration ecology and will inspire further research. We also hope to be able to show that it is worthwhile to link theoretical and conceptual aspects of ecology with more practical fields such as ecological restoration. In this regard, a multidisciplinary approach is indispensable, even if it sometimes seems to be extremely difficult to find agreement, for example, on the use of a common language. Ecologists and restoration practitioners come from a wide range of countries and scientific backgrounds. In compiling this book, it was interesting to see that even differences in English and scientific language usage among researchers from different English-speaking countries (let alone usage by nonnative English speakers) can cause confusion and require clarification for successful communication. Such differences in usage of specific words or approaches to science are also reflected in the many approaches to the study of ecological assembly or ecological restoration (see Chapter 3; also see Booth and Larson 1999). The fields of assembly and restoration ecology remain emerging disciplines, and as such they have often engendered heated debates among ecologists and restoration practitioners. Such debates are productive only as long as ideas keep moving forward and do not come to an impasse. It is important that we embrace such debates and differences, as complex issues require a diversity of approaches and cannot be resolved by going down one path alone. The aim of this book is to include a diversity of approaches to ecological assembly and restoration and to form a much-needed link between the two fields.

REFERENCES

Belyea, L. R.; and Lancaster, J. 1999. Assembly rules within a contingent ecology. *Oikos* 86:402–416.
Booth, B. D.; and Larson, D. W. 1999. Impacts of language, history, and choice of system on the study of assembly rules. In *Ecological Assembly Rules: Perspectives,*

Advances, Retreats, ed. E. Weiher and P. A. Keddy. 206–227. Cambridge: Cambridge University Press.

Hobbs, R. J.; and Harris, J. A. 2001. Restoration ecology: repairing the earth's ecosystems in the new millennium. *Restoration Ecology* 9(2): 239–246.

Hobbs, R. J.; and Norton, D. A. 1996. Towards a conceptual framework for restoration ecology. *Restoration Ecology* 4:93–110.

Keddy, P. A. 1992. Assembly and response rules: two goals for predictive community ecology. *Journal of Vegetation Science* 3:157–164.

Lockwood, J. L. 1997. An alternative to succession: assembly rules offer guide to restoration efforts. *Restoration and Management Notes* 15:45–50.

Pickett, S. T. A; and White, P. S. 1985. *The Ecology of Natural Disturbance and Patch Dynamics*. Chicago: Academic Press.

Urbanska K. M.; Webb, N. R.; and Edwards, P. J. 1997. *Restoration Ecology and Sustainable Development*. Cambridge: Cambridge University Press.

Weiher E.; and Keddy, P. A. 1999. *Ecological Assembly Rules: Perspectives, Advances, Retreats*. Cambridge: Cambridge University Press.

White, P. S.; and Jentsch, A. 2001. The search for generality in studies of disturbance and ecosystem dynamics. *Ecology* 62:399–450.

Young, T. P. 2000. Restoration ecology and conservation biology. *Biological Conservation* 92:73–83.

Young T. P.; Chase, J. M.; and Huddleston, R. T. 2001. Community succession and assembly. Comparing, contrasting and combining paradigms in the context of ecological restoration. *Ecological Restoration* 19(1): 5–18.

PART ONE

Assembly Rules and the Search for a Conceptual Framework for Restoration Ecology

Although the fields of assembly theory and restoration practice seem to be a natural fit, they actually have few linkages. The following three chapters explore the current state of affairs in these two areas to provide a background for the other chapters, which endeavor to remedy this situation. Chapter 2 makes a plea for developing a conceptual framework in restoration ecology. Chapter 3 brings us up to date with the many aspects of and perspectives on assembly theory. Chapter 4 draws linkages between specific assembly models and explores how these can be applied to the planning and evaluation of restoration projects.

Chapter 2

Advances in Restoration Ecology: Insights from Aquatic and Terrestrial Ecosystems

Stefan Halle and Marzio Fattorini

The term *ecological restoration* covers a wide range of activities involved with the repair of damaged or degraded ecosystems (Jordan et al. 1987, Berger 1990, Baldwin et al. 1994). Various terms have been used to describe these activities, including true restoration, remediation, rehabilitation, reclamation, reallocation, reconstruction, and replacement (Magnuson et al. 1980; Cairns 1982; Bradshaw 1992, 1997a; Gore and Shields 1995). Here we follow Hobbs and Norton (1996) and use the term *restoration* to refer broadly to activities that aim to repair damaged ecosystems. Ecological restoration was similarly defined in 1996 by the Board of the Society for Ecological Restoration as "the process of assisting the recovery and management of ecological integrity. Ecological integrity includes a critical range of variability in biodiversity, ecological processes and structures, regional and historical context, and sustainable cultural practices" (SER 2002).

In spite of the relatively short time span it covers, the history of ecological restoration has been quite ambiguous, which to some extent is due to an imprecise and confusing use of terms. When John Aber and Bill Jordan first introduced the term (Aber and Jordan 1985; later specified by Jordan et al. 1987), they did so in the tradition of Aldo Leopold's view of restoration in its literal sense; that is, to rebuild habitats and landscapes that have been lost. The North American prairie is probably the most prominent example of this approach; but later, the Madison Arboretum—which can be seen as the germ cell of ecological restoration—became a more extensive collection of small bits of otherwise rare or even lost habitat types. By using and redefining the term again and again, however, the common understanding of ecological restoration today has changed to recovery to any desired system state, even if there is no historical precursor. Also *ecological restoration* and *restora-*

tion ecology are often used as synonymous terms, although they were originally thought of as labeling quite different issues (Jordan et al. 1987; see
below).

Today the term *ecological restoration* covers a wide spectrum of quite different approaches. One end of this spectrum is marked by social aspects, an
idea that has been part of the field since its very early days and the Civilian
Conservation Corps. Leopold and such intellectual successors as Jordan
were concerned about ecosystems that were lost due to human activities.
Consequently, restoring the lost systems must include humans; otherwise,
the restored habitats will soon be lost again, since the very reason for the initial loss has not changed. Restoring nature—or at least parts of it—hence
will also require changing people's attitudes toward the nature that surrounds
them. This can be achieved by socioeconomic, educational, and cultural
efforts, including artwork and even religion. So, along the spectrum of
approaches, the first line of development of ecological restoration describes
a movement that aims at a harmonious coexistence of humankind and
nature.

The second line of development concerns practical aspects of restoration
projects. As Hobbs and Harris (2001) clearly stated, any action needs clearly
defined goals and agreements, including practical measures to assess whether
agreed-upon goals are achieved after a defined period of time and a structured overview of the available decision options during the restoration process. Decisions, however, will not be determined by rigorous scientific arguments or logic alone, because the political and economic interests of
different parties are involved. Therefore, the restoration ecologist must
engage in politics to promote the scientifically based options, which involves
tasks such as public relations work and lobbying, for which scientists traditionally are not well trained. Although scientists may regard the political
dimension of restoration projects as other people's business, the fulfillment-
oriented approach probably decides most of what actually is going to happen
when it comes to the restoration of degraded ecosystems.

To succeed more often than fail, restorationists must develop their tools.
To do this, current insights from basic ecology—and basic community ecology in particular—must be adapted and translated into practical management guidance. This third line, which considers ecological restoration as an
applied science, is probably the most widespread association with the term.
As current scientific discussion demonstrates, communities are challenging
enough to understand in existing systems; the task is much more demand-
ing for systems that have been lost. Nonetheless, ecological restoration is
even more than understanding community structure. One must be aware

not only of all the parts needed but also of how and in which order they have to be reassembled, since ecological restoration means that natural succession is skillfully manipulated to direct and speed up species turnover. The most essential difference from natural and unaided succession, however, is that its endpoint is not defined by a more or less stable state of species interactions but rather by the desired restoration goal, which often is—in a broad sense—a political compromise.

The other end of the spectrum of possible approaches is marked by the often-stated assertion that ecological restoration may serve as an "acid test" for theoretical ecology (Bradshaw 1987). Textbooks and journals provide a wide selection of well-established concepts for the community and ecosystem level, but we must always be aware that these are models. Probably the most powerful tests of such models are their use as tools in restoration projects, because the outcome will tell whether the model's assumptions were right or wrong and how far we have come in our understanding of nature. So in the fourth line of development, ecological restoration is seen primarily as an integrated part of basic ecology research (Jordan et al. 1987) in which the applied aspect is a welcome by-product.

The Present State

Although ecological restoration is a term that primarily describes techniques and practice, *restoration ecology* can be defined as the theoretical and empirical study of principles and theories concerning the development of degraded ecosystems; that is, the scientific background of ecological restoration. In its original definition (Jordan et al. 1987), restoration ecology was even thought of as a rigorous heuristic tool for a better understanding of basic ecology, with an underlying conceptual framework and research agenda (compare Cairns and Heckman 1996). Successful restoration obviously depends on the understanding of ecological principles, but restoring ecosystems is not limited to ecology, since it requires interdisciplinary approaches with other sciences, such as geography, chemistry, and physics. In addition, economics, sociology, and politics must be considered for successful restoration projects (Cairns and Heckman 1996; also see Chapter 5) but are beyond the scope of this overview.

When one looks at current contributions to journals and conferences in this field, it is obvious that the ambitious claim of an acid test of ecological theory and the recognition of restoration ecology as a basic science has not come to fruition. The typical restoration ecology paper is—still—a single case study, in which "tricks" are described to solve the main problems in a particular system. This is followed by species lists and management plans; that is, exactly the kind of descriptive approach that has long been criticized as not being helpful for a better understanding of the mechanisms behind systems (Harper 1987, Jordan et al. 1987). There have been initiatives to establish a theoretical framework for restoration ecology; for example, the standard paper on this issue, "Towards a conceptual framework for restoration ecology," by Hobbs and Norton (1996). However, in spite of the paper's inviting title, which opens a new realm just waiting to be explored, this paper reads as fresh as when it was printed more than six years ago.

This situation is particularly surprising because two fields that are intimately related to restoration ecology—that is, disturbance ecology and succession theory—both have well-developed conceptual backgrounds and are currently the subjects of lively debate. Ecological restoration in its strict sense (as reviewed by Cairns 1986) means that natural succession is used and manipulated to guide a system degraded by heavy disturbance back to its original state. Moreover, "designed disturbances" of smaller magnitude and impact than the original devastation are often used to direct or speed up succession; for example, to provide the required site conditions for establishment of desired species (Hobbs 1999). So a simpleminded and seemingly straightforward approach to a conceptual framework for restoration would be to combine extracts of disturbance and succession theory (see Chapter 4). Such an approach turned out not to be that simple, however, because restorationists almost never link disturbance and succession theory to approach restoration issues, nor has a consistent theory of restoration ecology developed on the basis of its two close relatives.

Predictive models and testable hypotheses are not common in restoration ecology (Cairns 1988, 1991), and statistical methods for testing assumptions are lacking (Cairns 1990). Where present, any hypotheses being tested are seldom formally stated or even explicitly recognized (Chapman 1999). It may be argued that a conceptual framework is not essential right from the beginning to establish a new branch of science and that the repeated demand for it only reflects the fashionable and pretentious attitudes of present basic ecology. This may be partly true, but the enormous progress in understanding ecological systems at all levels of complexity during the last decades was made possible only by the shift from descriptive, bottom-up natural history to theory-driven, top-down conceptions. It is just this crucial step that the subdiscipline of restoration ecology as a whole has not yet made.

This omission has severe consequences. First, restoration ecology in its present state is not compatible with the way of thinking and arguing characteristic of the field of basic ecology and is, therefore, not really part of it.

The lack of a theoretical basis prevents the mutual benefit that would develop from enduring exchange between conceptual frameworks and on-the-ground applications (see Chapter 4). Without a common platform of communication between theorists and empiricists, it is impossible to provide the restorationist with new ideas for more efficient restoration tools suggested by basic ecology theory, and there is no pathway for direct feedback from field experiences to verify ecological theory (Hobbs and Harris 2001).

Second, without a general theory, it is impossible to adapt knowledge and techniques that arise as experiences from other systems (Jordan et al. 1987, Hobbs and Norton 1996, Power 1999). It will never be possible to do all studies in all systems, and so there is an urgent need to extrapolate results from one system to another. In comparing ecosystems at different times or sites, it is extremely important to have a careful design, clear assumptions, and exact measurements; so reasonable standards of research and documentation are necessary (Higgs 1994). The lack of a conceptual framework that links different case studies together prevents restoration ecology from learning as a process of optimization. At present, restorationists often must encounter their system as a terra incognita and must try to solve the system-specific problems by trial and error. If successful, they become experts for their particular system with detailed knowledge and elaborated skills, but experiences from this one system only apply there and cannot be transferred to other situations.

So, at present, "restoration ecology" is mainly storytelling about case studies rather than conceptual analysis, not to mention rigid statistical or experimental testing. Only the implementation of a broadly accepted theoretical framework would make restorationists aware that they are actually doing similar things, irrespective of how different the species lists and abiotic peculiarities of their systems are. "Doing similar things" means more than just meeting at the same conferences and publishing in the same journals. As Alfred N. Whitehead put it in 1911, "The aim of science is to see what is general in what is particular," and in this sense, restoration ecology is still a technique, or even an art based on creativity, inspiration, and intuition (Joel Brown, personal communication), but not a science (compare Jordan et al. 1987).

More Weaknesses in Restoration Ecology

Apart from the major criticism of restoration ecology—that is, the lack of a general theory to allow the transfer of methodologies and knowledge from one situation to another—there are at least three more problems that are specific to this field and that must be solved to develop a sound scientific basis for ecological restoration.

Unclear Reference States

Because an ecosystem comprises an entire set of organisms and physical processes, a single-species or single-process approach to restoration is likely to fail. Most researchers agree that a restored ecosystem should be self-sustaining in the sense that no more management efforts are needed to keep the system state. For this purpose, ecological structure and function should be reestablished (Cairns 1988, 1991, 1995; Westman 1991; Wyant et al. 1995; Cairns and Heckman 1996; Hobbs and Norton 1996; Ehrenfeld and Toth 1997; Kauffman et al. 1997; Palmer et al. 1997; Parker and Pickett 1997; Heckman and Cairns 1998; Harker et al. 1999; Keddy 1999). Thus, an essential question at the core of restoration ecology is what frame of reference should be taken for the predisturbance condition in order to define the restoration goals.

We share the point of view that there is no one ideal or "natural" reference state for any type of community or ecosystem (Pickett and Parker 1994, Parker and Pickett 1997). Rather, the context and history of the site being restored must be considered to determine reasonable reference states. Ecosystems possess several "self-healing" mechanisms that enable them to overcome the effects of disturbance and very often include an activation of potentials that had not been revealed before (Dubos 1978). Restoration in the sense of active intervention is necessary only when a system has crossed a threshold of irreversibility or when the recovery processes are too slow to achieve management goals within a tolerable time frame (Hobbs 1999; also see Chapter 5). Nevertheless, one still must define the range of appropriate reference states, which is particularly difficult in intensely used cultural landscapes with considerable human impact over centuries.

Insufficient Understanding of Ecosystem Structure, Function, and Processes

It is also difficult to reestablish ecosystem structure or function because the structural and functional attributes of the ecosystem before a disturbance are most often not precisely documented (Cairns 1988, 1990, 1995; Westman 1991). In certain cases, it may be possible to restore the basic functions or ecosystem processes (for example, the flow of energy and matter, the water cycle, the production of standing crop), but to achieve the former structure in full (that is, a particular assemblage of species, the original soil profile, the age structure in a forest) may be more difficult (Bradshaw 1997a, Dobson et al. 1997). The restoration of ecosystem function may be regarded as the mayor step toward sustainability of restored areas (Bradshaw 1997a), but we must be aware that function and structure will often be intimately linked and in perpetual feedback. For example, one cannot expect that functions such as primary productivity levels, buildup of a litter layer, and water retention could be restored in a forest before at least some trees (that is, a structural component) are growing.

Although it is assumed that restoration of structure will automatically achieve restoration of function, this is not always the case, because these elements often develop at independent rates (Westman 1991). This has been well documented in wetland restoration (Simenstad and Thom 1996, Zedler 1996), where the establishment of appropriate plant populations does not necessarily result in the restoration of ecosystem processes. Relationships between ecological structure and function are still not well understood, even in undisturbed systems (Simenstad and Thom 1996). For instance, restoring soil health is made difficult by a limited understanding of many key processes that occur belowground, and neglecting the soil conditions is probably the most common cause of failure of sites to recover from disturbance (Allen et al. 1999).

Small and Site-Specific Restoration Studies

The most important level of biological organization in restoration ecology is the ecosystem, but only a few studies deal with a comprehensive set of ecosystem functions and structures (Cairns 1991, Kauffman et al. 1997). Important issues that need to be addressed include the control of ecosystem dynamics over time and the interchange of matter and energy with the surrounding landscape (Ehrenfeld and Toth 1997). Since ecosystems and communities are not isolated but are interconnected subsystems on a larger scale, restoration must consider the landscape context (Westman 1991, Naveh 1994, Wyant et al. 1995, Cairns and Heckman 1996, Hobbs and Norton 1996, Kauffman et al. 1997, Parker and Pickett 1997, Wissmar and Beschta 1998). Interconnections of system components must recover for the restored area to become a functioning part of the landscape (Heckman and Cairns 1998).

Unfortunately, in most cases, the focus of efforts is restricted to the attributes of specific sites. This problem is particularly obvious in stream restoration projects, which are often limited to the riverbed rather than considering the entire catchment area. Due to the upstream-downstream continuum, however, changes in any segment of a river are communicated throughout the system. In addition to the measures carried out on the channel, such parameters as sediment load, discharge, and land use of the surrounding terrestrial habitats need to be included (Muhar et al. 1995). Restoration would benefit from the principles, knowledge, and techniques of the disciplines that treat rivers and their floodplains (or streams and their riparian zones) as integral parts of a single ecosystem; that is, hydrology, fluvial geomorphology, and system ecology (National Resource Council 1992, Clark 1997). Only a landscape perspective of riparian ecosystems reveals how they may depend on other ecosystems and would provide essential information for developing catchment-wide restoration plans (Wissmar and Beschta 1998). It has been argued that the continuum of a river and its riparian environments are too extensive to allow universal solutions, so transferring approaches from one river corridor to another must be undertaken with caution (Schumm 1984). However, even if any step toward river restoration must take into account the unique properties of each riverine habitat, some generalizations seem to be possible, with particular respect to physiographic regions, river types, and developmental stages (Wiegleb 1991).

Lines of Progress

Especially during the last decade, several authors have contributed to the development of the science of restoration ecology and have attempted to circumvent its inherent weaknesses. Here we review some lines of progress that we feel are particularly promising. A special focus is on restoration in both aquatic and terrestrial ecosystems—which, in the past, have generally been treated separately. Direct comparisons between these two ecosystem types include food webs (Chase 2000), eutrophication (Smith et al. 1999), biodiversity and functional groups (Freckman et al. 1997), community structure (Cyr et al. 1997), and ecosystem response to stress (Rapport and Whitford 1999). These kinds of integrative approaches seem especially suitable to reveal general principles that are independent of the particular system.

Setting Clear Goals and Criteria

Lockwood and Pimm (1999) reviewed 87 projects that satisfied three criteria: (1) intended goals had been stated, (2) the general goal was to restore some portion of the former ecosystem, and (3) the systems had been subjected to at least some initial management. The duration of projects, from initial management to the date of latest publication, ranged from 1 to 53 years (average 6 years). Fifty-nine sites were of small size (1–10 ha), 20 of medium size (11–50 ha), and only 8 of large size (≥51 ha). Only 20 percent of the projects were considered successful (that is, all prescribed goals achieved and management ceased), whereas the great majority (60 percent) were only partially successful (that is, all goals reached but management still needed, or only some goals reached and management ceased). The remaining 20 percent failed to satisfy any of the criteria. Function and partial structure were restored 10 times more often than full structure (species-for-species reassemblage).

Clear and achievable restoration goals should focus on the desired characteristics for a system in the future and not on features of the present state (Hobbs and Harris 2001). Short-term as well as long-term restoration goals should be defined quantitatively whenever possible to allow analytical evaluation and statistical testing (Westman 1991, Grumbine 1994, Hobbs and Norton 1996, Harker et al. 1999). An explicit statement of criteria for success and failure, which should be established before the restoration project starts, permits rigorous examination of the restoration process and would help to identify necessary changes in strategy to reach the intended goals (Cairns and Heckman 1996).

Recovery times in aquatic systems are generally shorter than in terrestrial ones because (1) unaffected upstream and downstream areas as well as internal refugia serve as sources of organisms for recolonization, (2) life history traits of aquatic organisms allow rapid recolonization and favor flexibility and adaptability due to frequent natural disturbances, and (3) high flushing rates of lotic systems allow for fast dilution or replacement of polluted waters (Yount and Niemi 1990). If no toxic residuals impair environmental quality and if sources of colonizing species are nearby, merely restoring the physical habitat characteristics (river depth, velocity, sediment load and size, sinuosity, and so forth) may be enough for successful river restoration (Cairns 1995).

Because of its highly dynamic features, however, river restoration requires the definition of criteria for a desired flow pattern, which is challenging because the interactions of individual flow components with geomorphological and ecological processes are incompletely understood (Poff et al. 1997). Quantitative standards may be developed by reconstructing the river-specific natural flood regime (Richter et al. 1997). For instance, restoration of the Kissimmee River in Florida (70-km river channel and 11,000 ha of surrounding wetland) was based, among other things, on the reestablishment of five key parameters of hydrology characteristic of the state before channelization (Toth 1995).

Using Knowledge of Species Characteristics

Before using species in restoration projects, it is reasonable to first gather information on their ecological requirements (for example, Budelsky and

Galatowitsch 1999). For instance, it is important to determine which life stages are most sensitive to disturbance and, therefore, require most attention during restoration. A variety of physiological, morphological, and life history traits of species that need to be reintroduced are available in databases (for example, the Ecological Flora Database of the British Isles: http://www.york.ac.uk). For the European flora, specific databases on clonal plants (van Groenendael et al. 1996, Klimes and Klimesova 1999), seed banks (Thompson et al. 1997), and root systems (Kutschera and Lichtenegger 1982, 1992) were compiled, and an electronic database containing vegetative and generative life history traits of a relevant part of the flora of central Europe has been developed (Poschlod et al. 2000). In Canada, an identification database for seeds, with morphological and ecological information, covers the coastal lowlands of Manitoba (Chang et al. 2000), and in the United States, the Flora of North America Digital Library is in preparation at the Center for Botanical Informatics of the Missouri Botanical Garden.

Many studies of recovery following disturbances in streams are limited to genera or family taxa levels of insects instead of single species (Wallace 1990). Despite this constraint, a unified measure for the functional composition of invertebrate communities in European running waters has been created recently using the genus taxonomic level (Statzner et al. 2001). This tool was developed by considering general biological traits of organisms (for example, size, reproductive and dispersal potential, food and feeding habits) that indicate ecological functions. Such traits are comparable among communities, even across biogeographic regions that differ in their taxonomic composition. This kind of approach should also be applicable to other types of communities and ecosystems, provided that knowledge about the biological traits is extensive and well structured. This is the case for many terrestrial plants (for example, Grime et al. 1988), and so considering the functional composition of plant communities at large spatial scales should be possible.

The development of similar databases for other faunal groups such as soil arthropods would be helpful for a more concise planning of restoration steps. In soil, the existing community should be assessed because even a crude understanding of microarthropods, soil fungi, and other microorganisms can be helpful in designing restoration strategies (Allen et al. 1999; also see Chapter 10). For instance, large-spored species of mycorrhizal fungi disperse slowly back to disturbed sites, so the reintroduction of fungi and of animals that graze them would help to reestablish more stable cycling of soil nutrients.

Genetic differences within species, however, can also affect the restoration outcome (Seliskar 1995). When isolated from other populations, the introduction of a small initial population may induce genetic drift and a loss of genetic variation (that is, the founder effect, Charlesworth and Charlesworth 1987). More research is needed about how to select a founder population genetically and about how different mating systems would translate into different reestablishment strategies.

Planning a Restoration Time Table

Ecological systems are dynamic and open to extrinsic perturbations, so restoration of degraded systems will take time (Cairns 1990, Bradshaw 1997b, Harker et al. 1999). In addition, restoration is not a discrete event but a process in which different system components need different lengths of time to recover (Parker and Pickett 1997). A key to appropriate restoration actions, therefore, is an understanding of the ecosystem's natural disturbance regime and recruitment dynamics (see Chapter 17), which are coupled in many ecosystems. For instance, reintroduction of plants is more likely to be successful if the natural disturbance regime is mimicked and natural succession can proceed (see Chapter 4).

Seasonal variations also should be considered, especially in aquatic ecosystems where the wetland environment behaves as a sieve that permits the establishment of only certain species at any given time (van der Valk 1981). The characteristics of this environmental sieve can change in response to water level fluctuations. In addition, drift and aerial colonization are temporally varied, usually being highest in spring and autumn. Therefore, restoration efforts should concentrate on time periods with naturally high recolonization dynamics (Gore et al. 1995).

In terrestrial ecosystems, it is unlikely that colonizers will arrive on their own if sources of recolonizing animal species are too distant or the area between them is too hostile. Hence, the time scale for faunal colonization to create a functioning community may be unacceptably long, even if a site is prepared with substrate, hydrology, and vegetation virtually indistinguishable from those of a nondegraded site. Earthworms, for instance, often require assistance to reach degraded sites, because these sites may be surrounded by drainage ditches that are barriers to colonization. Soil development will often benefit from the presence of earthworms (Butt et al. 1995), but the timing of their reintroduction may need to coincide with the establishment of site drainage and the availability of suitable food sources.

Unifying Restoration Schemes, Studies, and Treatments

The successful transfer of information between basic ecology and ecological restoration is fundamental to the implementation of restoration schemes. To unify studies of restoration, Keddy (1999) suggested the application of tools that already exist; that is, (1) assembly rules with predictions based upon key environmental factors and the responses of species to those factors (see Chapter 3), and (2) indicators of ecosystem integrity. Although there will always be a need for indicators relevant to local communities, consensus on standard indicators for wide usage would also be desirable. Aronson et al. (1993) provided a list of "vital ecosystem attributes" for a quantitative evaluation of whole ecosystem structure, composition, and functional complexity over time, based on measures of current and potential future vegetation structure, faunal relationships, soil condition, water and nutrient availability, and microsymbiont efficiency. In addition, Aronson and LeFloc'h (1996) proposed 16 "vital landscape attributes" for monitoring the success of restoration using a landscape perspective.

Bradshaw (1983) proposed general treatments to overcome physical, nutritional, and toxicity problems that prevent satisfactory vegetation growth on mined land. Wherever possible, processes occurring in natural succession related to plant growth, organic matter, and nutrient accumulation should be used, because they are self-sustaining, do not require external inputs, and can provide long-term solutions (Bradshaw 1997b, 1997c). Especially in ecosystems with extreme environmental conditions, the creation of safe sites may promote the establishment of species in both terrestrial (Urbanska 1994, 1997) and aquatic ecosystems (Goodwin et al. 1997).

Controlled and replicated field experiments will make it possible to record what works and why. Experiments could also allow the assessment of whether particular restoration practices are likely to be successful at other sites. In particular, recognizing failures and their causes will allow the avoidance of further mistakes (Zedler 1995) and therefore will contribute to the optimization of restoration approaches. For future progress, restoration site design should include experimental testing of alternative approaches whenever possible (Zedler 2000, Fattorini 2001); for example, by using different genotypes of the same species (Seliskar 1995), different numbers and combinations of transplant species (Zedler 2001), or different herbicide and seeding treatments (Tyser et al. 1998).

Validating Success by Subsequent Monitoring

Most often, the budget for restoration projects covers only the management phase, not subsequent monitoring programs to ensure that the goals of restoration have been achieved. Powerful monitoring should consider (1) the adequacy of the ecological data upon which the judgment of success is based; (2) the duration of the observational period, which should permit a full display of natural variability, cyclic phenomena, and succession processes; and (3) the scale of the spatial information, which may vary from a small patch to a large drainage basin (Cairns 1991). Such programs are costly because they always imply long-term observation, but they are required if practical restoration approaches are to be optimized. Moreover, the common lack of scrutiny and validation is one of the most serious shortcomings of the science of restoration ecology, because this lack prevents learning, hypotheses testing, and, consequently, scientific progress.

A restored ecosystem should be evaluated in its entirety, and so a broadly representative range of assessment criteria must be used to reflect the complexity of food webs; habitat heterogeneity; and dynamic physical, chemical, and biological processes. For terrestrial systems, there is increasing awareness of the important role of positive plant interactions (Gigon and Ryser 1986, Callaway and Walker 1997, Urbanska 1997, Brooker and Callaghan 1998), mycosymbionts (E. B. Allen 1989, M. F. Allen 1989), decomposers (Edwards and Abivardi 1997), pollinators, and seed dispersers (Handel et al. 1994, Handel 1997, Majer 1997, Toh et al. 1999). In evaluating plant species performance on a restored site, for instance, percentage cover or flowering rates are commonly measured. For meaningful planning of future projects based on previous experience, however, it would be preferable to know why certain species did well and others failed. Such an approach is much more challenging, since the processes of germination, establishment, growth, and reproduction may depend on many abiotic and biotic factors, including microclimate, resources, safe sites, competitors, herbivores, pathogens, symbionts, genetic diversity, and population density. Unfortunately, animal communities are ignored in evaluating restoration success, but invertebrates in particular are essential to self-sustaining ecosystems (Wheater and Cullen 1997, Wheater et al. 2000) and provide useful measures of restoration success (Webb 1996, Majer 1997, Wheater et al. 2000). In riverine systems, the few long-term studies of recolonization of restored channels indicate a deterministic pattern of functional types of colonists, in which the arrival of predatory invertebrates indicates late succession stages (Gore and Shields 1995).

Indices that compare species composition at different sites provide a quantitative tool for comparing current site conditions to the baseline goal. Measures of community similarity can be used if the primary goal is to restore the pattern of species presence and absence rather than relative species abundance. There also have been advances in the analytical procedures of multivariate data sets to measure changes in organism assemblages in response to environmental changes (Bloom 1980, Clarke 1993). Although potentially powerful, such procedures have seldom been incorporated into restoration studies (for an exception, see Wheater et al. 2000). Ordinations and analyses that consider the abundance of individual species instead of species presence and absence data alone allow the use of natural species abundance as a gauge for restoration success (Chapman 1999). Bentham et al. (1992) suggested a method of habitat classification based on a three-dimensional ordination of soil microbiological properties (dehydrogenase activity, adenosine triphosphate, and ergosterol) to assess the restoration potential of soils. Although long-term monitoring of soil development is rare, the few available studies suggest that recovery of populations and communities is much faster than recovery of soil properties (Hart et al. 1989, Craft et al. 1999).

Theory suggests that temporal measures of recovery at a single site indicate that some state change has occurred, but not that this state change is attributable to a particular action. To understand the efficacy of management action fully, comparisons of a recovering ecosystem against nondegraded sites and degraded, nonrestored sites are required in addition to measures before and after restoration (Kondolf 1995, Power 1999). Information about the natural pattern of oscillation in ecosystem properties can be helpful to establish confidence intervals for variance around a mean expectation for parameters being measured during recovery (Westman 1991). Such thorough studies, however, will be possible only in a few exceptional cases, which will then have to serve as proxies for the management of other degraded sites. Thus, it will be necessary to transfer experiences across ecosystems and across a variety of spatial and temporal scales, which again demonstrates the urgent need for the connective platform of a conceptual framework.

Developing Models for Intermediate Evaluation

Final achievement of restoration goals may take a century or more. To allow for strategy corrections to restoration projects in progress, however, it is essential to estimate, on the basis of current performance, whether the restoration path is following the desired trajectory and to predict the time to complete recovery (Westman 1991). Four examples of tools that have been developed to achieve these goals may elucidate some general principles but also demonstrate some of the serious problems with such approaches.

Markov chain modeling expresses the probability of transition from one recovery state to the next. For example, this approach was followed by Kachi et al. (1986) and Turner (1987) with remote sensing data to track the development of forests. The Markov matrix can be extrapolated beyond the sampling period, assuming that future recovery rates will not change very much. However, since succession rates are often nonlinear (Westman 1985), the basic assumption for this family of models is questionable.

The expert system SUCCESS enables predictions of rates and directions of succession in disturbed terrestrial habitats of central Europe up to 50 years after disturbance (Prach et al. 1999). Information about geographical position, substratum, moisture, relief, nutrient content, surrounding vegetation, and size of the disturbed area is employed to predict site-specific succession. Such expert systems could be enlarged to apply to wider geographical areas, but this would require an enormous database.

A multivariate model based on canonical correspondence analysis (CANOCO, Ter Braak 1988) was outlined and tested on lowland heaths (Mitchell et al. 2000). In this model, data on species composition and abiotic and biotic factors are combined to measure direction and change through time against internally derived standards. The approach has proved useful for measuring the success of management and for deciding which succession stages are easiest to restore in degraded heathland.

The in-stream flow incremental methodology (IFIM) assumes that the distribution of lotic biota is controlled by the hydraulic conditions within the water column. Traditionally, IFIM has been used to predict minimum flow requirements to maintain biota after reservoir manipulation or irrigation diversion. The approach was modified in that channel morphology, flow characteristics, and biological preferences of target species were combined to predict gains or losses in physical habitat under new and modified flow regimes after placement of restoration structures in surface-mined rivers (Gore 1985, Gore et al. 1995). This approach was applied to benthic macroinvertebrates (Statzner et al. 1988) and allowed a current state evaluation of the system.

These four examples of models that were applied for intermediate evaluation of restoration projects are quite different in purpose but are not mutually exclusive approaches. They show that restoration ecology has reached a stage in which the use of predictive models is a complement to experimentation and that different systems and purposes require different modeling setups. Succession directions and rates can never be highly predictable because of stochastic effects (Pickett et al. 1987, Walker and Chapin 1987, Wyant et al. 1995, Hobbs and Norton 1996). However, even imprecise tools for predicting the time to achieve long-term restoration goals on the basis of current performances represent an enormous progress over current decision making, which is often based on gut feelings.

Toward a Unifying Framework for Restoration Ecology

For restoration ecology to be seen as a science rather than a technique, development of a common conceptual framework is an obvious prerequisite. As in many branches of basic ecology, scientific debate over alternative framework conceptions will increase our understanding. Suggestions for theoretical frameworks will also provide a research agenda for the field of restoration ecology because such suggestions outline questions of interest for stringent experimental testing of hypotheses. Despite the many obstacles that complicate the development of general models in restoration ecology, advances have recently been achieved because of increasing experience gained from the restoration of a variety of aquatic and terrestrial ecosystems and also because of new insights from theoretical considerations. Thus, some ideas for a theory of ecological restoration are now proposed and need to be discussed.

The apparent strength of general models—that is, their ability to describe a wide variety of different ecological situations with a limited set of model parameters—may also be seen as their most essential flaw. Generality is achieved by sacrificing many details specific to any particular system and by keeping only those features that are essential for the modeled processes. For example, in our dynamic environmental filter model (see Chapter 6), the interactions of established species are ignored and the ecosystem is seen as a black box in which the intrinsic processes at the community level are considered only insofar as they affect the probability of invasion by new species. Negating the intrinsic features of communities may stretch the usual need for simplification in modeling too far, because the representative characteristics of ecosystems are no longer recognizable. It may be argued that a model so general that it fits everything will not give a single hint at appropriate restoration actions in any particular system, and may, therefore, be regarded as a futile intellectual exercise. However, only by sacrificing the peculiarities of systems it is possible to make them directly comparable. This dilemma goes back to the three possible model types as defined by Levins (1966): models can be precise, realistic, or general. A model that is both precise (in the sense that system-specific predictions can be derived) and general (in the sense that any system may fit into its framework) is impossible by definition.

For any one system—or group of comparable systems—tailored models are helpful to choose among different restoration strategies and to enable intermediate evaluation. For this reason, they should always be part of any larger restoration project. Models with local applicability are much more suitable for giving direct advice to the restoration practitioner, but at the cost of a limited domain of validity. The conclusions derived will apply only to a certain set of systems with comparable characteristics, and hence this kind of model cannot be scaled up to develop a conceptual framework for the science of restoration ecology. We must accept that there are different layers of the same problem, each worthy of consideration in its own right, but that it will be impossible to find the one and only "all-purpose model" that covers all aspects equally well. Rather, every model will have a confined array of answerable questions and will be useful only for the layer for which it was designed.

If global models do not give any practical advice, what then is their pragmatic use in restoration ecology? Is it just nice to have one for theoretical reasons, or does a global model really contribute to the development as a branch of ecological science? This question was debated with respect to our dynamic filter model (see Chapter 6) at the Dornburg workshop. The heated discussion clearly verified one of the most significant benefits of a general model: everybody was challenged to argue about specific aspects of the model, but thoughts had to be put into the same rigid frame, irrespective of the highly diverse individual approaches and experiences. This forced the participants to take distinct positions, which worked out very clearly what people actually agreed and disagreed about. Such a sharp debate over restoration issues would not have been possible on the basis of case studies in particular systems, which would have addressed a totally different level of discussion; that is, questions on system details. No doubt, such a discussion would also have been valuable, but only for a handful of system experts concerned about comparable problems.

In addition to being a catalyst for substantial and system-independent scientific discussion, a conceptual model has further promising advantages. For instance, it helps to categorize case study examples. If successful as well as failed restoration projects—both being equally important as a scientific result—are pressed into the same restrictive framework, repeated patterns might occur with accumulating experience. Patterns will show up more clearly with a general model, since the degrees of freedom to describe cases and actions taken are very limited. Repeated patterns, which can be derived only from field experience, will indicate that under certain conditions some strategies will often lead to success, whereas others will tend to fail. These conclusions, in turn, will allow us to make predictions and formulate testable hypotheses—prerequisites for any scientific insight and progress. Moreover, without such patterns, it will never be possible to reach the really essential next step: to identify the processes that generate the patterns. Without a model general enough to fit all systems, the search for patterns in a myriad of case studies presented from different perspectives is difficult, if not impossible.

Acknowledgments

Susan Galatowitsch, Richard Hobbs, Patricia Holmes, Bill Jordan, Blandine Massonnet, Jörg Perner, and Vicky Temperton made constructive comments on earlier drafts of this chapter. Special thanks go to Tim Nuttle and John Sloggett who, in addition to substantial general comments, supplied linguistic improvements.

REFERENCES

Aber, J. D.; and Jordan, W. R. 1985. Restoration ecology: an environmental middle ground. *Bioscience* 35:399.

Allen, E. B. 1989. The restoration of disturbed arid landscapes with special reference to mycorrhizal fungi. *Journal of Arid Environments* 17:279–286.

Allen, M. F. 1989. Mycorrhizae and rehabilitation of disturbed arid soils: processes and practices. *Arid Soil Research* 3:229–241.

Allen, M. F.; Allen, E. B.; Zink, T. A.; Harney, S.; Yoshida, L. C.; Sigüenza, C.; Edwards F.; Hinkson, C.; Rillig, M.; Bainbridge, D.; Doljanin, C.; and MacAller, R. 1999. Soil microorganisms. In *Ecosystems of Disturbed Ground*, ed. L. R. Walker, 521–544. Amsterdam: Elsevier.

Aronson, J.; Floret, C.; LeFloc'h, E.; Ovalle, C.; and Pontanier, R. 1993. Restoration and rehabilitation of degraded ecosystems in arid and semiarid regions. I. A view from the South. *Restoration Ecology* 1:8–17.

Aronson, J.; and LeFloc'h, E. 1996. Vital landscape attributes: missing tools for restoration ecology. *Restoration Ecology* 4:377–387.

Baldwin, A. D. J.; De Luce, J.; and Pletsch, C. 1994. *Beyond Preservation: Restoring and Inventing Landscapes.* Minneapolis: University of Minnesota Press.

Bentham, H.; Harris, J. A.; Birch, P.; and Short, K. C. 1992. Habitat classification and soil restoration assessment using analysis of soil microbiological and physico-chemical characteristics. *Journal of Applied Ecology* 29:711–718.

Berger, J. J. 1990. *Environmental Restoration: Science and Strategies for Restoring the Earth.* Washington, D.C.: Island Press.

Bloom, S. A. 1980. Multivariate quantification of community recovery. In *The Recovery Process in Damaged Ecosystems*, ed. J. Cairns, 141–151. Ann Arbor: Ann Arbor Science Publishers.

Bradshaw, A. D. 1983. The reconstruction of ecosystems. *Journal of Applied Ecology* 20:1–17.

Bradshaw, A. D. 1987. Restoration: an acid test for ecology. In *Restoration Ecology: a Synthetic Approach to Ecological Research*, ed. W. R. Jordan, M. E. Gilpin, and J. D. Aber, 24–29. Cambridge: Cambridge University Press.

Bradshaw, A. D. 1992. The biology of land restoration. In *Applied Population Biology*, ed. S. K. Jain and L. W. Botsford, 25–44. Dordrecht: Kluwer.

Bradshaw, A. D. 1997a. What do we mean by restoration? In *Restoration Ecology and Sustainable Development*, ed. K. M. Urbanska, N. R. Webb, and P. J. Edwards, 8–14. Cambridge: Cambridge University Press.

Bradshaw, A. D. 1997b. The importance of soil ecology in restoration science. In *Restoration Ecology and Sustainable Development*, ed. K. M. Urbanska, N. R. Webb, and P. J. Edwards, 33–64. Cambridge: Cambridge University Press.

Bradshaw, A. D. 1997c. Restoration of mined lands using natural processes. *Ecological Engineering* 8:255–269.

Brooker, R. W.; and Callaghan T. V. 1998. The balance between positive and negative plant interactions and its relationship to environmental gradients: a model. *Oikos* 81:196–207.

Budelsky, R. A.; and Galatowitsch, S. M. 1999. Effects of moisture, temperature, and time on seed germination of five wetland Carices: implications for restoration. *Restoration Ecology* 7:86–97.

Butt, K. R.; Frederickson, J.; and Morris, R. M. 1995. An earthworm cultivation and soil inoculation technique for land restoration. *Ecological Engineering* 4:1–9.

Cairns, J. 1982. Restoration of damaged ecosystems. In *Research on Fish and Wildlife Habitat,* ed. W. T. Mason and S. Ilker, 220–239. Washington, D.C.: U.S. Environmental Protection Agency.

Cairns, J. 1986. Restoration, reclamation, and regeneration of degraded or destroyed ecosystems. In *Conservation Biology: the Science of Scarcity and Diversity,* ed. M. Soulé, 465–484. Sunderland, Mass.: Sinauer Associates.

Cairns, J. 1988. Restoration and the alternative: a research strategy. *Restoration and Management Notes* 6:65–67.

Cairns, J. 1990. Lack of theoretical basis for predicting rate and pathways of recovery. *Environmental Management* 14:517–526.

Cairns, J. 1991. The status of the theoretical and applied science of restoration ecology. *The Environmental Professional* 13:186–194.

Cairns, J. 1995. Restoration ecology: protecting our national and global life support system. In *Rehabilitating Damaged Ecosystems* ed. J. Cairns, 1–12. Boca Raton, Fla.: Lewis Publishers.

Cairns, J.; and J. R. Heckman. 1996. Restoration ecology: the state of an emerging field. *Annual Review of Energy and Environment* 21:167–189.

Callaway, R. M.; and Walker, L. R. 1997. Competition and facilitation: a synthetic approach to interactions in plant communities. *Ecology* 78:1958–1965.

Chang, E. R.; Dickinson, T. A.; and Jefferies, R. L. 2000. Seed flora of La Perouse Bay, Manitoba, Canada: a DELTA database of morphological and ecological characters. *Canadian Journal of Botany* 78:481–496.

Chapman, M. G. 1999. Improving sampling designs for measuring restoration in aquatic habitats. *Journal of Aquatic Ecosystem Stress and Recovery* 6:235–251.

Charlesworth, D.; and Charlesworth, B. 1987. Inbreeding depression and its evolutionary consequences. *Annual Review of Ecology and Systematics* 18:237–268.

Chase, J. M. 2000. Are there real differences among aquatic and terrestrial food webs? *Trends in Ecology and Evolution* 15:408–412.

Clark, M. J. 1997. The magnitude of the challenge: an outsider's view. In *Restoration Ecology and Sustainable Development,* ed. K. M. Urbanska, N. R. Webb, and P. J. Edwards, 353–377. Cambridge: Cambridge University Press.

Clarke, K. R. 1993. Non-parametric multivariate analyses of changes in community structure. *Australian Journal of Ecology* 18:117–143.

Craft, C.; Reader, J.; Sacco, J. N.; and Broome, S. W. 1999. Twenty-five years of ecosystem development of constructed *Spartina alterniflora* (Loisel) marshes. *Ecological Applications* 9:1405–1419.

Cyr, H.; Peters, R. H.; and Downing, J. A. 1997. Population density and community size structure: comparison of aquatic and terrestrial systems. *Oikos* 80:139–149.

Dobson, A. P.; Bradshaw, A. D.; and Baker, A. J. M. 1997. Hopes for the future: restoration ecology and conservation biology. *Science* 277:515–522.

Dubos, R. 1978. The resilience of ecosystems: an ecological view of environmental restoration. *The Reben G. Gustavson Memoria Lectures* 1:1–29.

Edwards, P. J.; and Abivardi, C. 1997. Ecological engineering and sustainable development. In *Restoration Ecology and Sustainable Development*, ed. K. M. Urbanska, N. R. Webb, and P. J. Edwards, 325–352. Cambridge: Cambridge University Press.

Ehrenfeld, J. G.; and Toth, L. A. (1997). Restoration ecology and the ecosystem perspective. *Restoration Ecology* 5:307–317.

Fattorini, M. 2001. Establishment of transplants on machine-graded ski runs above timberline in the Swiss Alps. *Restoration Ecology* 9:119–126.

Freckman, D. W.; Blackburn, T. H.; Brussaard, L.; Hutchings, P.; Palmer, M. A.; and Snelgrove, P. V. R. 1997. Linking biodiversity and ecosystem functioning of soils and sediments. *Ambio* 26:556–562.

Gigon, A.; and Ryser, P. 1986. Positive Interaktionen zwischen Pflanzenarten. *Veröffentlichungen des Geobotanischen Istitutes ETH Stiftung Rübel* 87:372–387.

Goodwin, C. N.; Hawkins, C. P.; and Kershner, J. L. 1997. Riparian restoration in the western United States: overview and perspective. *Restoration Ecology* 5:4–14.

Gore, J. A. 1995. Mechanisms of colonization and habitat enhancement for benthic macroinvertebrates in restored river channels. In *The Restoration of Rivers and Streams: Theories and Experience*, ed. J. A. Gore, 81–101. Boston: Butterworth.

Gore, J. A.; and Shields, F. D., Jr. 1995. Can large rivers be restored? *BioScience* 45:142–152.

Gore, J. A.; Bryant, F. L.; and Crawford, D. J. 1995. River and stream restoration. In *Rehabilitating Damaged Ecosystems*, ed. J. Cairns, 245–275. Boca Raton, Fla.: Lewis Publishers.

Grime, J. P.; Hodgson, J. G.; and Hunt, R. 1988. *Comparative Plant Ecology. A Functional Approach to Common British Species.* London: Unwin Hyman.

Grumbine, R. E. 1994. What is ecosystem management? *Conservation Biology* 8:27–38.

Handel, S. N. 1997. The role of plant-animal mutualisms in the design and restoration of natural communities. In *Restoration Ecology and Sustainable Development*, ed. K. M. Urbanska, N. R. Webb, and P. J. Edwards, 111–132. Cambridge: Cambridge University Press.

Handel, S. N.; Robinson, G. R.; and Beattie, A. J. 1994. Biodiversity resources for restoration ecology. *Restoration Ecology* 2:230–241.

Harker, D.; Libby, G.; Harker, K.; Evans, S.; and Evans, M. 1999. Principles and practices of ecological restoration. In *Landscape Restoration Handbook*, ed. D. Harker, S. Evans, M. Evans, and K. Harker, 63–73. Boca Raton, Fla.: Lewis Publishers.

Harper, J. L. 1987. The heuristic value of ecological restoration. In *Restoration Ecology: a Synthetic Approach to Ecological Research*, ed. W. R. Jordan, M. E. Gilpin, and J. D. Aber, 35–45. Cambridge: Cambridge University Press.

Hart, P. B. S.; August, J. A.; and West, A. W. 1989. Long-term consequences of topsoil mining on biological and physical characteristics of two New Zealand loessial soils under grazed pasture. *Land Degradation and Rehabilitation* 1:77–88.

Heckman, J. R.; and Cairns, J. 1998. Ecosystem restoration: a new perspective for sustainable use of the planet. In *Damaged Ecosystems and Restoration*, ed. B. C. Rana, 69–76. Vallabh Vidyanagar, India: Sardar Patel University.

Higgs, E. 1994. Expanding the scope of restoration ecology. *Restoration Ecology* 2:137–146.

Hobbs, R. J. 1999. Restoration of disturbed ecosystems. In *Ecosystems of Disturbed Ground*, ed. L. R. Walker, 673–687. Amsterdam: Elsevier.

Hobbs, R. J.; and Harris J. A. 2001. Restoration ecology: repairing the earth's ecosystems in the new millennium. *Restoration Ecology* 9:239–246.

Hobbs, R. J.; and Norton, D. A. 1996. Towards a conceptual framework for restoration ecology. *Restoration Ecology* 4:93–110.

Jordan, W. R.; Gilpin, M. E.; and Aber, J. D. 1987. *Restoration Ecology: A Synthetic Approach to Ecological Research.* Cambridge: Cambridge University Press.

Kachi, N.; Yasuoka, Y.; Totsuka, T.; and Suzuki, K. 1986. A stochastic model for describing revegetation following forest cutting: an application of remote sensing. *Ecological Modelling* 32:105–117.

Kauffman, J. B.; Beschta, R. L.; Otting, N.; and Lytjen, D. 1997. An ecological perspective of riparian and stream restoration in the western United States. *Fisheries* 22:12–24.

Keddy, P. 1999. Wetland restoration: the potential for assembly rules in the service of conservation. *Wetlands* 19:716–732.

Klimes, L.; and Klimesova, J. 1999. CLO-PLA2: a database of clonal plants in central Europe. *Plant Ecology* 141:9–19.

Kondolf, G. M. 1995. Five elements for effective evaluation of stream restoration. *Restoration Ecology* 3:133–136.

Kutchera, L.; and Lichtenegger, E. 1982. *Wurzelatlas mitteleuropäischer Grünlandpflanzen. Band 1, Monocotyledonae.* Stuttgart: Fischer.

Kutchera, L.; and Lichtenegger, E. 1992. *Wurzelatlas mitteleuropäischer Grünlandpflanzen. Band 2, Pteridophyta und Dicotyledonae (Magnoliopsida). Teil 1.*Stuttgart: Fischer.

Levins, R. 1966. The strategy of model building in population biology. *American Scientist* 54:421–431.

Lockwood, J. L.; and Pimm, S. L. 1999. When does restoration succeed? In *Ecological Assembly Rules: Perspectives, Advances, Retreats*, ed. E. Weiher and P. Keddy, 363–392. Cambridge: Cambridge University Press.

Magnuson, J. J.; Regier, H. A.; Christie, W. J.; and Sonzogi, W. C. 1980. To rehabilitate and restore Great Lakes ecosystems. In *The Recovery Process in Damaged Ecosystems*, ed. J. Cairns, 95–112. Ann Arbor: Ann Arbor Science Publishers.

Majer, J. D. 1997. Invertebrates assist the restoration process: an Australian perspective. In *Restoration Ecology and Sustainable Development*, ed. K. M. Urbanska, N. R. Webb, and P. J. Edwards, 212–237. Cambridge: Cambridge University Press.

Mitchell, R. J.; Auld, M. H. D.; Le Duc, M. G.; and Marrs, R. H. 2000. Ecosystem stability and resilience: a review of their relevance for the conservation management of lowland heaths. *Perspectives in Plant Ecology, Evolution and Systematics* 3:142–160.

Muhar, S.; Schmutz, S.; and Jungwirth, M. 1995. River restoration concepts: goals and perspectives. *Hydrobiologia* 303:183–194.

National Research Council 1992. *Restorations of Aquatic Ecosystems: Science, Technology, and Public Policy.* Committee on Restoration of Aquatic Ecosystems. Washington, D.C.: National Academy Press.

Naveh, Z. 1994. From biodiversity to ecodiversity: a landscape-ecology approach to conservation and restoration. *Restoration Ecology* 2:180–189.

Palmer, M. A.; Ambrose, R. F.; and Poff, N. L. 1997. Ecological theory and community restoration ecology. *Restoration Ecology* 5:291–300.

Parker, V. T.; and Pickett, S. T. A. 1997. Restoration as an ecosystem process: implications for the modern ecological paradigm. In *Restoration Ecology and Sustainable Devel-*

opment, ed. K. M. Urbanska, N. R. Webb, and P. J. Edwards, 17–32. Cambridge: Cambridge University Press.

Pickett, S. T. A.; Collins, S. L.; and Armesto, J. J. 1987. Models, mechanisms and pathways of succession. *The Botanical Review* 53:335–371.

Pickett, S. T. A.; and Parker, V. T. 1994. Avoiding the old pitfalls: opportunities in a new discipline. *Restoration Ecology* 2:75–79.

Poff, N. L.; Allan, J. D.; Bain, M. B.; Karr, J. R.; Prestegaard, K. L.; Richter, B. D.; Sparks, R. E.; and Stromberg, J. C. 1997. The natural flow regime: a paradigm for river conservation and restoration. *BioScience* 47:769–783.

Poschlod, P.; Kleyer, M.; and Tackenberg, O. 2000. Databases on life history traits as a tool for risk assessment in plant species. *Zeitschrift für Ökologie und Naturschutz* 9:3–18.

Power, M. 1999. Recovery in aquatic ecosystems: an overview of knowledge and needs. *Journal of Aquatic Ecosystem Stress and Recovery* 6:253–257.

Prach, K.; Pysek, P.; and Smilauer, P. 1999. Prediction of vegetation succession in human-disturbed habitats using an expert system. *Restoration Ecology* 7:15–23.

Rapport, D. A.; and Whitford, W. G. 1999. How ecosystems respond to stress: common properties of arid and aquatic systems. *BioScience* 49:193–203.

Richter, B. D.; Baumgartner, J. V.; Vigington, R.; and Braun, D. P. 1997. How much water does a river need? *Freshwater Biology* 37:231–247.

Schumm, S. A. 1984. River morphology and behavior: problems of extrapolation. In *River Meandering: Proceedings of the Rivers '83 Conference, 1983, New Orleans, Louisiana*, ed. C. M. Elliot, 16–29. New York: American Society of Civil Engineers.

Seliskar, D. 1995. Exploiting plant genotypic diversity for coastal salt marsh creation and restoration. In *Biology of Salt-Tolerant Plants*, ed. M. A. Khan and I. A. Ungar, 407–416. Karachi, Pakistan: University of Karachi, Department of Botany.

Simenstad, C. A.; and Thom, R. M. 1996. Functional equivalency trajectories of the restored Gog-Le-Hi-Te estuarine wetland. *Ecological Applications* 6:38–56.

Smith, V. H.; Tilman, G. D.; and Nekola, J. C. 1999. Eutrophication: impacts of excess nutrient inputs on freshwater, marine, and terrestrial ecosystems. *Environmental Pollution* 100:179–196.

Society for Ecological Restoration Science and Policy Working Group. 2002. *The SER Primer on Ecological Restoration.* www.ser.org/.

Statzner, B.; Gore, J. A.; and Resh, V. H. 1988. Hydraulic stream ecology: observed patterns and potential application. *Journal of the North American Benthological Society* 7:307–360.

Statzner, B.; Bis, B.; Dolédec, S.; and Usseglio-Polatera, P. 2001. Perspectives for biomonitoring at large spatial scales: a unified measure for the functional composition of invertebrate communities in European running waters. *Basic and Applied Ecology* 2:73–85.

Ter Braak, C. J. F. 1988. CANOCO: A FORTRAN *Program for Canonical Community Ordination by (Partial) (Detrended) (Canonical) Correlation Analysis, Principal Components Analysis and Redundancy Analysis*. Wageningen, The Netherlands: IWIS-TNO.

Toh, I.; Gillespie, M.; and Lamb, D. 1999. The role of isolated trees in facilitating tree seedling recruitment at a degraded sub-tropical rainforest site. *Restoration Ecology* 7:288–297.

Thompson, K.; Bakker, J. P.; and Bekker, R. M. 1997. *Methods of Seed Bank Analysis. The Soil Seed Bank of North West Europe: Methodology, Density and Longevity.* Cambridge: Cambridge University Press.

Toth, L. A. 1995. Principles and guidelines for restoration of river/floodplain ecosystems: Kissimmee River, Florida. In *Rehabilitating Damaged Ecosystems*, ed. J. Cairns, 49–73. Boca Raton, Fla.: Lewis Publishers.

Turner, M. G. 1987. Spatial simulation of landscape changes in Georgia: a comparison of 3 transition models. *Landscape Ecology* 1:29–36.

Tyser, R. W.; Asebrook, J. M.; Potter, R. W.; and Kurth, L. L. 1998. Roadside revegetation in Glacier National Park, U.S.A.: effects of herbicide and seeding treatments. *Restoration Ecology* 6:197–206.

Urbanska, K. M. 1994. Ecological restoration above the timberline: Demographic monitoring of whole trial plots in the Swiss Alps. *Botanica Helvetica* 104:141–156.

Urbanska, K. M. 1997. Safe sites: interface of plant population ecology and restoration ecology. In *Restoration Ecology and Sustainable Development*, ed. K. M. Urbanska, N. R. Webb, and P. J. Edwards, 81–110. Cambridge: Cambridge University Press.

van der Valk, A. G. 1981. Succession in wetlands: a Gleasonian approach. *Ecology* 62:688–696.

van Groenendael, J. M.; Klimes, L.; Klimesova, J.; and Hendriks, R. J. J. 1996. Comparative ecology of clonal plants. *Philosophical Transactions of the Royal Society London, Series B* 351:1331–1339.

Walker, L. R.; and Chapin, F. S. III. 1987. Interactions among processes controlling successional change. *Oikos* 50:131–135.

Wallace, J. B. 1990. Recovery of lotic macroinvertebrate communities from disturbance. *Environmental Management* 14:605–620.

Webb, N. R. 1996. Restoration ecology: science, technology and society. *Trends in Ecology and Evolution* 11:396–397.

Westman, W. E. 1985. *Ecology, Impact Assessment, and Environmental Planning.* New York: Wiley Interscience.

Westman, W. E. 1991. Ecological restoration projects: measuring their performance. *The Environmental Professional* 13:207–215.

Wheater, C. P.; and Cullen, W. R. 1997. Invertebrate communities of disused and restoration blasted limestone quarries in Derbyshire. *Restoration Ecology* 5:77–84.

Wheater, C. P.; Cullen, W. R.; and Bell, J. R. 2000. Spider communities as tools in monitoring reclaimed limestone quarry landforms. *Landscape Ecology* 15:401–406.

Wiegleb, G. 1991. Wissenschaftliche Grundlagen von Fliessgewässerrenaturierungskonzepten. *Verhandlungen der Gesellschaft für Ökologie* 19:17–24.

Wissmar, R. C.; and Beschta, R. L. 1998. Restoration and management of riparian ecosystems: a catchment perspective. *Freshwater Biology* 40:571–585.

Wyant, J. C.; Meganck, R. A.; and Ham, S. H. 1995. A planning and decision-making framework for ecological restoration. *Environmental Management* 19:789–796.

Yount, D. J.; and Niemi, G. J. 1990. Recovery of lotic communities and ecosystems from disturbance: a narrative review of case studies. *Environmental Management* 14:547–569.

Zedler, J. B. 1995. Salt marsh restoration: lessons from California. In *Rehabilitating Damaged Ecosystems*, ed. J. Cairns, 75–95. Boca Raton, Fla.: Lewis Publishers.

Zedler, J. B. 1996. Coastal migration in Southern California: the need for a regional restoration strategy. *Ecology* 6:84–93.

Zedler, J. B. 2000. Progress in wetland restoration ecology. *Trends in Ecology and Evolution* 15:402–407.

Zedler, J. B. 2001. *Handbook for Restoring Tidal Wetlands.* Marine Science Series. Boca Raton, Fla.: CRC Press.

Chapter 3

The Search for Ecological Assembly Rules and Its Relevance to Restoration Ecology

Vicky M. Temperton and Richard J. Hobbs

This book aims to bring together two fields of endeavor: restoration ecology and the search for assembly rules in ecosystems. By nature, these two areas of activity and research should complement each other well, but in practice people working in these fields have little contact. To improve this situation and begin a discourse between basic ecologists and restoration practitioners, it is necessary to provide the background for a constructive exchange of ideas. The chapters in Part I aim to provide such a background to the discourse, which we hope will ensue from this book.

Our goal in writing this chapter is not to present an extensive review of the assembly rules literature but to provide a synopsis of the current central approaches and tenets of assembly rules theory, where these tenets come from, and how relevant the theory might be for the practice of restoration ecology. For some more comprehensive reviews of assembly theory, see Lawton (1991), Keddy (1992), Belyea and Lancaster (1999), and Wilson (1999). As outlined in Chapter 1, there are certain long-debated questions in ecology that still fascinate ecologists, precisely because they are such simple yet big questions that are so difficult to answer. Two of these are: How do communities of organisms come to be the way they are? What are the constraints on membership in a community?

As defined by the Society for Ecological Restoration, "restoration ecology is an intentional activity that initiates or accelerates the recovery of an ecosystem with respect to its health, integrity and sustainability" (SER Science and Policy Working Group 2002). For restoration ecology to progress beyond treating each habitat or community as a singular case study (see Chapter 2), we must discover not only any rules governing how natural communities are structured but also how these rules, and hence community structure, change

34

over time. Young et al. (2001) have highlighted the need to extract and combine the most useful elements of various concepts (particularly succession and assembly) to guide restoration efforts. Thus, the framework of assembly rules theory is directly relevant to restoration ecology. There is a need for restoration when an ecosystem is stuck in an undesirable state, and the art and science of restoration is how best to move the system over any threshold holding it in that undesirable state (Hobbs and Norton 1996, Hobbs 1999). The rest of this chapter is an analysis of the different approaches, methodologies, and actual assembly rules involved in assembly theory, with a discussion of their applicability and relevance to restoration ecology.

General Approaches

The field of assembly rules is very broad and encompasses a variety of different approaches to finding rules that govern how ecological communities develop. J. Bastow Wilson (1999) defined assembly rules as "ecological restrictions on the observed patterns of species presence or abundance that are based on the presence or abundance of one or more other species or groups of species (not simply the response of species to the environment)." Belyea and Lancaster (1999) iterated that rules, as general principles, "constrain an action or procedure." They also emphasized that assembly rules arise from the interactions among living components of a system. This idea implies that assembly rules are about principles that constrain how a community is put together and functions. Another group of ecologists views assembly rules somewhat differently (see Figure 3.1 for a comparison of approaches). They say that assembly rules can include any constraint on the species pool, be it biotic or abiotic (Keddy 1992; Zobel 1992; Weiher and Keddy 1995; Díaz et al. 1998, 1999; and see, for example, Chapters 6, 14, and 16). In this approach, the interaction of the environment with the organisms of a community, as well as the interactions among organisms, restricts how a community is structured and develops. For example, Keddy (1992) stated that the objective of assembly rules is to predict, knowing a regional species pool and given a certain set of environmental conditions, which subset of organisms will occur in a habitat.

As yet, there is no real consensus (Booth and Larson 1999) about what assembly rules constitute, which brings us to ask: Which school of thought should be the focus of the search for a connection between assembly rules and restoration? The objective of this book is to be inclusive, by presenting an overview of the current status of a field that includes different approaches and conflicting opinions. As long as each school of thought clearly defines

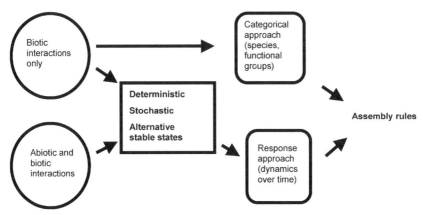

FIGURE 3.1 The different approaches to the search for assembly rules. There are two schools of thought about whether abiotic factors play a role: the "biotic interactions only" camp and the "abiotic and biotic interactions" camp. The "biotic interactions only" camp mainly uses the categorical approach (analyzing frequencies of species or functional groups) as a method for finding assembly rules. The "abiotic and biotic interactions" camp focuses primarily on the three main theories of community assembly over time (deterministic, stochastic, and alternative stable states) and uses a response approach based on dynamics over time to find assembly rules.

how it sees assembly rules, the debate will be able to progress accordingly and useful links to restoration practice will be forged. Of course, it would be better if we could at least agree on the level of abstraction at which we look for assembly rules. For now, this consensus is lacking, and perhaps this could turn out to be a strength, in that different perspectives and approaches are needed to understand the incredible complexity of ecological communities. It would be useful at this stage, however, if the different schools of thought could adopt specific names, thus enabling swift identification of whatever approach is meant (see the Methodological Issues section of this chapter for some suggested names).

Although the genesis of the assembly rules approach is usually attributed to the work of Jared Diamond on the distribution of bird species in an archipelago in Papua, New Guinea (1975), ecologists have been interested in how communities are structured for much longer than this. Plant ecologists in particular did some pioneering early work in this area. Clements (1916, 1936) worked in terrestrial ecosystems and developed the idea of a sequence of states in an ecosystem that gradually evolve into a final climax state (usually a forest) that functions as a whole unit (the "organism concept"). Gleason (1917, 1926) formulated the opposing "individualistic concept," which states that a community does not function as a whole but is merely the sum of its individual parts. The development of these ideas is discussed more fully by Young et al. (2001), who focus on the differences between succession and assembly theory and their relevance to restoration ecology. Harper's work (1977) on safe sites and the invasion of plants into communities forms a more recent approach to the study of development of communities. Much of this early work on succession and community development is still relevant today, since the three main theories of assembly derive directly from this research (see the next section). This legacy of early work is also particularly applicable as a conceptual framework for the field of restoration ecology because it relates to one of the fundamental questions in restoration ecology (Young et al. 2001): Do disturbed communities tend to repair themselves and return to the predisturbance state, or do historical events in a particular community produce a variety of alternative stable states?

Exchange of information and active debate among ecologists studying different kingdoms of the living world has often been lacking. When the term *assembly rule* was first coined by Diamond in 1975, most vertebrate ecologists involved in the subsequent debate seemed to be ignoring the legacy of relevant groundbreaking work by plant ecologists. Botanists and invertebrate zoologists have tended to look for the connection between physical factors and their effect on individual species distribution, whereas vertebrate zoologists commonly look at competition as a defining force to explain species distribution. This dichotomy is currently disappearing, since much of the microcosm work on community assembly includes organisms from more than one of the five kingdoms (Fraser and Keddy 1997, Buckland and Grime 2000), not just those of the Plantae or Animalia kingdoms. Some authors (Booth and Larson 1999) even go as far as to say that the current study of assembly rules is a rehashing of old knowledge and, furthermore, sometimes does not even take this previous knowledge into account. In this book, we therefore strive to acknowledge the pioneering work of early ecologists, while also moving forward to new ideas.

Deterministic Versus Stochastic Versus Alternative Stable States

There are three major models of community assembly: the deterministic, the stochastic, and alternative stable states (ASS) (see Figure 3.1). In the deterministic model, a community's development is seen as the inevitable consequence of physical and biotic factors. Clements (1916) believed in the importance of the interaction of organisms and in ecosystems reaching equilibrium. Believers in the stochastic or carousel model (Gleason 1917, 1926;

Van der Maarel and Sykes 1993) state that community composition and structure is essentially a random process; it depends only on the availability of vacant niches and the order of arrival of organisms (hence the carousel image, swinging around and letting random organisms on board fly off into the surroundings). This theory contends that the complexity of nature notoriously evades simple description and is primarily defined by chance events, which means that communities do not follow any specific trajectory over time. Sutherland (1974) formulated the alternative stable states model (hereafter, ASS), which is intermediate between the first two, and asserts that communities are structured and restricted to a certain extent but can develop into numerous stable states because of an element of randomness inherent in all ecosystems (see chaos theory for a similar concept; for example, Gleick 1987).

Put in simple terms, if natural and degraded communities follow the deterministic model of assembly, then reassembly of a community after disturbance should occur along predictable lines of development (provided that environmental conditions can be restored to a state similar to that existing before the major disturbance) (Figure 3.2b). Ecologists restoring a site would therefore do best to focus on achieving the required environmental conditions for the desired community and then let natural succession do the rest of the work.

If communities follow the stochastic model, then there is no reason to expect exact reassembly of the community after a major disturbance. Any number of new states could develop, depending on the availability of organisms for invasion, on environmental conditions, and on historical events. Successful restoration would depend mainly on clearly defining a goal-community (whether in structure and function or just in function; see Lockwood and Pimm 1999) and making sure the desired species are introduced in as many different ways as possible, to increase the likelihood that one's goal-community will be reached.

If communities follow the ASS model, then one would expect the recovery of a degraded system to follow one of several possible trajectories, depending on historical events, the availability and order of arrival of organisms, and the element of randomness inherent in all systems (Figure 3.2c). Successful restoration in this case would entail knowing the history of the degradation and how best to restore the functions of the system, before or concomitant with the introduction of target species. Given the element of randomness, which would inevitably be part of the assembly process, species introductions should include a wide spectrum of appropriate species for the given habitat (because the establishment success of organisms depends on many factors,

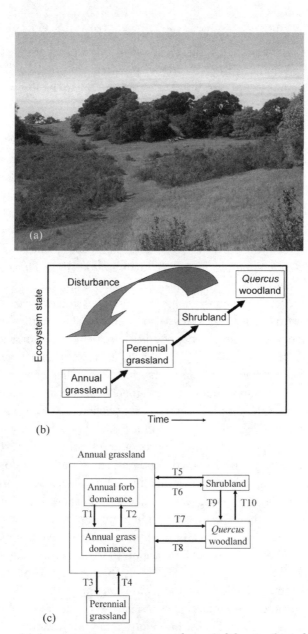

FIGURE 3.2 (a) A vegetation mosaic in northern California. (b) A deterministic model for the dynamics of the mosaic shown in (a). (c) An alternative stable states model for the dynamics of the mosaic shown in (a). The deterministic model assumes a linear development of vegetation from annual grassland
to woodland, with disturbance resetting the system to annual grassland. The
alternative stable states model assumes that the vegetation can occur in a
number of different states, with transitions depending on the occurrence of
disturbance and climatic variation. (See Chapter 5 for more details.)

including weather conditions and safe site availability for plants; see Chapters 13 and 14). Essentially, the ASS model predicts that one may be able to reestablish a group of species, but these species may not be the species one desires to have in such a habitat (Lockwood and Pimm 1999).

Currently, we think that most ecologists would tend to agree that the ASS model of assembly is likely to be the one that reflects reality in nature most closely. For example, Anthony Bradshaw, who has long-standing experience in restoration ecology and is one of the main proponents of using ecological theory in restoration (Bradshaw 1983, 1987), sees the establishment of species on degraded land as the outcome of deterministic processes "complicated by stochastic events" (Bradshaw 1983). All communities fall within a general framework established by the environment and the biota, but the order of arrival of organisms (so-called priority effects; Wilbur and Alford 1985, Drake 1991), as well as an element of randomness, can play an important role in determining which of several stable states is reached. A number of studies have shown that the order of arrival of different species can drastically change the trajectory of development of a community. These studies have used organisms such as spiders (Ehmann and MacMahon 1996), aquatic microorganisms (Drake 1991), acacia ants (Palmer et al. 2002), and nurse plants (see Chapter 14). This idea, of course, is linked to an earlier succession theory, that of Connell and Slatyer (1977), in which early colonists have a positive (facilitative) or negative (inhibitive) effect on later colonists. The tolerance model of Connell and Slatyer does not have its equivalent in the concept of priority effects, however.

Disturbances form an additional important factor in the development of communities, being "ubiquitous, inherent and unavoidable, causing both stability and change" (A. Jentsch, personal communication). Disturbance is increasingly seen not only as a negative phenomenon (in the case of human degradation of ecosystems, devastating hurricanes, volcanic eruptions, and so forth) but also as a positive one that helps to shape community structure and development and maintains biodiversity (Sousa 1984, White and Pickett 1985, Menge and Sutherland 1987, Chesson and Huntly 1997, Bradt et al. 1999, White and Jentsch 2001; see also Part V, "Disturbance and Assembly"). As such, disturbance must be incorporated into restoration projects more than it has been. This is a fast developing field, which will no doubt provide many useful links to practical ecological restoration.

Pattern and Process

Gilpin (1993) asked whether one can deduce the factors determining the composition of ecological communities by simply looking at the spatial distributions of species. Wilson and coworkers consider a pattern found in nature as an assembly rule and assert that environmental factors are not part of the assembly rule (see Wilson 1999), whereas Weiher and Keddy (1999) and Belyea and Lancaster (1999) insist that a rule is not known until the mechanism behind a pattern is known. According to Belyea and Lancaster (1999), finding a nonrandom pattern in nature is potentially a signpost signaling that there could be some rule in the system, but essentially this is a "far cry from defining the rules which generated that pattern." They conclude that assembly rules are "general and mechanistic, and operate within the case-specific constraints imposed by colonization sequence and environment." Drake et al. (1999) see assembly not as a mechanism but as a mechanic. The pieces that make up a system are affected by the action of mechanisms (for example, processes such as facilitation and inhibition or interactions such as competition, mutualism, or predation). These mechanisms, in turn, are regulated by assembly mechanics, which set the possible trajectories a system can go through.

Of course, patterns are found more often than the mechanisms behind them. Can one infer processes from patterns in systems as complex and multifactorial as ecological communities? Does one have good enough data to show unequivocally which factors are driving community assembly? The interplay of various abiotic factors makes it difficult to determine whether alternative stable states are due to species interactions, abiotic factors, or chance alone (or a combination of all three). The following section on methodology addresses this issue further, covering a big debate in assembly rules theory pertaining to the important differences between pattern and process.

Despite the lack of consensus in the debate over the relative merits of finding patterns or processes, restoration practitioners—who are faced with the job of trying to produce a certain type of ecosystem—need guidelines to follow, and if possible, knowledge of mechanisms as well as the patterns involved in community development. The more generally applicable any such guidelines or rules are, the more useful they will be to restoration ecologists working in different ecosystems.

The trick seems to be to find rules that are general enough to be universally applicable but specific enough to not be useless. Belyea (Chapter 7) advocates using systems models as a complement to filter models of assembly (see Part II), since they describe assembly in "terms of higher-level characteristics of community structure and ecosystem function." It is these higher-level characteristics that potentially form the most useful theoretical background to the practice of restoration ecology. Without any guidelines or rules, the process of ecosystem restoration will remain rather vague and similar to the British legal system, in which a precedent case is taken as a guide to the future. There is no written constitution on which to base one's decisions.

Lockwood and Pimm (1999) presented a review of the success of restoration ecology projects and found that although ecosystem function and partial structure could be restored to human-damaged systems relatively easily, restoring species composition and diversity (complete structure) was far more difficult. This finding potentially implies a certain functional redundancy within communities, because a subset of the original species present before degradation is sufficient to maintain ecosystem functions. Species within one functional group (such as nitrogen fixers or secondary consumers) could be interchangeably replaceable without having a major effect on ecosystem functioning. Thus, if a species from a functional group goes extinct, the consequences may not be so serious, since another species in the same group could fulfill the function for the community. However, some species may not be replaceable or have functional equivalents; such species have been called keystone species (see Power et al. 1996), ecological engineers (Jones, Lawton, and Shachak 1994, 1997), ecosystem drivers (Walker 1992, 1995), and nexus species (see Chapter 4). Identification of such irreplaceable species, or at least categories of them, is an important task for basic and restoration ecologists, if success rates in restoration projects are to be improved. Much of the current research in biodiversity and ecosystem functioning also addresses the issue of how replaceable species might be (Chapin et al.1997, Loreau et al. 2002; see also the Jena Biodiversity Project, www. uni-jena.de/biologie/ecology/biodiv). However, care must be applied to the question of species redundancy and replaceability; this concept needs to be considered over time, and system resilience may depend on the abilities of different species to cope with, for instance, different climatic or disturbance events.

The concept of ecological filters forms one of the main approaches in assembly rules theory (Keddy 1992, Kelt et al. 1995, Zobel 1997, Díaz et al. 1998). Out of a total species pool of potential colonists, only those that are adapted to the abiotic and biotic conditions present at a site will be able to establish themselves successfully. Both the abiotic and biotic conditions are seen as "filters" that allow or disallow membership in a community. A process of deletion takes place, analogous to a filtering out of those organisms not adapted to the habitat conditions (see Part II, "Ecological Filters as a Form of Assembly"). This approach focuses on the end product of numerous interactions between a colonist and the ecosystem components (a top-down approach) and treats individual interactions as a black box. Its potential use for restoration ecology lies in its application at the beginning of restoration projects, to ascertain what factors may be limiting membership in a community (see Chapter 6 for a conceptual model potentially applicable to restoration sites and Chapter 7 for critical discussion).

A crucial question concerns the level of abstraction to use to look for rules. Assembly rules research encapsulates the usual conflicts between ecologists, such as whether it is best to study nature via the top-down or bottom-up approach (note: trophic web dynamics are not meant here). In general, research on assembly follows a top-down approach to the natural world (Drake et al. 1999); it looks at whole systems and extracts only the minimum of detail from the multiple interactions within the system. On the other hand, the search for assembly rules can also be approached as bottom-up research. To understand the development of the community, one needs to know a certain amount about the autecology of its organisms (their growth, nutrition, and reproductive requirements) and the interactions among the organisms that make up the system (see Chapter 7). This is especially true if one intends to categorize them into functional groups (Keddy 1992, Hobbs 1997).

As yet, there are few experimental or descriptive data relating ecological assembly to symbioses or mutualisms, although research on such interactions is now coming into its own (Bertness and Callaway 1994, Vosátka and Dodd 1998, Malcov et al. 1999, Malcov et al. 2001). There is now evidence of the importance of mycorrhizal and bacterial ecology at all stages of a community's development (van der Heijden et al. 1998, Buscot et al. 2000; also see Chapter 10). Scientists from socialist countries have historically placed more emphasis on mutualism and symbiosis than their counterparts in capitalist countries (Martin Zobel, personal communication). This historical bias is now diminishing, but it is food for thought for ecologists living in a traditionally capitalist-driven system, where competition can all too easily seem the driving force behind the behavior of living things. Insights from the application of mycorrhizae inocula to degraded sites in restoration projects (for example, Vosátka and Dodd 1998) need to be used to test theories of community assembly (see Chapter 10).

Methodological Issues in the Search for Assembly Rules

There are two main methodological approaches to studying assembly; Drake (1999) calls them the categorical and the topological (or response) (see Figure 3.1). The former involves categories (such as species, functional groups, guilds) and the study of the extant community (a snapshot in time), whereas the latter involves changes in the community and in expression of assembly rules over time. Wilson and coworkers (Wilson and Roxburgh 1994, Wilson et al. 1996, Mucina and Bartha 1999, Wilson and Gitay 1999; working on plants) and Diamond and coworkers (Diamond 1975; Gilpin and Diamond 1982; Brown and Bowers 1984; Fox 1987, 1999; Fox and Brown 1993; working on vertebrates) belong to the "biotic interactions only" and the "categorical" camp (see Figure 3.1). They use the idea of guild proportionality, in which species from similar guilds or functional groups compete more with one another, because they use resources similarly (the ecological concept of "limiting similarity," whereby too much similarity between species hinders their ability to coexist). Hence, one would expect a relatively constant proportion of species from each guild within a community. Guild proportions or other patterns found in communities are compared with appropriate statistical tests to a simulation of random distributions of the variable in question (the null model). This allows one to test whether the found frequency of a pattern could have been generated by chance (Fox 1999) or the pattern is significantly different from a random generation of patterns. This is an interesting approach that is often used in searching for assembly rules, but it is again weakened because it looks only at patterns in communities and tells us little about whether such patterns really arise from competition or whether other processes are more important. In addition, it is based on the assumption that competition for the same resources is higher within, not between, functional groups (Belyea and Lancaster 1999). It is difficult at present to know just how correct this assumption is. Work has also been done on other pairwise interactions (such as plant-herbivore, herbivore-carnivore, mutualists, pathogens) (Lawton 1984; see also Belyea and Lancaster 1999 for examples), asking questions such as whether proportions of trophic levels are relatively constant over succession (see Chapter 9).

Ecologists who focus on the dynamics of communities over time (the "response" and generally also the "biotic interactions only" approach; see Figure 3.1) emphasize the importance of looking at more than the extant (current) community in the search for assembly rules (Drake 1991, Drake et al. 1993, Chesson and Huntly 1997, Lockwood et al. 1997; also see Chapter 6 for an example of the "response" and "abiotic and biotic interactions" approach). One of the main outcomes of a growing body of research on species invasion into ecological communities (Harper 1977; Johnstone 1986; Drake et al.1993; Lockwood et al. 1997, 1999; Tilman 1997; Davis et al. 2000) is the realization that time scale is as important as spatial scale for the outcome of invasion scenarios. Drake et al. (1991, 1993) found patterns in artificial aquatic landscapes to have been generated by the interplay among processes occurring on several levels of spatial and temporal scale. Furthermore, different time intervals between successive invasions changed the assembly trajectory entirely (Drake 1991). Lockwood et al. (1997) modeled the development of simple communities; interestingly, they also found that rapid invasion (in relation to life span of the organisms) allowed the formation of different assemblages over time. This is reminiscent of the carousel model (van der Maarel and Sykes 1993), in which species turnover is high and each species had its chance to form part of the community. With infrequent invasions, on the other hand, invaders were more successful and persisted longer in the community (Lockwood et al. 1997). Drake (1991) made a plea for more consideration of the history of community development, since the forces that constrain an extant community may not be the same ones that constrained its development to that point. He even went so far as to say that the "sequence of invasions determines the set of rules that are possible and likely to operate," even when the species pool remains the same (a reiteration of the priority effect). An example is the invasion of a herbivore, which enables the subsequent invasion of a new plant species; without the herbivore invasion, the community would have developed in a quite different manner. These results highlight the importance of studying assembly at many scales of space and time and over whole communities, since mere pairwise interactions (for example, the categorical approach) form only a part of the dynamics that constrain community membership.

Restoration ecology is centered on the recovery of ecosystems, which inherently focuses on the dynamics of systems over time. It follows, therefore, that the categorical approach to assembly rules has limited value for direct application to restoration projects, whereas the response approach is a much more appropriate one.

The results of assembly experiments using the response approach could have important ramifications for restoration ecology, where the successful invasion (or not) of desired or undesired species is often a key issue. A test of the Drake (1991) or Lockwood et al. (1997) models, using a comparison of several large restoration projects and different timings of species introductions, would enable a direct test of whether such differences in invasion success also apply to (degraded) ecosystems in the real world and not just to simulation models or microcosms. Differences in invasion success related to timing and turnover of invasion would have far-reaching effects on future methods in restoration ecology.

After Diamond coined the first "assembly rule" in 1975, a heated debate over the interpretation of his data ensued. He collected data on the presence or absence of bird species on an archipelago and formulated assembly rules in terms of permissible and forbidden combinations of species. He attributed the checkerboard patterns he found to competition between species and concluded that interspecific competition played an important role in structuring the communities he studied. Similarly, since then, other vertebrate ecologists (Fox 1987, Kelt and Brown 1999; hereafter called the guild-rule camp) have looked at guild proportionality in small mammals and claim to have found some assembly rules. This was disputed by another camp of ecologists (for example, Connor and Simberloff 1979, Simberloff et al. 1999; hereafter called the null-model camp), who claimed that finding a pattern does not allow one to ascribe the pattern directly to a specific mechanism (for example, competition), constraining which organisms can coexist in a community. For example, Connor and Simberloff (1979) countered Diamond's competition assembly rule with the claim that much the same pattern could emerge even if colonization had occurred independently of which species were present on an island or in a habitat. The null-model camp also had issues with the null models used by Gilpin and Diamond (1982, 1984), and Brown and Bowers (1984), stating that they were not true null models, since they were manipulated to contain too much biological reality (such as maintaining the same number of species in the null models as found in the field, which biases the results). The null-model camp claimed that although one might find a pattern in nature, one cannot unequivocally prove what causes this pattern. It could have arisen at random, or it could be the result of biotic (and abiotic) interactions; that is, some form of rule. One could infer certain processes as being dominant, but there is still a lack of data to show clearly that, say, competitive exclusion is behind the structure of a community, even though single experiments can show that competition does occur between organisms in the community. The guild-rule camp countered this argument by stating that the null models proposed by the null-model camp were too far from biological reality to be of any relevance. The null models were entirely random, with no consideration of biological restrictions such as the number of species a habitat can hold. They also pointed out that no counterarguments are provided by the null-model camp as to why patterns of coexistence found by the guild-rule camp seem to obey assembly rules predicted on the basis of competition (for example, see guild-based assembly rules in rodent communities found by Fox 1987, Kelt and Brown 1999).

The lack of alternative mechanisms to competition put forward by the null-model camp to explain patterns in nature could be explained because (1) good enough field data are still lacking to invoke competition unequivocally as the driving force in assembly of certain communities; and (2) the study of other potentially defining factors in community assembly are equally difficult, if not harder, to quantify than interspecific competition; such factors include mutualisms (Bertness and Callaway 1994; Callaway and Walker 1997), regeneration (Grubb 1977, Brand and Parker 1995), effects of herbivory (Buckland and Grime 2000), and safe sites (Jumpponen et al. 1999).

The guild-rule versus null-model controversy consumed a lot of "time, energy, resources and paper" (Fox 1999:27) and essentially reached an impasse in the 1980s. There are certainly lessons to be learned from this controversy, however—not least for restoration ecologists, who can focus on what the impasse in the debate signifies in terms of potentially useful approaches to assembly theory. Essentially, one camp found a pattern for assembly and assigned it to a single mechanism of ecological interaction for a specific ecosystem, ignoring the complexity of possible driving forces behind assembly. This was an attempt to delve into the black box of possible interactions between organisms in a system (see Chapter 20) and extract one defining driving force behind community development. The guild-rule versus null-model controversy ended up in a set of arguments and counterarguments, which seemed to put the potential marriage between assembly rules theory and restoration ecology on hold. As discussed earlier in this chapter, however, certain approaches to assembly theory are more constructive as a guide to restoration ecology than others. The main lesson one could learn from the debate, therefore, would be that it is insufficient to infer and investigate only one possible mechanism behind the assembly of a community. A combination of filter (treating the black box as one entity) and systems approaches (investigating the interactions in the black box) seems likely to be the way ahead.

Assembly Rules

In their book on ecological assembly rules, Weiher and Keddy (1999) call not only for continuing the search for assembly rules but also for now finding some rules, focusing especially on how they vary along gradients and between taxa. Few authors state hypothesized rules as to how a community develops (Grime 1979, Post and Pimm 1983, Tilman 1982, Fox 1987), and even fewer have come up with explicit rules with underlying mechanisms (Wilson and Whittaker 1995, Fox 1987, and see Belyea and Lancaster 1999). And yet when one walks in a forest or near a river, repeatable collections of species are clearly visible in nature. Bradshaw (1983) states that "deterministic, autogenic and facilitation components of succession are stronger than we sometimes consider."

Certain assembly rules are immediately obvious and seem trivial, such as the observation that predators without prey will starve; others go into more interesting aspects of interactions, such as what abundance of prey is necessary for a new predator to enter a community (Morin 1999). Rules found by Fox (1987) or Wilson and coworkers (for example, Wilson and Whittaker 1995, Wilson and Gitay 1999) apply to specific ecosystems and specific organisms. Grime (1979) formulated a set of plant functional groups based on a species's response to the environment (such as stress and disturbance tolerators) or its ability to forage for resources and compete with other species (competitors). Southwood (1988) provided a similar approach based on evolutionary strategies. Evolutionary approaches of this kind do not involve short-term responses to any particular disturbance. Many of the planet's current degraded ecosystems do not have any natural analogs, given the scale and the rapidity of the disturbances involved. It is questionable whether evolutionary strategies and templates would apply in degraded ecosystems in which the extant species have not had the time to evolve a strategy to deal with the new conditions. On the other hand, the natural selection of certain traits, such as metal tolerance in plants, evolves from normal plant populations exposed to metal contamination (Bradshaw 1983). It is rare to find metal-tolerant plant species in natural or seminatural habitats. This suggests that species do have the ability to evolve adaptive strategies even in systems that are suddenly subjected to extreme disturbances without any natural analogs. Such adaptability could become a crucial characteristic in a world affected by global climate change.

How general should assembly rules be? Should they apply to all trophic levels or within trophic levels, or is it enough to find rules for different kinds of habitat? Kelt and Brown (1999) conclude that the generality of the rules they found for North American desert rodents is not always transferable to rodents in other deserts, let alone to other organisms. This kind of assembly rule will not be directly applicable to restoration projects, unless the ecosystem in question is situated in the same habitat as that in which the rule was found to hold true. If the search for assembly rules is only able to elucidate rules of very specific applicability, then this could perpetuate the current problem in restoration ecology; namely, the preponderance of treating each case study as unique without a guiding conceptual framework (see Chapter 2). For this reason, we advocate working on assembly theories that are as universal as possible and that include the dynamics of a system (not just extant categories). The higher-level characteristics of systems are those that are most likely to yield directly useful results for restoration (see Chapter 7) rather than rules pertaining to pairwise interactions of specific organisms in specific habitats. In addition, we make a plea to include the role of the environment in the search for assembly rules, since the living and nonliving aspects of an ecosystem are so inextricably linked.

Conclusions

Assembly rules theory encompasses a wide array of different approaches (see Figure 3.1), including

1. Different models of community assembly: the deterministic, the stochastic, and the alternative stable states (ASS) model. The ASS model seems to be the most promising model for restoration practitioners to use, since its incorporation of both randomness and determinism best reflects ecological reality.
2. Investigating patterns and processes found in ecological communities. Rules of most use to restoration practitioners are those that provide not just a pattern but a mechanism behind the pattern.
3. Filter models and systems models offer different approaches to assembly, but should complement each other well when applied to restoration projects. The filter model is probably best applied at the beginning of a project, in order to ascertain what is constraining membership of certain species from a system.

Methodologies used in the search for assembly rules are either categorical (using species, functional groups, guilds) or response-based (dynamic). Since restoration ecology is concerned with the recovery of a habitat over time, theories derived from the response approach will be most useful as a framework for restoration. For there to be a more fruitful exchange between basic ecologists and restoration practitioners, assembly rules need to be as general as possible and include system dynamics over time as well as environmental factors.

Acknowledgments

Thanks to Lisa Belyea and Julie Lockwood for providing very useful comments on earlier drafts of the manuscript. We would like to thank the *Deutsche Forschungsgemeinschaft* (DFG; German Research Council) for providing generous funding for the Graduate Training Group (Graduiertenkolleg; Project GRK266) on "Functioning and Regeneration of Degraded Ecosystems," within which Vicky Temperton had the luxury, as an integrating postdoc, of spending quality time researching and thinking about ecological assembly and restoration ecology. In addition, DFG funding for invited speakers provided the framework for the close collaboration with Richard Hobbs in Australia.

REFERENCES

Belyea, L. R.; and Lancaster, J. 1999. Assembly rules within a contingent ecology. *Oikos* 86:402–416.

Bertness, M. D.; and Callaway, R. 1994. Positive interactions in communities. *Trends in Ecology and Evolution* 9:191–193.

Booth, B. D.; and Larson, D. W. 1999. Impact of language, history and the choice of system on the study of assembly rules. In *Ecological Assembly: Perspectives, Advances, Retreats*, ed. E. Weiher and P. Keddy. Cambridge: Cambridge University Press.

Bradshaw, A. D. 1983. The reconstruction of ecosystems: presidential address to the British Ecological Society, December 1982. *Journal of Applied Ecology* 20(1): 1–17.

Bradshaw, A. D. 1987. Restoration: an acid test for ecology. In *Restoration Ecology: A Synthetic Approach to Ecological Research*, ed. W. R. I. Jordan, M. E. Gilpin, and J. D. Aber, 23–30. Cambridge: Cambridge University Press.

Bradt, P.; Urban, M.; Goodman, N.; Bissell, S.; and Spiegel, I. 1999. *Stability and resilience in benthic macroinvertebrate assemblages*. Hydrobiologia 403:123–133.

Brand, T.; and Parker, V. T. 1995. Scale and general laws of vegetation dynamics. *Oikos* 73:375–380.

Brown, J. H.; and Bowers, M. A. 1984. Patterns and processes in three guilds of terrestrial vertebrates. In *Ecological Communities*, ed. D. R. Strong, D. Simberloff, L. G. Abele, and A. B. Thistle, 282–297. Princeton: Princeton University Press.

Buckland, S. M.; and Grime, J. P. 2000. The effects of trophic structure and soil fertility on the assembly of plant communities: a microcosm experiment. *Oikos* 91:336–352.

Buscot, F.; Munch, J. C.; Chacosset, J. Y.; Gardes, M.; Nehls, U.; and Hampp, R. 2000. Recent advances in exploring physiology and biodiversity of ectomycorrhizas highlight the functioning of these symbioses in ecosystems. *FEMS Microbiology Reviews* 699:1–14.

Callaway, R. M.; and Walker, L. R. 1997. Competition and facilitation: a synthetic approach to interactions in plant communities. *Ecology* 78:1958–1965.

Chapin, F. S.; Sala, O. E.; and Burke, I. C. 1997. Biotic control over the functioning of ecosystems. *Science* 277:500–504.

Chesson, P.; and Huntly, N. 1997. The roles of harsh and fluctuating conditions in the dynamics of ecological communities. *American Naturalist* 150(5): 520–534.

Clements, F. E. 1916. *Plant Succession*. Publication No. 242. Washington, D.C.: Carnegie Institution.

Clements, F. E. 1936. The nature and structure of the climax. *Journal of Ecology* 22:9–68.

Connell, J. H.; and Slatyer, R. O. 1977. Mechanisms of succession in natural communities and their role in community stability and organization. *American Naturalist* 111:1119–1144.

Connor, E. F.; and Simberloff, D. 1979. The assembly of species communities: chance or competition? *Ecology* 60:1132–1140.

Davis, M. A.; Grime, J. P.; and Thompson, K. 2000. Fluctuating resources in plant communities: a general theory of invasibility. *Journal of Ecology* 88:528–534.

Diamond, J. M. 1975. Assembly of species communities. In *Ecology and Evolution of Communities*, ed. M. L. Cody and J. M. Diamond, 342–444. Cambridge, Mass.: Harvard University Press.

Díaz, S.; Cabido, M.; and Casanoves, F. 1998. Plant functional traits and environmental filters at a regional scale. *Journal of Vegetation Science* 9(1): 113–122.

Díaz, S.; Cabido, M.; and Casanoves, F. 1999. Functional implications of trait-environment linkages in plant communities. In *Ecological Assembly Rules—Perspectives, Advances and Retreats*, ed. E. Weiher and P. Keddy, 338–363. Cambridge: Cambridge University Press.

Drake, J. A. 1991. Community-assembly mechanics and the structure of an experimental species ensemble. *American Naturalist* 137(1): 1–26.

Drake, J. A.; Flum, T. E.; Witteman, G. J.; Voskuil, T.; Hoylman, A. M.; Creson, C.; Kenny, D. A.; Huxel, G. R.; Larue, C. I.; and Duncan J. R. 1993. The construction and assembly of an ecological landscape. *Journal of Animal Ecology* 62:117–130.

Drake, J. A.; Zimmermann, C. R.; Parucker, T.; and Rojo, C. 1999. On the nature of the assembly trajectory. In *Ecological Assembly: Advances, Perspectives, Retreats*, ed. E. Weiher and P. Keddy, 233–250. Cambridge: Cambridge University Press.

Ehmann, W. J.; and MacMahon J. A. 1996. Initial tests for priority effects among spiders that co-occur on sagebrush shrubs. *Journal of Arachnology*, 24:173–185.

Fox, B. J. 1987. Species assembly and the evolution of community structure. *Evolutionary Ecology* 1:201–213.

Fox, B. J. 1999. The genesis and development of guild assembly rules. In *Ecological Assembly: Advances, Perspectives, Retreats*, ed. E. Weiher and P. Keddy, 23–57. Cambridge: Cambridge University Press.

Fox, B. J.; and Brown J. H. 1993. Assembly rules for functional groups of North American desert rodent communities. *Oikos* 67:358–370.

Fraser, L. H.; and Keddy, P. 1997. The role of experimental microcosms in ecological research. *Trends in Ecology and Evolution* 12:478–481.

Gilpin, M. E. 1987. Experimental community assembly: competition, community structure and the order of species introductions. In *Restoration Ecology: A Synthetic Approach to Ecological Research*, ed. W. R. I. Jordan, M. E. Gilpin, and J. D. Aber, 151–161. Cambridge: Cambridge University Press.

Gilpin, M. E.; and Diamond, J. M. 1982. Factors contributing to non-randomness in species co-occurrences on islands. *Oecologia* 52:75–82.

Gilpin, M. E.; and Diamond, J. M. 1984. Are species co-occurrences on island non-random, and are null hypotheses useful in community ecology? In *Ecological Communities: Conceptual Issues and the Evidence*, ed. D. R. Strong, D. Simberloff, L. G. Abele, and A. B. Thistle, 297–315. Princeton: Princeton University Press.

Gleason, H. A. 1917. The structure and development of the plant association. *Bulletin of the Torrey Botany Club* 44:463–481.

Gleason, H. A. 1926. The individualistic concept of the plant association. *Bulletin of the Torrey Botany Club* 53:7–26.

Gleick, J. 1987. *Chaos: Making a New Science*. New York: Penguin.

Grime, P. 1979. *Plant Strategies and Vegetation Processes*. Chichester, England: John Wiley and Sons.

Grubb, P. J. 1977. The maintenance of species-richness in plant communities: the importance of the regeneration niche. *Biological Review* 52:107–145.

Harper, J. L. 1977. *Population Biology of Plants*. London: Academic Press.

Hobbs, R. J. 1997. Can we use plant functional types to describe and predict responses to environmental change? In *Plant Functional Types*, ed. T. M. Smith, H. H. Shugart, and F. I. Woodward, 66–91. Cambridge: Cambridge University Press.

Hobbs, R. J. 1999. Restoration of disturbed ecosystems. In *Ecosystems of Disturbed Ground*, ed. L. R. Walker, 673–687. Amsterdam: Elsevier.

Hobbs, R. J.; and Norton, D. A. 1996. Towards a conceptual framework for restoration ecology. *Restoration Ecology* 4:93–110.

Johnstone, I. M. 1986. Plant invasion windows: a time-based classification of invasion potential. *Biological Review* 61:369–394.

Jones, C. G.; Lawton, J. H.; and Shachak, M. 1994. Organisms as ecosystem engineers. *Oikos* 69:373–386.

Jones, C. G.; Lawton, J. H.; and Shachak, M. 1997. Positive and negative effects of organisms as physical ecosystem engineers. *Ecology* 78:1946–1957.

Jumpponen, A.; Väre, H.; Mattson, K. G.; Ohtonen, R.; and Trappe, J. M. 1999. Characterization of "safe sites" for pioneers in primary succession on recently deglaciated terrain. *Journal of Ecology* 87:98–105.

Keddy, P. 1992. Assembly and response rules: two goals for predictive community ecology. *Journal of Vegetation Science* 3:157–164.

Kelt, D. A.; and Brown, J. 1999. Community structure and assembly rules: confronting conceptual and statistical issues with data on desert rodents. In *Ecological Assembly: Advances, Perspectives, Retreats*, ed. E. Weiher and P. Keddy, 75–108. Cambridge: Cambridge University Press.

Kelt, D. A.; Taper, M. L.; and Meserve, P. L. 1995. Assessing the impact of competition on community assembly: a case study using small mammals. *Ecology* 76(4): 1283–1296.

Lawton, J. H. 1984. Non-competitive populations, non-convergent communities, and vacant niches: the herbivores of bracken. In *Ecological Communities*, ed. D. R. Strong, D. Simberloff, L. G. Abele, and A. B. Thistle, 67–99. Princeton: Princeton University Press.

Lawton, J. H. 1991. Are there assembly rules for successional communities? In *Colonization, Succession and Stability*, ed. A. J. Gray, M. J. Crawley, and P. J. Edwards, 225–244. London: Blackwell Scientific.

Lockwood, J. L.; Moulton, M. P.; and Balent, K. L. 1999. Introduced avifaunas as natural experiments in community assembly. In *Ecological Assembly: Advances, Perspectives, Retreats*, ed. E. Weiher and P. Keddy, 108–125. Cambridge: Cambridge University Press.

Lockwood, J. L.; and Pimm, S. L. 1999. When does restoration succeed? In *Ecological Assembly: Advances, Perspectives, Retreats*, ed. E. Weiher and P. Keddy, 363–392. Cambridge: Cambridge University Press.

Lockwood, J. L.; Powell, R. D.; Nott, M. P.; and Pimm, S. L. 1997. Assembling ecological communities in time and space. *Oikos* 80:549–553.

Loreau, M.; Naeem, S.; and Inchausti, P. 2002. *Biodiversity and Ecosystem Functioning. Synthesis and Perspectives.* Oxford: Oxford University Press.

Malcov, R.; Albrechtov, A.; and Vosátka, M. 2001. The role of extraradical mycelium network of arbuscular mycorrhizal fungi on the establishment and growth of *Calamagrostis epigejos* in industrial waste substrates. *Applied Soil Ecology* 18:129–142.

Malcov, R.; Vosátka, M.; and Albrechtov, J. 1999. Influence of arbuscular mycorrhizal fungus *Glomus mosseae* and simulated acid rain on the coexistence of the grasses *Calamagrostis villosa* and *Deschampsia flexuosa*. *Plant and Soil* 207:45–57.

Menge, B. A.; and Sutherland, J. P. 1987. Community regulation: variation in disturbance, competition, and predation in relation to environmental stress and recruitment. *American Naturalist* 130:730–757.

Morin, P. J. 1999. *Community Ecology.* Malden, Mass.: Blackwell Science.

Mucina, L.; and Bartha, S. 1999. Variance in species richness and guild proportionality in two contrasting dry grassland communities. *Biologia Bratislava* 54:67–75.

Palmer, T. M.; Young, T. P.; and Stanton, M. L. 2002. Burning bridges: priority effects and the persistence of a competitively subordinate acacia-ant in Laikipia, Kenya. *Oecologia* 133:372–379.

Post, W. M.; and Pimm, S. L. 1983. Community assembly and food web stability. *Mathematical Biosciences* 64:169–192.

Power, M. E.; Tilman, D.; Estes, J. A.; Menge, B. A.; Bond, W. J.; Mills, L. S.; Daily, G.; Castilla, J. C.; Lubchenco, J.; and Paine, R. T. 1996. Challenges in the quest for keystones. *Bioscience* 46(8): 609–620.

Simberloff, D.; Stone, L.; and Dayan, T. 1999. Ruling out a community assembly rule: the method of favored states. In *Ecological Assembly: Perspectives, Advances, Retreats*, ed. E. Weiher and P. Keddy, 58–74. Cambridge: Cambridge University Press.

Society for Ecological Restoration Science and Policy Working Group. 2002. *The SER Primer on Ecological Restoration*. www.ser.org/.

Sousa, W. P. 1984. The role of disturbance in natural communities. *Annual Review of Systematics* 15:353–391.

Southwood, T. R. E. 1988. Tactics, strategies and templates. *Oikos* 52:3–18.

Sutherland, J. P. 1974. Multiple stable states in natural communities. *American Naturalist* 108(964): 859–873.

Tilman, D. 1982. *Resource Competition and Community Structure*. Monographs in Population Biology, vol. 17. Princeton: Princeton University Press.

Tilman, D. 1997. Community invasibility, recruitment limitation, and grassland biodiversity. *Ecology* 78(1): 81–92.

Van der Heijden, M. G. A.; Kironomos, J. N.; Ursic, M.; Moutoglis, P.; Streitwolf-Engel, R.; Boller, T.; Wiemken, A.; and Sanders, I. R. 1998. Mycorrhizal fungal diversity determines plant biodiversity, ecosystem variability and productivity. *Nature* 396:69–72.

Van der Maarel, E.; and Sykes, M. T. 1993. Small-scale plant species turnover in a limestone grassland: the carousel model and some comments on the niche concept. *Journal of Vegetation Science* 4:179–188.

Vosátka, M.; and Dodd, J. C. 1998. The role of different arbuscular mycorrhizal fungi in the growth of *Calamagrostis villosa* and *Deschampsia flexuosa*, in experiments with simulated acid rain. *Plant and Soil* 200:251–263.

Walker, B. H. 1992. Biodiversity and ecological redundancy. *Conservation Biology* 6:18–23.

Walker, B. H. 1995. Conserving biological diversity through ecosystem resilience. *Conservation Biology* 9:747–752.

Weiher, E.; and Keddy, P. 1995. The assembly of experimental wetland communities. *Oikos* 73:323–335.

Weiher, E.; and Keddy, P. 1999. *Ecological Assembly Rules: Perspectives, Advances, Retreats*. Cambridge: Cambridge University Press.

White, P. S.; and Jentsch, A. 2001. The search for generality in studies of disturbance and ecosystem dynamics. *Annals of Botany* 62:399–450.

White, P. S.; and Pickett, S. T. A. 1985. Natural disturbance and patch dynamics: an introduction. In *The Ecology of Natural Disturbance and Patch Dynamics*, ed. S. T. A. Pickett and P. S. White, 3–13. New York: Academic Press.

Wilbur, H.; and Alford, R. A. 1985. Priority effects in experimental pond communities: responses of *Hyla* to *Bufo* and *Rana*. *Ecology* 73:1106–1114.

Wilson, J. B. 1999. Assembly rules in plant communities. In *Ecological Assembly: Advances, Perspectives, Retreats*, ed. E. Weiher and P. Keddy, 130–164. Cambridge: Cambridge University Press.

Wilson, J. B.; and Gitay, H. 1999. Alternative classifications in the intrinsic guild struc-
    ture of a New Zealand tussock grassland. *Oikos* 86:566–572.
Wilson, J. B.; and Roxburgh, S. H. 1994. A demonstration of guild-based assembly rules
    for a plant community, and determination of intrinsic guilds. *Oikos* 69:267–276.
Wilson, J. B.; Wells, T. C. E.; Trueman, I. C.; Jones, G.; Atkinson, M. D.; Crawley, M.
    J.; Dodd, M. E.; and Silvertown, J. 1996. Are there assembly rules for plant species
    abundance? An investigation in relation to soil resource and successional trends. *Jour-
    nal of Ecology* 84:527–538.
Wilson, J. B.; and Whittaker, R. J. 1995. Assembly rules demonstrated in a saltmarsh
    community. *Journal of Ecology* 83:801–807.
Young,, T. P.; Chase, J. M.; and Huddleston, T. 2001. Community succession and assem-
    bly: comparing, contrasting and combining paradigms in the context of ecological
    restoration. *Ecological Restoration* 9:5–18.
Zobel, M. 1992. Plant species coexistence: the role of historic, evolutionary and ecolog-
    ical factors. *Oikos* 65:314–320.
Zobel, M. 1997. The relative role of species pools in determining plant species richness:
    an alternative explanation of species coexistence. *Trends in Ecology and Evolution*
    12(7): 266–269.

Chapter 4

Assembly Models and the Practice of Restoration

JULIE L. LOCKWOOD AND COREY L. SAMUELS

The central theme of this volume is how better to incorporate assembly rules into restoration ecology. This is a long overdue effort. It is hampered, however, by the plurality with which both assembly and restoration are considered. There are several ways to think about assembly rules, including whether they include biotic or abiotic limiting factors and how these factors are themselves affected over time and across space (see Chapter 3). Restoration ecologists are faced with a bewildering array of ecological starting conditions, including everything from reclaiming a spoil heap to reintroducing a disturbance regime into a mostly intact ecosystem (Cairns 1989). We discuss computer algorithms and controlled experiments that seek to understand how biotic interactions shape communities over time. This approach is distinct from several other conceptual models of the assembly process in that these models do not include abiotic factors and are carried out under the most controlled of conditions (that is, computers or microcosms). Clearly, these are not the same conditions that restoration practitioners experience. However, the simplicity of the method allows us to see complex temporal dynamics that are extremely difficult to capture in more realistic settings. We will first review the methods and primary results from assembly models and then relate these results to the practice of restoration. We will end by identifying ways to push these models toward greater realism, especially in terms of their contribution to understanding restoration.

Assembly Models

What do we mean by assembly models? Assembly models are either computer algorithms or microcosm experiments that typically attempt to document important temporal events that affect community metrics such as species richness, stability (often measured as invasion resistance), and compositional similarity. We consider computer algorithms and microcosm experiments as models (being abstractions) of the assembly process, and will use that term for both approaches throughout the remainder of this chapter.

Historically, most assembly computer algorithms have used Generalized Lotka-Volterra (GLV) equations to represent the interactions among species that coexist in some species pool of a fixed size (that is, number of species). The effect of these species on one another was partially preset by the investigator (for example, determining the number of predators and prey to be included) and was also determined through random assignment of interaction coefficients (that is, a_{ij} values of the community matrix) drawn from a uniform random distribution. Species were drawn randomly from this pool and placed in the assembling community. Time passes by iterating the community matrix for some predetermined number of time steps. In the early models (for example, Drake 1990, Case 1990), several primary consumers were initially placed in the community. Then, one at a time, other species from the pool were randomly picked and seeded into the community at some arbitrarily low population density. Whether or not these species stayed in the community, and thus did not become extinct, was determined solely by their interactions with existing community members. Because these interactions were prescribed by GLV equations, they included indirect competition and predation only.

Because the generation of these model communities is quite easy, researchers typically produced several of them at once, thus allowing comparison of their species compositions over time. Later incarnations of these computer algorithms relaxed several of these protocols, including allowing more than one species to attempt to colonize (or invade) the community at a time, seeding species into the community at higher initial densities, and assessing the successful entry of a species into the community using methods other than determining the resultant communities' mathematical stability (Hewitt and Huxel 2002; Lockwood et al. 1997; Law and Morton 1993, 1996).

Microcosm experiments were developed alongside computer algorithms and became a complementary, and somewhat more realistic, way of verifying the patterns that emerged from computer simulations. Almost all of these microcosm experiments used aquatic systems. The aquatic microcosms were often nothing more than simple food webs kept in fish tanks or flasks (Drake et al. 1996). However, because it is possible to maintain several such microcosms in the lab and under controlled conditions, researchers could add dif-

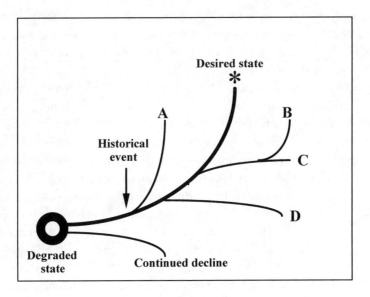

Time

FIGURE 4.1 The sequence in which species attempt to colonize an assembling community produces alternative pathways. Thus, development is a nonlinear process containing several threshold events. This suggests that restoration will require incorporation of historical events in order to drive the community along the pathway that leads to the desired state or, possibly, toward higher levels of complexity.

ferent sequences of species over time and readily observe the result of their interactions across multiple communities. Interactions were again limited mostly to competition and predation, and species were drawn from a large, unchanging species pool.

Perhaps the most interesting result of these assembly models is evidence that there can be alternative community trajectories (that is, the sequence of species additions and subtractions) and thus alternative stable endpoints (for example, Robinson and Edgemon 1988; Wilson 1992; Drake et al. 1993, 1997; Law and Morton 1993, 1996). These stable endpoints are defined as being resistant to invasion by all the species that remain in the species pool. In some assembly models, it was possible to identify the specific event during assembly that triggered a switch to a different developmental pathway (Figure 4.1). Two communities assembling in tandem (A and Desired) from a common species pool may be the same at the beginning, but a difference in one historical event (arrow) can produce two very different community compositions. Thus, assembly models indicate that the act of building communities from scratch is a highly contingent process whereby the successful establishment of one species depends on a series of events that happened well before it attempted to colonize. This view of community assembly emphasizes the role of history in determining community patterns and dynamics. The species present in the past, even if they are no longer extant, and the sequence in which they arrived, made the community what it is today. This history matters in at least three ways, which we describe below.

First, the order in which species attempt colonization tremendously influences final community composition and richness (Robinson and Edgemon 1988, Drake 1990). A variety of mechanisms, all operating at the population level, determine whether or not the species attempting colonization will establish. For example, competitive exclusion between two or more species may determine which one(s) successfully establish and at what time during the assembly process. The outcome of this interaction is itself highly influenced by order of arrival, termed a priority effect by Wilbur and Alford (1985). In this way, small differences in the timing of colonization sets the stage for the future final community composition. These stochastic differences in colonization accumulate over time, driving the community toward one of many possible alternative stable states. A relevant corollary to this result is that it is not possible to reassemble a particular community composition using only the species present in that final steady-state community (termed the "Humpty-dumpty effect" by Pimm 1991).

Second, one species can play an inordinately important role in determining which of the alternative stable states the community may eventually reach. Ecologists have long understood this concept, but usually only in terms of the effects of keystone species (Paine 1966). Keystone species are represented in the steady-state community and play a significant role in maintaining a particular species composition. In assembly models, species that play this keystone role come and go during different stages of community development. In this case, they are termed "nexus" species (Drake et al. 1997). Nexus species determine final community composition through their effect on other contemporary species in the community, but at some point in the assembly process, they themselves go extinct. Thus, nexus species are a special and pronounced case of the effect of colonization sequence described above. Unlike keystone species, they may not be readily identifiable. Keystone species are defined by their unexpectedly large influence given their relatively limited abundance or distribution (Mills et al. 1993). Nexus species may show no unusually strong influence on other concurrent species. Yet these nexus species are the principal driving forces that determine which of the many alternative developmental pathways the community will enter.

Finally, how often species attempt establishment, and at what densities they are seeded into the community, can have significant effects on community dynamics (Hewitt and Huxel 2002, Lockwood et al. 1997). In nature, communities may not be built one species at a time, and when species attempt colonization, they may not always arrive as small propagule populations. The early incarnations of assembly algorithms and experiments incorporated both of these assumptions (for example, Drake 1990). However, achieving a stable end community composition, in which none of the remaining species in the pool can colonize the community, is apparently a special outcome of these protocols.

When the models make species colonization attempts more common, community composition changes continuously, never reaching a stable composition (Lockwood et al. 1997, Hewitt and Huxel 2002). Chance still determines colonization order, although the effect of this sequence of events on community composition is less pronounced. A greater subset of the species in the pool eventually colonizes, but many are ephemeral. Thus, these model communities are generally richer in species than those created by more orderly introductions (Lockwood et al. 1997). Similarly, when species are allowed to attempt colonization at higher densities than in the original algorithm protocols, invasion resistance is never achieved. These models thus mimic results obtained by the random process model (see Chapter 3), but there are still some species from the pool that are always excluded. Which species those are, and how many are excluded, seems to depend on the number of species that try to establish simultaneously, their initial invasion density, and the composition of the species pool (Hewitt and Huxel 2002).

Relevance of Assembly Model Results to Restoration Practice

Community assembly models indicate that it is impossible to see all the pertinent events of development by observing reference or predisturbance ecosystems. Assembly is a nonlinear process, and quirks of history force a community along a variety of alternative developmental pathways. Thus, what we observe as *natural* depends on the ongoing history of invasion and extinction. Restoring nature, therefore, requires recreating elements of this history. Most of the communities we seek to restore have developed over millennia (Cairns 1989). We cannot, and will not, know the detailed history of any such system, nor can we afford millennia to recreate it. Yet somehow our prospects for successful restoration must be less bleak and patchy than is our knowledge of the history of a community's development.

The assembly principles that make species-for-species restoration appear improbable may serve as a bare-bones guide to increasing success at attaining desired species, higher species richness, or more ecosystem functions (Palmer et al. 1997). For example, assembly models tell us that restoration must somehow involve more species than are found in the predisturbance or reference community. Also, there are identifiable events, such as the presence of nexus species, that play a strong role in determining community composition. If these few events can be built into restoration protocol, the group of species and functions restored may be more precisely controlled. Using these approaches, it may be possible to guide restoration along the one pathway that will lead toward the desired state or along the few pathways (A, B, or C) that will lead to substantial increases in complexity (see Figure 4.1).

Although some restoration ecologists may have attempted to incorporate similar steps into their protocols, it is unlikely that they considered their results in the context of assembly models (Pritchett 1997). Some, however, have used what is called a succession-based approach, and most recent succession models suggest a similar approach (Packard 1994). Next we review some of these efforts, as we believe that these approaches lay the groundwork that will help us see where assembly-based steps will come into play.

Passive Restoration

The most critical first step in restoration is to halt degradation activities that prevent recovery (Hobbs and Norton 1996, Kauffman et al. 1997). Depending on the system, this may be sufficient to restore desired composition and function (Kauffman et al. 1997). Understanding the system-specific steps in community recovery can inform management decisions. For example, many western riparian zones may be capable of rapid recovery after the cessation of grazing because the biota has evolved to survive frequent natural disturbance events (Kauffman et al. 1997). A minimal amount of monitoring and observation of the natural recovery process can tell us whether passive restoration might work. Beyond saving a phenomenal amount of time and money, this approach can inform management decisions if active restoration methods are required (Kauffman et al. 1997, Yates and Hobbs 1997).

Monitoring unaided recovery also informs an assembly-based approach to active restoration. Identifying obstacles that hinder natural recovery leads to insight into when and where transitions onto alternative pathways might have taken place (Hobbs and Norton 1996, Yates and Hobbs 1997). If active restoration is to proceed, incorporating these insights will help determine which steps in community development are most crucial (Yates and Hobbs 1997).

Interseeding

Packard (1994, 1997) developed a protocol for prairie restoration based on a very simple insight derived from observations of natural community development: "The seeds of prairies most often take root not on bare soil but among other established plants." This interseeding protocol involves sowing selected prairie species in with existing turf. The turf may be a prairie remnant or composed of exotic grasses. In his ongoing restorations, Packard essentially allows nature to initiate succession and then he takes the next logical step in the process: adding missing species. This simple insight improves the chances of incorporating rare species into the restored prairie. In turn, the prairies produced by an interseeding protocol dramatically improve a site's quality and potential to contribute to biodiversity conservation as compared to restoration protocols that seed all plants into the site at once (Packard 1997).

This example highlights how an understanding of the implementation of assembly steps can contribute to management decisions if active restoration is required. Midwestern grassland restoration is often limited by failure to recruit prairie species. Packard's restoration protocol illustrates the second fundamental jump in incorporating assembly principles: recognizing where the degraded community is along a continuum of possible states (Hobbs and Norton 1996).

Packard used traditional succession theory to reach his conclusions. Because assembly experiments suggest that community development may not be as linear and deterministic as classic succession theory dictates, we must use caution in determining where a community stands among the plethora of alternative assembly pathways. State-transition protocols first developed for range management (Westoby et al. 1989) but since converted into a restoration framework (Hobbs and Norton 1996, Yates and Hobbs 1997) may provide much-needed help. These models allow for alternative stable states and nonlinear, or threshold, dynamics, and they provide a framework for placing the degraded system among the possible alternative pathways. For example, Yates and Hobbs (1997) formulated a state-transition model of *Eucalyptus* woodlands in Australia. They used expert knowledge of the surrounding landscape to determine the range of states possible and to identify grazing (disturbance) thresholds. Within this framework, hypotheses about the ease and effectiveness of particular restoration protocols could be generated and tested.

Reintroduction of Nexus Species

Reintroductions of species must be well planned in accordance with community development processes. For example, reintroducing beaver into riparian zones when suitable woody species have reestablished can dramatically accelerate community development (Naiman et al. 1988, Kauffman et al. 1997). The beavers influence biotic structure by improving salmonid habitat and species composition as well as by improving abiotic characteristics such as hydrology and wetland extent (Naiman et al. 1988; Kauffman et al. 1997). If this same reintroduction of beaver is done too early, however, it can do more harm than good by preventing woody species from becoming sufficiently established to be resilient to beaver herbivory (Kauffman et al. 1997).

That the timing of introduction of a keystone species is vital is another relatively simple insight gained from common knowledge of riparian dynamics and the habitat requirements of beavers. It is also something we can deduce from an examination of a reference watershed. Nexus species may have similar effects on composition by precipitating beneficial, neutral, or catastrophic long-term results. Nexus species, by definition, will not be observed in reference sites or before degradation begins. Thus, identifying these species or knowing when they will do more harm than good is not as easy as it is in the case of beaver reintroductions. Sources of information are likely to be found by a close examination of similar habitats in the surrounding landscape. Early-colonizing ruderal species may once have been important components of the development process (Packard 1988), and in this regard considerable insight may be gleaned from studies of primary successions (Ash et al. 1994; also see Chapter 14). In sum, accounting for nexus species means that an assembly-based protocol will include species that we fully expect to phase out over time. These species may provide the key to selecting the developmental pathway with the most desired outcome (Hobbs and Norton 1996).

Timing Isn't Everything

From a close look at assembly models, we can produce a fairly unique set of suggestions for the restoration practitioner. These suggestions are not comprehensive, however, and are derived from the least realistic of assembly rule investigations: computer models and aquatic microcosms. Notably, most of the published assembly models we reviewed in the first section are aging. Surely we can push these models beyond their original purposes and beyond results that simply relax original assumptions. The failure to follow up has left the role of assembly models in the revival of assembly rule investigations undefined, especially as they relate to restoration ecology. Therefore, we will end by suggesting four ways for assembly models to proceed in ways that continue to contribute to our understanding of community ecology and restoration.

Incorporation of Abiotic Conditions

There is no need to exclude abiotic conditions from these models. Although several investigations into assembly rules explicitly consider abiotic elements (usually as filters; see, for example, Chapter 6), it remains unclear how sequence effects and alternative endpoints stand up to variation within the abiotic environment. These constraints could severely limit the number of possible alternative stable states by homogenizing assembly pathways (Weiher and Keddy 1995a, 1995b; Lord et al. 2000). On the other hand, several succession studies indicate that the interplay between abiotic and biotic factors significantly shapes the outcome of species interactions (Tilman 1982, 1985; Connell and Slatyer 1977; McCook 1994; McIntosh 1981; Berlow 1997). Thus, it is possible that small differences in the outcome of interactions that are driven by abiotic factors could further magnify the importance of historical events.

The addition of abiotic factors to assembly models could occur in one of two ways. The first, and simplest, is to fix the sequence of colonization attempts and allow one environmental variable to change. The environmental constraint could be varied either continuously or categorically, and its effect on community composition could be observed easily by documenting differences in colonization success between assembly "runs." Such an approach is similar to the one taken by Keddy et al. (1994), in which they contrasted the competitive rankings of wetland plants across two levels of environmental conditions (that is, mesic, infertile versus flooded, fertile). The principal addition to such protocols that we recommend is to set these communities into temporal motion, adding and subtracting species in a prescribed fashion across all environmental levels. This approach could be mimicked within computer-generated communities by using niche-based competitive algorithms in which species are responding to levels of one resource gradient (for example, Tilman and Karieva 1997).

The second, and more complex, approach is to vary abiotic conditions through space while simultaneously varying which species attempt to invade over time. Weiher and Keddy (1995a, 1995b; 1998) have performed roughly equivalent experiments using wetland plants sowed into experimental ponds that differ in depth and thus in flooding regime. We suggest that instead of introducing all species into the pool only once at the beginning of the experiment (as did Weiher and Keddy), the species should be added selectively over time and their sequence of introduction altered between experimental units. This approach introduces a higher level of complexity and requires the investigator to keep close track of community composition over time. Initially, the task may be handled most appropriately by using microcosms instead of the more natural ponds of Weiher and Keddy (1995a, 1995b). Given the backdrop of results obtained from simpler designs, however, we should be able to narrow the number of effects we hope to document and place these results into a broader context.

Incorporation of Starting Condition

All restorations do not start from the same point (see Part V, "Disturbance and Assembly"). Restoration is typically not construction in the sense that we are starting from scratch and building a community; and in that sense, assembly models are missing their mark. In real restorations, the preexisting community has suffered some degree of disturbance that has set back the assembly process, and it is the goal of restoration to reassemble the pieces. This implies that we need to know how communities *dis*assemble, and then how they *re*assemble from each unique starting point. To some extent, existing studies of assembly incorporate this perspective in their investigations of how communities respond after natural disturbance (for example, see Chapter 17). However, Beylea and Lancaster (1999) correctly note that communities may disassemble in ways that existing assembly rules fail to capture.

Lord et al. (2000) and Wilson et al. (2000) have considered the question of reassembly (from scratch) by examining the ease with which British plant communities have been reassembled in New Zealand. Their results indicate that reassembly is possible if environmental conditions in the donor (or predisturbance) and recipient (or restored) community are well matched. Beyond these empirical investigations of single events, however, to our knowledge no community assembly models have explicitly disassembled a community in an effort to see how variation in the starting point affects the ease with which the original community may be reassembled. How the disassembly and reassembly process affects the number of possible alternative stable endpoints must be considered. Because GLV equations are deterministic, computer algorithms that explore this process must incorporate a stochastic approach to the determination of community membership at each time step (for example, Donalson and Nisbet 1999). However, microcosm experiments seem ideally suited to such investigations because it is relatively easy to remove species from such systems and then reintroduce them in systematic ways across the various reassembly runs.

Incorporation of Spatial Scale

Most assembly models have a static species pool from which potential colonizers are drawn. All members of this pool have equal access to the community in the sense that they all attempt to invade an approximately equal number of times and are seeded into the community at the same initial population densities. Thus, it is as if the assembling community sat adjacent to the original species pool and there were no limitations on recruitment of each species in this pool. This approach does not reflect the realities of restoration. The spatial disconnect between the appropriate species pool and the community to be restored varies tremendously in actual restorations. Among restoration ecologists, this variation has spawned a growing literature on how to overcome recruitment limitation by varying landscape components (MacMahon and Holl 2001). Models of assembly are ideal for understanding the effects of spatial issues on assembly outcomes for the same reason they are useful for understanding the role of temporal history; the underlying methods are drastic simplifications of community interactions that allow more complex patterns to be observed clearly (Donalson and Nisbet 1999).

At its simplest, the incorporation of space could begin by following the development of island biogeography theory. In this context, past assembly models represent the mainland-island scenario, in which the mainland serves as the species pool for the island and all mainland species have equal access to the island (Hewitt and Huxel 2002; also see Figure 4.2). It would be simple enough to extend these models to reflect a stepping-stone scenario whereby the probability of invasion is determined by the spatial location of each island relative to all others (Figure 4.2). The island nearest the mainland receives potential colonists from the entire mainland pool, and each species has an equal probability of attempting colonization. The next island in the chain receives its colonists only from the first island, and so on, until the last island receives the fewest species from the mainland pool. There are many more scenarios that one could imagine, and although it would be

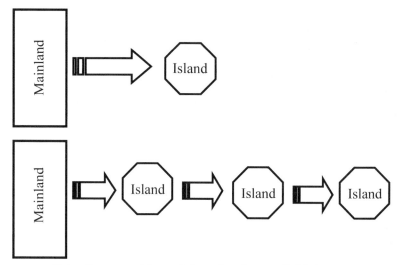

FIGURE 4.2 Extending assembly models so that they explicitly incorporate space can follow the development of island biogeography theory. In the top panel, assembly is modeled as a mainland-island system. This representation characterizes most existing assembly models. The bottom panel represents a simple extension of this pattern whereby the assembling communities are viewed as stepping-stone islands.

interesting to document the effects of each, their benefits to restoration follow from how closely the modeled scenarios match the actual conditions under which most restorations proceed (Hewitt and Huxel 2002).

Incorporation of Indices of Compositional Variability

If community assembly is constrained by a set of rules, we should be able to set compositional bounds, or indices of variability, on the result of any assembly process. Depending on how one views the assembly process, the outcome can be deterministic, random, or somewhere in between (that is, having many alternative stable states). If assembly is deterministic, the rules that govern community development must be quite rigid and unresponsive to seemingly stochastic events. In this case, the concept of compositional bounds is meaningless; no matter how the assembly proceeds, the end result is always the same and predictable. However, this view of community assembly is almost certainly too simplistic (Young et al. 2001). Instead, the process of community assembly responds to several factors that will push the final species composition in one of many directions (see Chapter 11). If we were to use our understanding of the assembly process as a guide to restoration practice, it would be very helpful to describe how much variability in species composition is reduced by each of the rules we identify.

There are two relatively novel ideas embedded in this recommendation. First, instead of viewing the outcome of assembly as a categorical variable (that is, deterministic, random, in-between), we should view it is a continuous variable. The number of alternative states that a community may achieve is thus determined by the strength of the rule(s) that guide the assembly process, such that the stronger the rule (or as the number of rules incorporated goes up), the fewer resulting states are possible. Second, and following from this, the achievement of any one community composition becomes a probabilistic event, hopefully one that can be measured.

This approach is typified in recent work by Samuels, who began to explore a variation of Markovian models, called Markov set-chains, that allow for ranges of outcomes of assembly (Samuels and Lockwood 2002). In this approach, Markov transition probabilities are specified as ranges instead of single values (for example, species A replaces B between 40 and 60 percent of the time instead of species A replacing B with a probability of .45). The model then predicts the final community composition, also expressed as a range. For example, this method allows us to make statements such as "After 10 years, between 45 and 60 percent of the community will be made up of species A." Unlike stochastic approaches to modeling community dynamics (see Chapter 11), Samuels's algorithms are optimization routines that explore community composition by following the extreme outcomes of species interactions at each time step (Samuels and Lockwood 2002). This approach (as well as a stochastic one if couched properly) provides a new type of information, especially in terms of restoration. By predicting a range of possible assembly results, it provides insight into how one could narrow the window of possible outcomes by adding assembly-based rules to the restoration protocol. If fully developed, this suggestion could be directly useful for restoration practitioners by allowing them to judge how important it is to control their initial planting scheme or to incorporate one species versus another.

Conclusions

Assembly models were not developed with restoration in mind. They provided unexpected insights into the role of historical contingencies on community composition and directly challenged the prevailing view of community development as a deterministic process (Drake 1990). Because early assembly model investigations coincided with the rise of restoration ecology, however, it did not take long for the obvious parallels between the two to be drawn (Pimm 1991, Lockwood and Pimm 1999). The results of assembly models changed the way most academic restoration ecologists perceived their endeavors (for example, Hobbs and Norton 1996, Palmer et al. 1997, Young et al. 2001; also see Chapter 2). Here, we reviewed these models in the context of how they would inform basic restoration protocols. It is difficult to achieve a full translation between abstract models and hard reality, and there is considerably more to do in this regard. However, in the context of our review, and especially in the context of the remaining chapters in this volume, it becomes obvious that assembly models have not attained their full potential in terms of what they can provide to restoration ecology. Their principal benefit lies in their ability to explore the full array of possible outcomes of community assembly quickly and under controlled conditions. Although the temporal dynamics of species interactions are clearly an appropriate avenue for such exploration, so too are several other lines of inquiry. We therefore end with a call for such research.

Acknowledgments

We thank Tim Nuttle, Gary Huxel, Truman Young, Vicky Temperton, Anke Jentsch, Daniel Simberloff, Jeffrey Duncan, Katherine Fenn, and Stuart Pimm for development of ideas and for comments on the various stages of this manuscript.

REFERENCES

Ash, H. J.; Gemmell, R. P.; and Bradshaw, A. D. 1994. The introduction of native plant species on industrial waste heaps: a test of immigration and other factors affecting primary succession. *Journal of Applied Ecology* 31:74–84.
Berlow, E. L. 1997. From canalization to contingency: historical effects in a successional rocky intertidal community. *Ecological Monographs* 67:435–460.
Cairns, J., Jr. 1989. Restoring damaged ecosystems: is predisturbance condition a viable option? *The Environmental Professional* 11:152–159.
Case, T. J. 1990. Invasion resistance arises in strongly interacting species-rich model competition communities. *Proceedings of the National Academy of Sciences* 87:9610–9614.
Connell, J. H.; and Slatyer, R. O. 1977. Mechanisms of succession in natural communities and their role in community stability and organization. *American Naturalist* 111:1119–1144.
Donalson, D. D.; and Nisbet, R. M. 1999. Population dynamics and spatial scale: effects of system size on population persistence. *Ecology* 80:2492–2507.
Drake, J. A. 1990. The mechanics of community assembly and succession. *Journal of Theoretical Biology* 147:213–234.

Drake, J. A.; Flum, T. E.; Witteman, G. J.; Voskuil, T.; Hoylman, A. M.; Creson, C; Kenny, D. A.; Huxel, G. R.; LaRue, C. S.; and Duncan, J. R. 1993. The construction and assembly of an ecological landscape. *Journal of Animal Ecology* 62:117–130.

Drake, J. A.; Hewitt, C. L.; and Huxel, G. R. 1996. Microcosms as models for generating and testing community theory. *Ecology* 77(3): 670–677.

Drake, J. A.; Hewitt, C. L.; Huxel, G. R.; and Kolasa, J. 1997. Diversity at higher levels of organization. In *Biodiversity: A Biology of Numbers and Rarity*, ed. K. Gaston. Boca Raton, Fla.: Blackwell Scientific.

Hewitt, C. L.; and Huxel, G. R. 2002. Invasion resistance and community resistance in single and multiple invasion models: do the models support the conclusions? *Biological Invasions* 4:263–272.

Hobbs, R. J; and Norton, D. A. 1996. Toward a conceptual framework for restoration ecology. *Restoration Ecology* 4(2): 93–110.

Kauffman, J. B.; Beschta, R. L.; Otting N.; and Lytjen, D. 1997. An ecological perspective of riparian and stream restoration in the western United States. *Fisheries* 22(5): 12–24.

Keddy, P. A.; Twolanstrutt, L.; and Wisheu, I. C. 1994. Competitive effect and response rankings in 20 wetland plants: are they consistent across 3 environments. *Journal of Ecology* 82:635–643.

Law, R.; and Morton, R. D. 1993. Alternative permanent states of ecological communities. *Ecology* 74:1347–1361.

Law, R.; and Morton, R. D. 1996. Permanence and the assembly of ecological communities. *Ecology* 77:762–775.

Lockwood, J. L.; and Pimm, S. L. 1999. When does restoration succeed? In *Assembly Rules: Perspectives, Advances and Retreats*, ed. E. Weiher and P. Keddy. Cambridge: Cambridge University Press.

Lockwood, J. L.; Powell, R.; Nott, M. P.; and Pimm, S. L. 1997. Assembling ecological communities in time and space. *Oikos* 80:549–553.

Lord, J. M.; Wilson, J. B.; Steel, J. B.; and Anderson, B. J. 2000. Community reassembly: a test using limestone grassland in New Zealand. *Ecology Letters* 3:213–218.

MacMahon, J. A.; and Holl, K. D. 2001. Ecological restoration: a key to conservation. In *Conservation Biology: Research Priorities for the Next Decade*, ed. M. E. Soule and G. H. Orians, 245–270. Washington, D.C.: Island Press.

McCook, L. J. 1994. Understanding ecological community succession: casual models and theories, a review. *Vegetatio* 110:115–147.

McIntosh, R. P. 1981. Succession and ecological theory. In *Forest Succession*, ed. D. C. West, H. H. Shugart, and D. B. Botkin. New York: Springer-Verlag.

Mills, L. S.; Soule, M. E.; and Doak, D. F. 1993. The keystone-species concept in ecology and conservation. *BioScience* 43(4): 219.

Naiman, R. J.; Williams, J. D.; and Williams, J. E. 1988. Alteration of North American streams by beaver. *BioScience* 38:753–762.

Packard, S. 1988. Just a few oddball species: restoration and the rediscovery of the tallgrass savanna. *Restoration and Management Notes* 6:13–22.

Packard, S. 1994. Successional restoration: thinking like a prairie. *Restoration and Management Notes* 12:32–39.

Packard, S. 1997. Interseeding. In *The Tallgrass Restoration Handbook: For Prairies, Savannas and Woodlands*, ed. S. Packard and C. F. Mutell, 163–192. Boca Raton, Fla.: CRC Press.

Paine, R. T. 1966. Food web complexity and species diversity. *American Naturalist* 100:65–75.

Palmer, M. A.; Ambrose, R. F.; and Poff, N. L. 1997. Ecological theory and community restoration. *Restoration Ecology* 5(4): 291–300.

Pimm, S. L. 1991. *The Balance of Nature?* Chicago: Chicago University Press.

Pritchett, D. A. 1997. Maybe, sometimes: a practitioner replies. *Restoration and Management Notes* 15(1): 51.

Robinson, J. V.; and Edgemon, M. A. 1988. An experimental evaluation of the effect of invasion history on community structure. *Ecology* 69:1410–1417.

Samuels, C. L.; and Lockwood, J. L. 2002. Weeding out surprises: incorporating uncertainty into restoration models. *Ecological Restoration* 20(4): 262–269.

Tilman, D. 1982. Resource competition and community structure. Princeton: Princeton University Press.

Tilman, D. 1985. The resource ratio hypothesis of succession. *American Naturalist* 125:827–852.

Tilman, D.; and Karieva, P. M. 1997. Spatial ecology: the role of space in population dynamics and interspecific interactions. Princeton: Princeton University Press.

Weiher, E.; Clarke, G. D. P; and Keddy, P. A. 1998. Community assembly rules, morphological dispersion, and the coexistence of plant species. *Oikos* 81:309–322.

Weiher, E.; and Keddy, P. A. 1995a. The assembly of experimental wetland plant communities. *Oikos* 73:323–335.

Weiher, E.; and Keddy, P. A. 1995b. Assembly rules, null models and trait dispersion: new questions from old patterns. *Oikos* 74:159–163.

Westoby, M.; Walker, B.; and Noy-Meir, I. 1989. Opportunistic management for rangelands not at equilibrium. *Journal of Range Management* 42:266–274.

Wilbur, H.; and Alford, R. A. 1985. Priority effects in experimental pond communities: responses of *Hyla* to *Bufo* and *Rana*. *Ecology* 73:1106–1114.

Wilson, D. S. 1992. Complex interaction in metacommunities, with implications for biodiversity and higher levels of selection. *Ecology* 73:1984–2000.

Wilson, J. B.; Steel, J. B.; Dodd, M. E.; Anderson, B. J.; Ullmann, I.; and Bannister, P. 2000. A test for community reassembly using the exotic communities of New Zealand roadsides in comparison to British roadsides. *Journal of Ecology* 88:757–764.

Yates, C. J.; and Hobbs, R. J. 1997. Woodland restoration in the western Australian wheatbelt: a conceptual framework using a state and transitions model. *Restoration Ecology* 5(1): 28–35.

Young, T. P.; Chase, J. M.; and Huddleston, R. T. 2001. Community succession and assembly. *Ecological Restoration* 19(1): 5–18.

PART TWO

Ecological Filters as a Form of Assembly Rule

A specific approach to assembly rules theory is the concept of ecological filters. The following three chapters explore this concept from various perspectives, focusing on its application for restoration. Chapter 5 considers the interaction of filtering effects, environmental gradients, and thresholds and their importance in determining how well restoration projects succeed. Chapter 6 develops a general filter model that has potential applications for planning and evaluating all kinds of restoration projects. Chapter 7 extends the ecological filter concept by advocating a two-pronged strategy for restoration: the use of a filter approach (focusing on constraints to community membership) in conjunction with a systems approach (focusing on species traits).

Chapter 5

Ecological Filters, Thresholds, and Gradients in Resistance to Ecosystem Reassembly

RICHARD J. HOBBS AND DAVID A. NORTON

The way in which biological communities are assembled from the regional pool of biodiversity has received increasing study, and the question of whether recognizable "assembly rules" exist has been examined (see Chapter 3). There is now evidence to suggest that assembly rules may be found for both experimental and natural communities. However, there is also evidence to suggest that different assemblages result from different starting conditions, order of species arrival or introduction, and type and timing of disturbance or management.

Other chapters in this book discuss the ongoing debate over how the biotic elements of ecosystems assemble, particularly following disturbance, and examine whether this process is relevant to ecological restoration. In Hobbs and Norton (1996), we proposed some general principles that might help build a broad framework for restoration ecology. In this chapter, we explore various aspects of ecosystem dynamics and restoration to develop such a framework further. Specifically, we look at how a general framework can serve as a basis for better understanding what can be achieved in restoration at a particular site. We start by reviewing recent developments in ecological concepts and examining how they radically alter the way in which we should consider ecosystem dynamics and restoration. In particular, we reflect on the importance of alternative stable states in ecosystems and the drivers that force transitions between states. We look at these state and transition approaches in a restoration context, particularly in relation to restoration thresholds. We then consider the concept of biotic and abiotic filters and examine whether it meshes with the ideas of states, transitions, and thresholds to assist in developing a comprehensive framework for ecosystem restoration.

72

New Ecosystem Paradigms

Ecological concepts have changed dramatically over the past 20 years (Pickett and Ostfeld 1995, Pickett et al. 1992). It is now increasingly recognized that ecosystems are complex, dynamic entities that vary at multiple scales in both space and time. The complex dynamic behavior of ecosystems often makes it difficult to understand how they function and to predict the outcomes of any given event or management activity (Hobbs 1998, Hobbs and Morton 1999, Pahl-Wostl 1995). Following Hobbs and Morton (1999), we now outline some of the main changes in ecological thinking that are relevant to understanding ecological assembly and restoration.

The Flux of Nature

Ecologists historically viewed the natural world as a fundamentally stable place in which each species had its ordered position within a community and in which any disturbance would result in an ordered successional progression leading back to the original climax state (Christensen 1988). Ecological communities were considered to be organized, patterned collections of coevolved species that incompatible species could not easily penetrate (Simberloff 1982). Ecologists now speak of this view as the *equilibrium paradigm*. In recent years, we have seen this notion of organization and stability give way to the view that the natural world is dynamic. In this paradigm, ecosystems are characterized more by instability than by permanence. Ecosystems are also marked by frequent disturbances that continually push them in alternative directions instead of causing them to return to their original condition. We now see ecosystems as characterized by more or less unpredictable individualistic species responses rather than by predictable and correlated species responses.

The *nonequilibrium paradigm* just described recognizes that the natural world is an uncertain place in which disturbances are constantly causing alterations in the composition of assemblages and in the spatial patterns of the environment (Fiedler et al. 1997, Pickett et al. 1992, Pickett and White 1985, Sousa 1984). This view does not suggest that ecological equilibria do not exist but rather that they are scale-dependent and embedded in nonequilibrial conditions. Nevertheless, the nonequilibrium paradigm does imply that predictable endpoints to the successional process following disturbance are rare, that multiple stable states may exist, and that some quasi-stable states can persist for long periods.

Multiple Stable States

Disturbance inevitably sets in motion some form of succession. Current discussion of ecosystem development following disturbance asks whether concepts of individualistic assembly and multiple endpoints are more appropriate than classical successional theory (Young et al. 2001). The course of succession can be difficult to predict, because the direction the ecosystem or assemblage takes is contingent upon the particular circumstances of the disturbance and the nature of the biophysical conditions that precede and follow it. The notion of contingency implies that history matters in patterns and processes of community change. As a consequence, the endpoint of a successional process is not a predictably uniform outcome; instead, several states are possible, depending on the contingent circumstances (Hobbs 1994, Noble and Slatyer 1980). A general feature of many ecosystems is the potential for the system to exist in a number of different states, depending on both past and present biotic and abiotic factors.

Patchiness and Landscape Ecology

Recognition of the importance of spatial and temporal variability, together with the increased availability of suitable tools for analyzing it, is at the root of the emerging discipline of landscape ecology. The understanding and management of the natural world depends as much on the analysis of flows of resources across ecosystems as it does on the detailed study of individual places. One of the principal issues underpinning landscape ecology is recognition of the vital importance of patchiness (Levin 1989, Ostfeld et al. 1997, Turner and Gardner 1991). Patchiness focuses on the spatial matrix of ecological processes and emphasizes the fluxes of materials and organisms within and between different parts of the landscape.

Thresholds in Restoration

The recent changes in the way ecological systems are viewed have a number of lessons for restoration ecology. If ecosystems are nonequilibrium, patchy, and liable to exist in a number of different states, then restoration goals need to take these facts into account. Aiming for a single endpoint may not be valid or may constrain the restoration endeavor too much. Furthermore, it seems likely that for many ecosystems, restoration thresholds exist as a result of human activities that prevent the system from returning to a less degraded state without the input of management, restoration, and aftercare effort (Aronson et al. 1993a; Hobbs and Harris 2001, Whisenant 1999).

Whisenant (1999) has recently suggested that two main types of such thresholds exist, one caused by biotic interactions and alterations and the other by abiotic alterations, transformations, or inherent limitations (Figure 5.1). If the system in question has been degraded mainly as a result of biotic changes (such as grazing-induced changes in vegetation composition), restoration efforts should focus on biotic manipulations that remove the degrading factor (for example, the grazing animal) and adjust the biotic composition (for example, replanting lost species). If, on the other hand, the system has become degraded as a result of changes in abiotic features (such as soil erosion or changed hydrology), restoration efforts must focus first on removing the degrading factor and repairing the physical and/or chemical environment (for example, to reinstate a particular hydrological regime). There is little point in focusing exclusively on biotic manipulation and ignoring the abiotic problems. In other words, system functioning should be corrected or maintained before questions of biotic composition and structure are considered. Taking system function into account provides a useful framework for initial assessment of the state of the system and subsequent selection of repair measures (Ludwig et al. 1997, Tongway and Ludwig 1996). Where function is not impaired, restoration can then focus on composition and structure.

Note that the threshold model (see Figure 5.1) suggests that the biota simply respond passively to the abiotic environment; that is, biotic thresholds always come after abiotic ones. In some situations, however, the presence of a keystone species or "ecosystem engineers" may be required to change the abiotic environment. Hence, biotic manipulation may be required in some cases to overcome an abiotic threshold.

Filters on Ecosystem Assembly

Several authors have explored the idea that the process of community assembly involves a series of filters that sift species out of the regional pool (Díaz et al. 1998, 1999a; Weiher and Keddy, 1995). In restoration ecology, we are also concerned with community assembly; we start with a set of individual species as the raw ingredients and try to build them into a community. In some cases, we have an in situ assemblage; in others (for example, mine-site restoration and wetland creation), we have no in situ assemblages at all. We suggest that there are a number of different filters that vary in their importance along relatively easily defined gradients. The resistance of degraded systems to restoration (that is, the degree of effort needed to restore the system to a particular state) will vary along these gradients. The idea of resistance to restoration is embodied in the idea of restoration thresholds, discussed above.

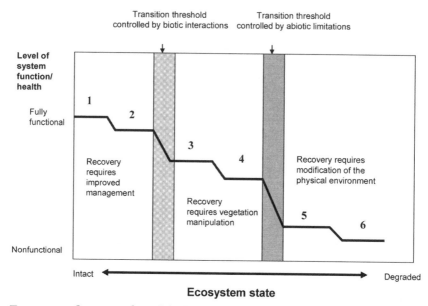

FIGURE 5.1 Conceptual model of transitions between undegraded and degraded ecosystem states and the presence of biotic and abiotic thresholds (after Whisenant 1999).

Further, we suggest that the difficulty in finding generalizable principles for restoration ecology (as discussed by Hobbs and Norton 1996) may be due, in part, to a failure to view ecosystems in this broader context. Using the concepts of filters and gradients in ecosystem resistance may help to provide a useful framework for linking and comparing restoration efforts in widely differing systems.

The approach taken to restoration in any given situation will be the product of a number of different filters reflecting the different factors that influence the course and success of the restoration. Restoration will, of course, also inevitably be approached in differing ways as a function of available money, manpower, and legislation or contractual guidelines to be respected. Previous studies have identified the potential for different types of filter to be important. A differentiation is usually made between biotic and abiotic filters, and some authors also suggest that disturbance may be considered as a distinct filter (Díaz et al. 1999b, Woodward and Diament 1991). It has also been recognized that the action of filters is likely to change over both space and time (Woodward and Diament 1991); for example, different filters may be important at different stages of restoration. In a way, environmental filters simply represent ecological variables and processes that are likely to be important in any given situation. However, by formalizing our understanding of the probable importance of a range of these variables, we can highlight those that are likely to be important determinants of restoration success in any given place. The following filters can be considered in restoration.

Abiotic Filters

- Climate: Rainfall and temperature gradients, especially species tolerance to local conditions such as frost, rather than regional effects.
- Substrate: Fertility (relative abundance and diversity of decomposers, nutrient cycling), soil water availability, toxicity.
- Landscape structure: Landscape position, previous land use, patch size, and isolation.

Biotic Filters

- Competition: With preexisting and potentially invading species and between planted or introduced species.
- Predation-trophic interactions: From preexisting and potentially invading species, and predation between reintroduced animal species.
- Propagule availability (dispersal): Bird perches, proximity to seed sources, presence of seed banks.
- Mutualisms: Mycorrhizae, rhizobia, pollination and dispersal, defense, and so forth.
- Disturbance: Presence of previous or new disturbance regimes.
- Order of species arrival and successional model: Facilitation, inhibition, and tolerance; priority effects; lottery effects.
- Current and past composition and structure (biological legacy): How much original biodiversity and original biotic and abiotic structure remains.

The importance of these various abiotic and biotic filters is considered in Chapter 6.

Socioeconomic Filters

In addition to the biotic and abiotic filters, a set of socioeconomic filters should also be considered in a restoration context.

- What the community wants: The goals and aspirations of local communities, landowners, and other stakeholders (for example,

alternative goals relating to biodiversity conservation, watershed
protection, soil conservation, or maximizing production values).
• What the community can afford: The costs of achieving stated goals
relating to production, planting, maintenance, and overcoming
potential threats and existing thresholds.

Both of these socioeconomic considerations may act to reinforce or
reduce the effects of the biotic and abiotic filters. For instance, if the costs
to overcome particular abiotic filters are too great to allow the project to
achieve a particular goal, the project may need to be redesigned to achieve
a different goal. Alternatively, social considerations may specifically exclude
particular species in restoration projects (for example, top carnivores such
as wolves).

It is likely that the importance of different filters will vary from place to
place and over time through restoration. For instance, the importance of
rainfall in determining the resistance to restoration is likely to increase as
rainfall decreases. Rainfall amount and distribution are much more critical
to the success of a restoration effort in the low-rainfall areas of the southwestern Australian wheat belt than they are in the high-rainfall areas of southwestern New Zealand (see "Case Studies" later). Similarly, invasions are
likely to be a more important filter in heavily invaded systems (for example,
parts of the New Zealand high country) than in systems with few invaders.
Although some abiotic filters (for example, fertility or soil moisture) are likely
to be more important early in restoration, others (for example, dispersal) are
likely to be more important later.

Our list of filters could be viewed as simply a catalog of every conceivable
factor that could affect organisms and ecological systems. Given this mas-
sive list, how do we go about formalizing our understanding of the likely
importance of a range of these variables? The filter concept may be useful
only in the very limited case of identifying factors that prevent establishment
of particular species. Trying to imagine landscape structure as a filter with a
dynamic mesh size that varies spatially and temporally may be difficult, and
there may be no point in doing so unless landscape structure constrains
ecosystem assembly in some way. However, we may gain some insight into
whether dispersal represents a filter on establishment of a particular species
if we consider landscape structure as a constraint around the degraded site—
for instance, by preventing movement of key dispersal agents. We cannot
examine all constraints on a system, so we must look at our measures of function and structure and then identify the constraints that "resist" transition
from one state to another along the trajectory.

The combination of key filters will determine the resistance to restoration in any given place and hence will determine the steps to be taken and the level of investment needed to overcome that resistance. Generalizing these relationships can provide a powerful tool for determining the most desirable and appropriate approach to restoration in any particular situation, recognizing that particular systems embody much specificity. It is clear that the effects of environmental filters influence the array of species that may arrive or survive in any given situation; hence, these effects should guide how the restoration project is approached. Because filters are dynamic, there may be little point in aiming to put back a particular suite of species. Although it may be of some value to guide restoration efforts in some situations by using reference ecosystems (Aronson et al. 1995, Moore et al. 1999, Parsons et al. 1999), the dynamic nature of the filters acting on species reassembly makes it highly unlikely that the exact composition of the reference system can be re-created, except perhaps in the very simplest of systems. The fact that future conditions are highly unpredictable further reinforces the unlikeliness of re-creating a particular reference state. However, such states do have value as benchmarks or generalized goals, especially for purposes of communication within and outside the group of people participating in a restoration.

Nevertheless, restoration efforts need not passively depend on the results of the filtering process but can instead try to modify the effects of filters to allow desired species in and prevent the establishment of undesired species (Figure 5.2). Indeed, restoration aims to speed up or direct the natural recolonization process where it is too slow or is impeded in some way. Modification of abiotic filters includes such activities as modifying the nutrient status or toxicity of the substrate, modifying soil structure or microtopography to improve water infiltration and retention, and shoring up gullies that are created by surface erosion that lead to silting and mudslides. Modification of biotic filters includes such activities as overcoming dispersal barriers by introducing species propagules to the site, preventing grazing, and controlling weeds. Overcoming dispersal barriers also may include increasing the overall species pool by transporting species from place to place and providing the opportunity for species from other regions to establish. This approach is often essential in the most modified systems. Reinstating lost disturbance regimes is a particularly important component of restoration in some situations (for example, Norton and de Lange 2002). It is therefore important to understand the key filters acting in any given situation when determining what the appropriate restoration activities might be.

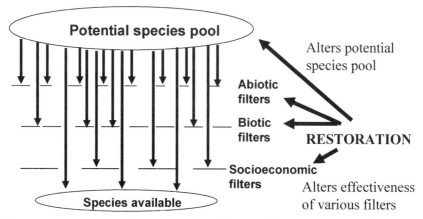

FIGURE 5.2 Conceptual view of biotic, abiotic, and socioeconomic filters as they affect species inclusion from the regional species pool into a local community and the ways in which restoration can alter both the availability of species and the impacts of filters.

Linking Filters, Thresholds, and Assembly Rules

We can consider restoration a conscious effort to manipulate filters to arrive at a desired species composition (and hence state) at any given site, coupled with an attempt to speed up the recolonization process by recognizing and dealing with potential restoration thresholds. Can we consider restoration thresholds simply as special cases of filters in which the mesh size is unacceptably small? We would argue that although this may be possible, it is still useful to consider thresholds separately: they represent discontinuities, whereas the actions of filters vary more continuously across different environments and create gradients of resistance to restoration.

A conceptual map of the relationships among assembly rules, filters, and thresholds is shown in Figure 5.3. We are ultimately interested in the composition (and function) of the ecosystem being restored. What influences how species reassemble in a particular location? As we have discussed, there will be varying degrees of resistance to restoration, depending on the action of a number of biotic and abiotic filters. These filters, in turn, can be considered as embodying a set of rules for which species can and cannot arrive and survive on the basis of key abiotic, biotic, and disturbance factors. In addition to the set of filters that are important in any given situation, there may also be important threshold effects that need to be considered. These thresholds may act to alter the filtering process, but they can also lead to the establishment or perpetuation of alternative states. The transition from a

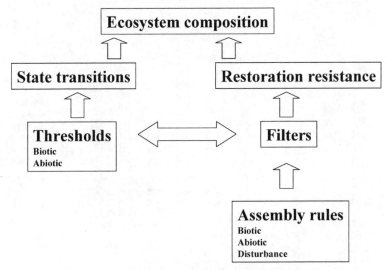

FIGURE 5.3 Interrelationships among the concepts of assembly rules, filters, and thresholds as they relate to ecosystem composition, via the effects of transitions between states and resistance to restoration.

more degraded to a less degraded state may involve manipulation of the biotic or abiotic components of the system, or both. Figure 5.3 is presented as one way of conceptualizing the linkages among thresholds, filters, and assembly rules, but it is by no means the final word on the matter—indeed, we welcome others to argue with us and develop these ideas further.

Case Studies

In this section, we examine a set of case studies in which the concepts put forward earlier are explored in specific contexts. The case studies provide examples of systems in which species assembly varies spatially and temporally as a result of thresholds and/or filters. In several of these cases, hypotheses have been formulated concerning the possible range of ecosystem states, the likely transitions between them, potential system thresholds, and the implications of all these factors for restoration.

Recolonization of Disturbed Areas in California Serpentine Grassland

Serpentine grassland in northern California consists predominantly of native annual forb species. The small-scale spatial patterning of this grass-

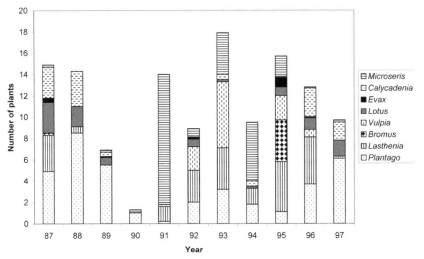

FIGURE 5.4 Mean numbers of plants of different species growing in ten 10-cm by 10-cm quadrats located on bare soil mounds formed the preceding year by gophers (*Thomomys bottae*) within annual grassland on serpentine soil in Jasper Ridge Biological Preserve, northern California. (Data from Hobbs and Mooney 1995 and R. J. Hobbs, unpublished.)

land is linked to localized disturbance by gophers, burrowing rodents that dig tunnels underground and excavate soil to the surface, where it covers existing vegetation and forms areas of bare soil. These bare areas have to be recolonized, mostly by seed, the following season (Hobbs and Mooney 1985). The dynamics of the grassland are also affected by climatic variations, particularly the highly variable rainfall (Hobbs and Mooney 1991, 1995). Figure 5.4 shows the relative abundances of species recolonizing gopher mounds formed each year over the period 1987 to 1997. The diagram indicates a considerable degree of variation in both the total number of plants recolonizing from year to year and the relative importance of different species. No clearly repeatable pattern of recolonization is apparent over all years, and some years show markedly different patterns. Part of this variation can be explained by a species's individual response to rainfall amounts and timing (Hobbs and Mooney 1995) (Figure 5.5), but obviously other factors are also involved. Hence, in this case, there is a clear impact of disturbance as a filter on composition (although the disturbance agent is biotic). Rainfall is also a strong abiotic filter and interacts with disturbance to produce considerable variation in year to year recolonization dynamics.

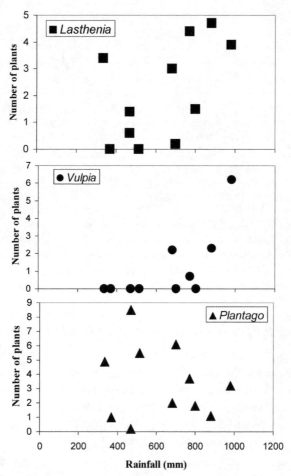

Figure 5.5 Number of plants established on bare soil mounds formed the previous year, plotted against annual rainfall amount (July–June), for *Lasthenia californica*, *Vulpia microstachys*, and *Plantago erecta* (R. J. Hobbs, unpublished data).

This case clearly indicates the importance of recognizing that the same disturbance (by gopher diggings) will not result in reassembly of the same suite of plants every time. Here, the results are contingent on the impacts of year-to-year variations in rainfall and how these variations affect the population dynamics of the plant species present in the vicinity. By analogy, therefore, we should expect results of particular restoration projects to vary, also depending on contingent effects acting to modify the filtering process.

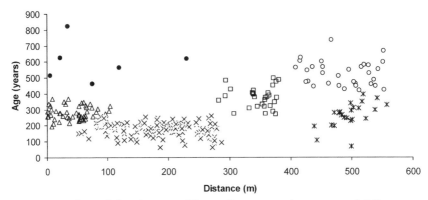

FIGURE 5.6 Spatial distribution of *Dacrydium cupressinum* trees of different ages grouped into age cohorts (same symbols) along a 600-m by 10-m transect in southwestern New Zealand lowland podocarp forests. (Data from B. Cornere, unpublished master's thesis, University of Canterbury.)

Disturbance Responses in New Zealand Mixed Podocarp Forests

Forests comprising mixtures of conifer species in the family Podocarpaceae occur in high-rainfall areas of New Zealand. In these forests, the scale and type of abiotic disturbance is the major factor determining, or filtering, forest canopy composition (Ogden and Stewart 1995). For example, after catastrophic river flooding, *Dacrycarpus dacrydioides* typically dominates the regenerating forest canopy (Duncan 1993). In the absence of such large-scale disturbance, however, other conifer species (especially *Dacrydium cupressinum* and *Prumnopitys ferruginea*) dominate in the smaller gaps created by windthrow. These two species are also differentiated by their establishment site requirements: *D. cupressinum* establishes on more elevated sites than *P. ferruginea*. In addition, the same area of forest may have been affected by several different past disturbance events, giving rise to overlapping cohorts of trees in both space and time (Figure 5.6; see also Allen and Norton 2000). The complexity of spatial patterns in terms of species composition and even the age of different cohorts of the same tree species highlight the complex responses of species to an abiotic filter such as disturbance. Furthermore, the transition between different forest states can be either gradual (in the absence of disturbance) or rapid (when disturbance occurs). Again, it is important to recognize that the same disturbance at different times or in different locations is likely to result in complex and variable biotic responses.

Annual grassland

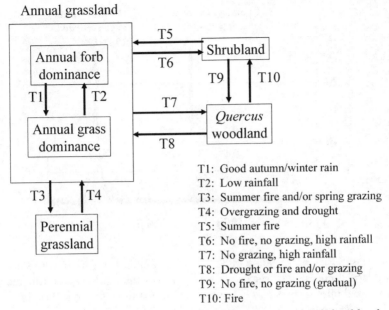

T1: Good autumn/winter rain
T2: Low rainfall
T3: Summer fire and/or spring grazing
T4: Overgrazing and drought
T5: Summer fire
T6: No fire, no grazing, high rainfall
T7: No grazing, high rainfall
T8: Drought or fire and/or grazing
T9: No fire, no grazing (gradual)
T10: Fire

FIGURE 5.7 Alternative states in a northern California grassland-shrubland-woodland mosaic and the likely drivers of transitions between states. (After George et al. 1992.)

Grassland-Shrubland-Woodland Mosaics in Northern California

Figure 5.7 presents a proposed conceptual framework for understanding the dynamics of vegetation in coastal northern California, developed from studies by George et al. (1992), Huntsinger and Bartolome (1992), and Marañon and Bartolome (1993). The area consists of a mosaic of various vegetation types including annual grassland, perennial grassland, shrubland, and oak woodland. It is thought that perennial grassland was originally much more widespread than it is at present but was transformed in the eighteenth century into annual grassland dominated by invasive grass and forb species predominantly from the Mediterranean Basin, as a result of fire and/or excessive grazing. There is ongoing interest in how this process occurred and whether it can be reversed (Bartolome et al. 1986, Hamilton et al. 1999, Menke 1992, Nelson and Allen 1993). The vegetation mosaic has been considered to be relatively stable, but there is evidence that transitions between different states are possible. For instance, Figure 5.8 illustrates a rapid transition from annual grassland to shrubland, as described in Williams et al. (1987), which

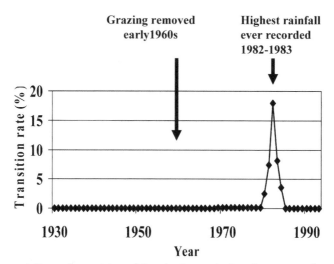

FIGURE 5.8 Rate of transition of 3-m by 3-m subplots from annual grassland to shrubland dominated by *Baccharis pilularis* ssp. *consanguinea*, as estimated from aerial photographs and field observations within a grid of 1100 subplots in Jasper Ridge Biological Preserve, northern California. (After Williams et al. 1987.)

corresponds to transition 6 in Figure 5.7. This transition requires the absence of grazing and fire; the site had certainly not been burned for some time, and grazing by cattle ceased in the 1960s. The trigger for invasion by shrubs, however, was unusually high rainfall, which allowed shrub seedlings to estab-lish and survive in competition with the annual grasses (Williams and Hobbs 1989).

In this vegetation mosaic, the factors controlling transitions between dif-ferent states are predominantly grazing, fire, and climate variables, particu-larly rainfall. It can be assumed that these factors are important in deter-mining the relative performance of individual plant species in any given area. It can also be assumed that seed supply is not a major factor affecting the abil-ity of species to invade different vegetation states, although seed supply will obviously vary from place to place (Marañon and Bartolome 1989, 1993). In management and restoration terms, if the goal is to reduce the area domi-nated by nonnative annual grassland, an understanding of the factors caus-ing transitions between grassland and other states indicates that control of grazing and facilitation of shrub establishment (by control of herbaceous vegetation) are likely to be needed.

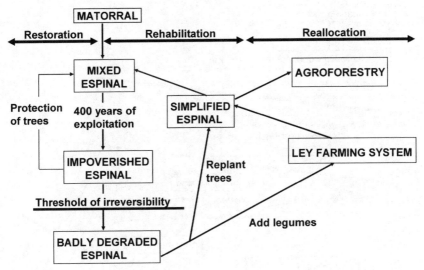

FIGURE 5.9 States and transitions for central Chilean ecosystems. (After Aronson et al. 1993b.)

Matorral Ecosystems in Chile

In Figure 5.9, states and transitions for central Chilean ecosystems are illustrated, as described by Aronson et al. (1993b). The relevant historical reference in this case is considered to be matorral, a Mediterranean-type shrubland-woodland that is affected by human use but retains high functional group diversity and overall complexity. The probable preexisting woodland is considered too long removed to be used as a reference. Overuse through grazing and other activities leads to the degradation of this system to a more open, clearly anthropogenic savanna-like system known as espinal (Aronson et al. 2000). Mixed espinal retains a reasonable degree of canopy species and structural diversity, but both decline rapidly with intensified exploitation and increasing degradation. A threshold appears to have occurred as a result of continued overuse of impoverished espinal, resulting in badly degraded espinal. A number of options are available for the restoration of degraded espinal. One is recovery to a simplified espinal ecosystem, with the potential for continued recovery back to mixed espinal. Alternatively, reallocation to a ley or mixed farming system using annual legumes, or to an array of agroforestry systems, is possible. Note that in this case, complete restoration to matorral is not considered a viable option if the system is going to continue to be used for economic purposes, with a high density and low prevailing economic levels of local human population. Note also that a

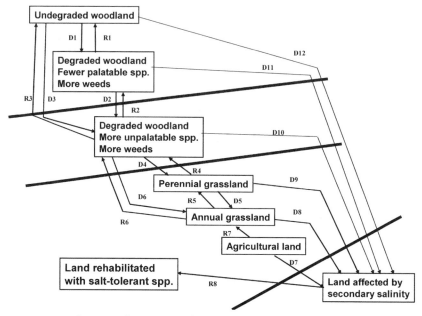

FIGURE 5.10 States and transitions for woodland ecosystems in Western Australia. See text for details. (After Yates and Hobbs 1997.)

reallocation can be employed as an intermediate measure in the effort to achieve rehabilitation or restoration.

Woodlands in Southwestern Australia

Figure 5.10 illustrates states and transitions for woodland ecosystems in Western Australia, as described by Yates and Hobbs (1997). In this case, the undisturbed state is open woodland dominated by *Eucalyptus salmonophloia*, with a shrubby understory. With degradation, the understory is lost, as are canopy mistletoes (Norton et al. 1995). In addition, soil structure declines and weed invasion occurs. With continued degradation, loss of adult trees and lack of recruitment eventually lead to the development of a secondary grassland system (annual or perennial). Any of these systems can degrade still further to salt land if rising water tables bring soil-stored salt to the surface. Transitions D1–D9 represent system degradation, whereas transitions R1–R8 represent system repair. Several hypothesized thresholds occur, and active restoration management may be required to force transitions back to less degraded states. For instance, removal of livestock grazing does not result in the return of degraded woodland to a less degraded state because soil compaction by

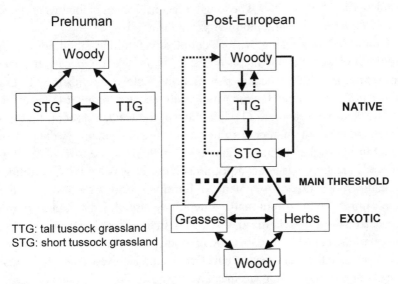

FIGURE 5.11 Prehuman and post-European settlement vegetation change in the New Zealand high country. Solid lines indicate transitions that are reasonably easy to make; dotted lines indicate transitions that are more difficult because of thresholds that act as barriers. (D. A. Norton, unpublished data.)

livestock causes reduced water infiltration and prevents tree and shrub regeneration (Yates et al. 2000a, Yates et al. 2000b). Effective treatment such as soil ripping overcomes this threshold.

Tussock Grasslands in New Zealand

The last example comes from the New Zealand high country: the rain-shadow ranges of the eastern South Island, which are now used for extensive pastoralism (Norton 1991). This area was predominantly wooded before human settlement, although infrequent fire would have maintained a dynamic mosaic of woody and herbaceous communities (Figure 5.11; also see McGlone 2001). Since early human settlement (some 700 to 800 years ago), unsustainable fire and grazing regimes have degraded this system to tall tussock grassland and subsequently to short tussock grassland (Connor 1964; also see Figure 5.11). These tussock grasslands are successional communities that are maintained by the cultural practices of pastoralism. If seed sources are available or restoration can return lost species, it may be possible to reverse the changes that have occurred. The tussock grasslands themselves, however, albeit cultural ecosystems, are highly valued by the local people

(Ashdown and Lucas 1987). The long-term conservation of the tussock grasslands will require a continuation of pastoralism.

Continued overgrazing by stock coupled with a high rabbit population and the invasion of exotic plants (grass, herb, and woody species) has seen some of these grasslands degraded even further, effectively crossing a threshold to a new set of stable states in which exotic species dominate (Walker and Lee 2000; also see Figure 5.11). For many of these areas, it seems that simply removing the original perturbing factors (fire and grazing) will be insufficient to return the system to the previous state. In some parts of the high country, in fact, the species composition of tussock grasslands has continued to change independently of grazing or burning history, which suggests that factors unrelated to current management have also played a significant role in grassland dynamics (Duncan et al. 2001). In these highly degraded ecosystems, because of widespread invasion of exotic plant species, it may be impossible to reverse the changes that have occurred. Clearly, we need to be realistic about our long-term goals in such areas.

Conclusions

In this chapter, we have highlighted the importance of recent shifts in ideas about how ecosystems work in determining how we should think about, plan, and carry out ecological restoration. In any one place, the species present in the ecosystem will be a result of the complex interaction of a number of abiotic and biotic filters, the effects of which change over space and time. A restoration project following a disturbance of any kind will be able to return the same set of species only by replicating the exact combination of filtering effects that created the preexisting ecosystem. Even then, it appears that ecosystems are complex dynamic systems with the propensity to exist in different states. Transitions between these states may be rapid, and there may be thresholds that, once crossed, render movement back to previous states difficult or perhaps even impossible. The existence of alternative ecosystem states and hence ecosystem trajectories is an important concept for restoration ecology: it means that we must assess the goals we set for restoration and how stringently we are going to demand the return to a particular state, in relation to our ability to force the system over any thresholds that may be present.

The case studies highlight the importance of recognizing where thresholds are likely to occur or have occurred. Threshold crossings theoretically prevent ecosystems from recovering without significant management input or long periods of time (see Figure 20.2 in Aronson et al. 2002). Thus, in such cases, major effort should be devoted to identifying clear goals about the desired outcome of restoration. Thresholds may be considered as brakes on the assembly process that may prevent successful restoration; hence, they are worthy of considerably more research.

How do these considerations relate to the practicality of deciding what to do in particular restoration projects? Figure 5.1 shows a system's trajectory through different states. The first difficulty in practical application is assigning variables to the *x*-axis and *y*-axis that can be defined operationally. How do we measure "level of system function/health" or "ecosystem state" in particular cases? If we cannot measure these things, how do we identify thresholds and how do we evaluate progress? The difficulty of defining operational variables reflects both scientific and socioeconomic issues. To use Figure 5.1 operationally, one must identify measures of system function and structure that are important from a scientific point of view and that also permit definition of a restoration goal (which may conflict with socioeconomic constraints). Reference sites provide an opportunity to define realistic goals if they are chosen on the basis of operational measures of system function and structure (note that species composition may not come into play at all). An assessment of restoration progress then becomes the distance in state space between degraded and reference sites. The identification of thresholds requires some knowledge of intermediate states. Although our goal is to develop a general framework, deep understanding of particular systems is still required.

Does all this help? Can filters, gradients in restoration resistance, and thresholds increase our understanding of and ability to restore degraded ecosystems? That, we suggest, is up to the readers of this book to judge!

Acknowledgments

We thank Lisa Belyea and James Aronson for constructive comments on the draft manuscript.

REFERENCES

Allen, R. B.; and Norton, D. A. 2000. Disturbance ecology: a basis for sustainable management. In *Sustainable Management of Indigenous Forest. Proceedings of Symposium of Southern Connections Congress III*, 16–23. Christchurch: Wickliffe Press.

Aronson, J.; Dhillion, S.; and Le Floc'h, E. 1995. On the need to select an ecosystem of reference, however imperfect: a reply to Pickett and Parker. *Restoration Ecology* 3:1–3.

Aronson. J.; Floret, C.; Le Floc'h, E.; Ovalle, C.; and Pontanier, R. 1993a. Restoration and rehabilitation of degraded ecosystems in arid and semiarid regions. I. A view from the South. *Restoration Ecology* 1:8–17.

Aronson, J.; Le Floc'h, E.; Floret, C.; Ovalle, C.; and Pontanier, R. 1993b. Restoration and rehabilitation of degraded ecosystems in arid and semiarid regions. II. Case studies in Chile, Tunisia and Cameroon. *Restoration Ecology* 1:168–187.

Aronson, J.; Le Floc'h, E.; and Ovalle, C. 2002. Semi-arid woodlands and desert fringes. In *Handbook of Ecological Restoration*, vol. 2., ed. M. Perrow and A. Davy, 466–485. Cambridge: Cambridge University Press.

Ashdown, M.; and Lucas, D. 1987. *Tussock Grasslands: Landscape Values and Vulnerability*. Wellington: New Zealand Environment Council.

Bartolome, J. W.; Klukkert, S. E.; and Barry W. J. 1986. Opal phytoliths as evidence for displacement of native California grassland. *Madroño* 33:217–222.

Christensen, N. L. 1988. Succession and natural disturbance: paradigms, problems, and preservation of natural ecosystems. In *Ecosystem Management for Parks and Wilderness*, ed. J. K. Agee and D. R. Johnson, 62–86. Seattle: University of Washington Press.

Connor, H. E. 1964. Tussock grassland communities in the Mackenzie Country, South Canterbury, New Zealand. *New Zealand Journal of Botany* 2:325–351.

Díaz, S.; Cabido, M.; and Casanoves, F. 1998. Plant functional traits and environmental filters at a regional scale. *Journal of Vegetation Science* 9:113–122.

Díaz, S.; Cabido, M.; and Casanoves, F. 1999a. Functional implications of trait-environment linkages in plant communities. In *Ecological Assembly Rules: Perspectives, Advances, Retreats*, ed. P. Weiher and P. Keddy, 338–362. Cambridge: Cambridge University Press.

Díaz, S.; Cabido, M.; Zak, M.; Martínez Carretero, E.; and Aranibar, J. 1999b. Plant functional traits, ecosystem structure and land-use history along a climatic gradient in central-western Argentina. *Journal of Vegetation Science* 10:651–660.

Duncan, R. P. 1993. Flood disturbance and the coexistence of species in a lowland podocarp forest, south Westland. *New Zealand. Journal of Ecology* 81:403–416.

Duncan, R. P.; Webster, R. J.; and Jensen, C. A. 2001. Declining plant species richness in the tussock grasslands of Canterbury and Otago, South Island, New Zealand. *New Zealand Journal of Ecology* 25:35–47.

Fiedler, P. L.; White, P. S.; and Leidy, R. A. 1997. The paradigm shift in ecology and its implications for conservation. In *The Ecological Basis of Conservation: Heterogeneity, Ecosystems and Biodiversity*, ed. S. T. A. Pickett, R. S. Ostfeld, M. Shachak, and G. E. Likens, 83–92. New York: Chapman and Hall.

George, M. R.; Brown, J. R.; and Clawson, W. J. 1992. Application of non-equilibrium ecology to management of Mediterranean grassland. *Journal of Range Management* 45:436–440.

Hamilton, J. G.; Holzapfel, C.; and Mahall, B. E. 1999. Coexistence and interference between a native perennial grass and non-native annual grasses in California. *Oecologia* 121:518–526.

Hobbs, R. J. 1994. Dynamics of vegetation mosaics: can we predict responses to global change? *Écoscience* 1:346–356.

Hobbs, R. J. 1998. Managing ecological systems and processes. In *Ecological Scale: Theory and Applications*, ed. D. Peterson and V. T. Parker, 459–484. New York: Columbia University Press.

Hobbs, R. J.; and Harris, J. A. 2001. Restoration ecology: repairing the Earth's ecosystems in the new millennium. *Restoration Ecology* 9:239–246.

Hobbs, R. J.; and Mooney, H. A. 1985. Community and population dynamics of serpentine grassland annuals in relation to gopher disturbance. *Oecologia* 67:342–351.

Hobbs, R. J.; and Mooney, H. A. 1991. Effects of rainfall variability and gopher disturbance on serpentine annual grassland dynamics. *Ecology* 72:59–68.

Hobbs, R. J.; and Mooney, H. A. 1995. Spatial and temporal variability in California
annual grassland: results from a long-term study. *Journal of Vegetation Science*
6:43–57.

Hobbs, R. J.; and Morton, S. R. 1999. Moving from descriptive to predictive ecology. In
Agriculture as a Mimic of Natural Ecosystems, ed. E. C. Lefroy, R. J. Hobbs, M. H.
O'Conner, and J. S. Pate, 43–55. Dordrecht, The Netherlands: Kluwer.

Hobbs, R. J.; and Norton, D. A. 1996. Towards a conceptual framework for restoration
ecology. *Restoration Ecology* 4:93–110.

Huntsinger, L.; and Bartolome, J. W. 1992. Ecological dynamics of *Quercus* dominated
woodlands in California and southern Spain: a state-transition model. *Vegetatio*
99/100:299–305.

Levin, S. A. 1989. Challenges in the development of a theory of community and ecosystem structure and function. In *Perspectives in Ecological Theory*, ed. J. Roughgarden,
R. M. May, and S. A. Levin, 244–255. Princeton: Princeton University Press.

Ludwig, J.; Tongway, D.; Freudenberger, D.; Noble, J.; and Hodgkinson, K., eds. 1997.
*Landscape Ecology, Function and Management: Principles from Australia's Range-
lands.* Melbourne: CSIRO Publishing.

Marañon, T.; and Bartolome, J. W. 1989. Seed and seedling populations in two contrasted communities: open grassland and oak (*Quercus agrifolia*) understory in California. *Acta Oecologica/Oecologia Plantarum* 10:147–158.

Marañon, T.; and Bartolome, J. W. 1993. Reciprocal transplants of herbaceous communities between *Quercus agrifolia* woodland and adjacent grassland. *Journal of Ecology*
81:673–681.

McGlone, M. S. 2001. The origin of the indigenous grasslands of southeastern South
Island in relation to pre-human woody ecosystems. *New Zealand Journal of Ecology*
25:1–15.

Menke, J. W. 1992. Grazing and fire management for native perennial grass restoration
in California grasslands. *Fremontia* 20:22–25.

Moore, M. M.; Covington, W. W.; and Fulé, P. Z. 1999. Reference conditions and ecological restoration: a southwestern ponderosa pine perspective. *Ecological Applications* 9:1266–1277.

Nelson, L. L.; and Allen, E. B. 1993. Restoration of *Stipa pulchra* grasslands: effects of
mycorrhizae and competition from *Avena barbata*. *Restoration Ecology* 1:40–50.

Noble, I. R.; and Slatyer, R. O. 1980. The use of vital attributes to predict successional
changes in plant communities subject to recurrent disturbances. *Vegetatio* 43:5–21.

Norton, D. A. 1991. Conservation of high country landscapes. *The Landscape*
1991:15–18.

Norton, D. A.; and de Lange, P. J. 2003. Fire and vegetation in a temperate peat bog:
implications for threatened species management. *Conservation Biology* 17:138–148.

Norton, D. A.; Hobbs, R. J.; and Atkins, L. 1995. Fragmentation, disturbance and plant
distribution: mistletoes in woodland remnants in the Western Australian wheatbelt.
Conservation Biology 9:426–438.

Ogden, J.; and Stewart, G. H. 1995. Community dynamics of the New Zealand conifers.
In *Ecology of the Southern Conifers*, ed. N. J. Enright and R. S. Hill, 81–119. Washington, D.C.: Smithsonian Institute Press.

Ostfeld, R. S.; Pickett, S. T. A.;, Shachak, M.; and Likens, G. E. 1997. Defining the sci-
entific issues. In *The Ecological Basis of Conservation: Heterogeneity, Ecosystems and
Biodiversity*, ed. S. T. A. Pickett, R. S. Ostfeld, M. Shachak, and G. E. Likens, 3–10.
New York: Chapman and Hall.

Pahl-Wostl, C. 1995. *The Dynamic Nature of Ecosystems. Chaos and Order Entwined.*
Chichester, England: John Wiley and Sons.

Parsons, D. J.; Swetnam, T. W.; and Christensen, N. L. 1999. Uses and limitations of historical variability concepts in managing ecosystems. *Ecological Applications*
9:1177–1178.

Pickett, S. T. A.; and Ostfeld, R. S. 1995. The shifting paradigm in ecology. In *A New Century for Natural Resources Management*, ed. R. L. Knight and S. F. Bates, 261–278.
Washington, D.C.: Island Press.

Pickett, S. T. A.; Parker, V. T.; and Fiedler, P. 1992. The new paradigm in ecology: impli-
cations for conservation biology above the species level. In *Conservation Biology: The
Theory and Practice of Nature Conservation*, ed. P. Fielder and S. Jain, 65–88. New
York: Chapman and Hall.

Pickett, S. T. A.; and White, P. S., eds. 1985. *The Ecology of Disturbance and Patch
Dynamics.* New York: Academic Press.

Simberloff, D. 1982. A succession of paradigms in ecology: essentialism to materialism
and probabilism. In *Conceptual Issues in Ecology*, ed. F. Saarinen, 63–99. Boston:
Reidel.

Sousa, W. P. 1984. The role of disturbance in natural communities. *Annual Review of
Ecology and Systematics* 15:353–391.

Tongway, D. J.; and Ludwig, J. A. 1996. Rehabilitation of semi-arid landscapes in Australia. I. Restoring productive soil patches. *Restoration Ecology* 4:388–397.

Turner, M. G.; and Gardner, R. H., eds. 1991. *Quantitative Methods in Landscape Ecology. The Analysis and Interpretation of Landscape Heterogeneity*, vol. 82. New York:
Springer-Verlag.

Walker, S.; and Lee, W. G. 2000. Alluvial grasslands in south-eastern New Zealand: vegetation patterns, long-term and post-pastoral change. *Journal of the Royal Society of
New Zealand* 30:72–103.

Weiher, E.; and Keddy, P. A. 1995. Assembly rules, null models, and trait dispersion: new
questions from old patterns. *Oikos* 74:159–164.

Whisenant, S. G. 1999. *Repairing Damaged Wildlands: A Process-Orientated, Landscape-
Scale Approach.* Cambridge: Cambridge University Press.

Williams, K.; and Hobbs, R. J. 1989. Control of shrub establishment by springtime soil
water availability in an annual grassland. *Oecologia* 81:62–66.

Williams, K.; Hobbs, R. J.; and Hamburg, S. 1987. Invasion of annual grassland in northern California by *Baccharis pilularis* ssp. *consanguinea*. *Oecologia* 72:461–465.

Woodward, F. I.; and Diament, A. D. 1991. Functional approaches to predicting the ecological effects of global change. *Functional Ecology* 5:202–212.

Yates, C. J.; and Hobbs, R. J. 1997. Woodland restoration in the Western Australian wheatbelt: a conceptual framework using a state and transition model. *Restoration Ecology*
5:28–35.

Yates, C. J.; Norton, D. A.; and Hobbs, R. J. 2000b. Grazing effects on plant cover, soil
and microclimate in fragmented woodlands in south-western Australia: implications
for restoration. *Austral Ecology* 25:36–47.

Yates, R. J.; Hobbs, R. J.; and Atkins, L. 2000a. Establishment of perennial shrub and tree species in degraded *Eucalyptus salmonophloia* remnant woodlands: effects of restoration treatments. *Restoration Ecology* 8:135–143.

Young, T. P.; Chase, J. M.; and Huddleston, R. T. 2001. Community succession and assembly: comparing, contrasting and combining paradigms in the context of ecological restoration. *Ecological Restoration* 19:5–8.

Chapter 6

The Dynamic Environmental Filter Model: How Do Filtering Effects Change in Assembling Communities after Disturbance?

Marzio Fattorini and Stefan Halle

The idea of community assembly in plant ecology is much older than the emergence of such terms as *restoration ecology* and *assembly rules* (see Booth and Larson 1999; also see Chapter 3) and dates back to the early twentieth century (Clements 1916, 1936; Gleason 1917, 1926). Environmental filters as a metaphor useful to describe the order of invasion and establishment of species was explicitly formulated by Mueller-Dombois and Ellenberg (1974) and by other authors in the following decades (for example, van der Walk 1981, Drake 1990, Poff and Ward 1990). According to this concept, large-scale processes such as speciation, migration, and dispersal determine how many and which species are "waiting at the entrance" to a particular community, thus forming the local species pool (in the sense of Zobel 1997). Biotic species-species interactions and the environmental conditions together function as a sieve, removing all species lacking specific combinations of traits (Keddy 1992, Weiher et al. 1998, Keddy and Weiher 1999).

Factors regulating plant species pools are resources (for example, light, nitrogen, water, space) and conditions (for example, pH, soil type, salinity, organic matter, flooding, grazing). This simple view of the world will not reveal community assembly, however, because the establishment of species may be mediated by additional factors, both in the present and historically (Grace 2001a). Grazing history, like other important historical processes, means that resources and conditions may have been different in the past, and the present community structure can be understood only in light of this information. Ecosystem engineers (in the sense of Jones et al. 1994, 1997) profoundly change environmental constraints, with marked consequences for physical conditions, coexisting species, and future colonists. *Sphagnum* mosses, for example, alter soil chemistry and hydrology to such an extent that many previously extant species are excluded (van Breemen 1995). To increase prediction in stream ecology, Poff (1997) proposed the identification of species traits that are sensitive to habitat characteristics at different hierarchical scales, ranging from microhabitats to watersheds or basins. Díaz et al. (1998, 1999) recognized filters acting at different scales in terrestrial ecosystems.

This chapter deals with a recurring question within the long debate on assembly in ecosystem ecology: Knowing the local species pool, is it possible to predict the future composition of a community at smaller spatial scales? To answer this question, we studied unaffected regeneration processes at local scales in two ecosystems in eastern Germany: a grassland and a small river that were both degraded by decades of heavy pollution.

The Jena Graduate Research Group

The Graduate Research Group (Graduiertenkolleg) on Functioning and Regeneration of Degraded Ecosystems at the Friedrich-Schiller-University in Jena, Germany, has been funded since 1996 by the *Deutsche Forschungsgemeinschaft* (DFG, German Research Council). It is a long-term interdisciplinary research project with a total duration of 9 years (1996–2005) and integrates the work of 42 doctoral students and 5 postdocs, supervised by 18 faculty researchers from 8 institutes in the fields of geography, chemistry, and biology. The major discipline, however, is ecology, which covers three aspects in the project: (1) change in community structure as an indicator for a degraded system state; (2) the "machinery" of the recovery process, driven by the species remaining after disturbance and their interactions; and (3) restoration in the sense of speeding up or directing recovery, which involves manipulation of succession pathways.

The primary objective of the research group is to study two degraded ecosystems in parallel. The grassland was covered with dust emission from a fertilizer factory for more than three decades (see Chapter 13), resulting in a soil pH of up to 10.1; toxic concentrations of phosphate, cadmium, and sodium; and an almost complete dieback of vegetation except for the grass *Puccinellia distans*, a halophytic specialist. The river suffered from eutrophication due to input from arable farmland and untreated industrial and communal wastewater, resulting in frequent blooms of *Cladophora* green algae. In addition, the river continuum has been, and still is, interrupted by more than 60 weirs over a total river length of 135 km, which hinders upstream movements of aquatic animals and changes the natural flood regime considerably. In both cases the aim is to define the state of the systems and to identify the functional relationships that were significantly affected by the degradation.

Although the two systems are completely different and probably do not have one macroscopic species in common, their degradation history is comparable. Both systems were degraded by a detrimental oversupply of nutrients, and in both systems the input of the pollutant stopped at a rather distinct point in time. Due to the reunification of East Germany and West Germany and the resulting dramatic changes in the economy of eastern Germany, the fertilizer factory was shut down in 1990. Also, water quality of the river improved considerably in the early 1990s because of changed agricultural practices and the processing of an increasing proportion of wastewater in modern and efficient sewage treatment plants. The most important similarity for our project, however, is that both systems have been going through unassisted recovery since then.

So, in a strict sense, our project is not about restoration but instead focuses on how ecosystems respond to relaxed environmental stress and disturbance. We do not look at our systems as two independent case studies, however; rather, we are searching for general features that apply to both systems—and that may also be applicable to other degraded systems. To avoid the pitfall of a merely descriptive approach, we implemented ecological modeling right from the beginning to translate the current state of understanding into a more general abstraction (see Chapter 2). Our aim was to develop from that abstraction a general model that describes the common features of (re)colonization and establishment of invading species after disturbance in both aquatic and terrestrial ecosystems.

As a result of this approach, we formulated the dynamic filter model, which—for the first time—provides structurally consistent categories to classify different restoration management actions in all kinds of degraded systems. In this global model, particular attention is given to the idea of filtering effects (previously proposed by van der Walk 1981, Poff and Ward 1990, Keddy 1992, Zobel 1997, Booth and Larson 1999, Díaz et al. 1999, and others) and to how these effects change over time. A literature review on this subject revealed that several researchers had already discussed similar ideas. This finding was encouraging and motivated us to go a step further; that is, to try our model as a suggestion for a conceptual framework for restoration ecology.

The Dynamic Environmental Filter Model

The dynamic environmental filter model is supposed to be general enough to apply to all kind of organisms in all systems. In terrestrial systems, one

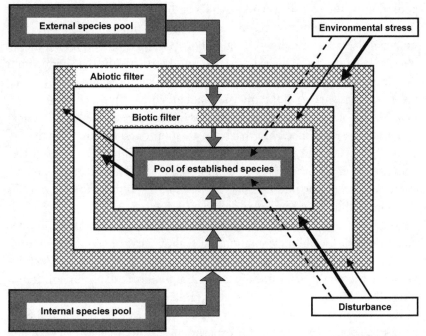

FIGURE 6.1 Schematic diagram of the dynamic environmental filter model. The pool of established species represents the present state of the community structure at a site. New species may invade by dispersal from the surroundings (external species pool) or from an organism bank (internal species pool), but only if they are able to pass the abiotic filter (description of the fundamental niche edges; that is, physical or chemical site conditions) and the biotic filter (description of the realized niche edges; that is, interactions with other species). The filter mesh sizes are constantly readjusted by internal feedback loops and by external cues (environmental stress and disturbance), depicted as black arrows. Thick arrow lines indicate a strong effect, thin arrow lines a weak effect, dashed arrow lines an eventual effect of severe stress or disturbance.

may initially think about plants; in aquatic systems, about freshwater invertebrates. Species composition varies either abruptly or continuously through space, so that it is necessary to arbitrarily choose a region to define the pool of established species of the community of interest (Figure 6.1). To become part of the community, new species must reach the site and must be able to establish. By establishment, we mean that new species are able to cope with abiotic conditions at the site, are able to penetrate the competitive barrier of the currently present species, and thus are persistent at the site.

There are two possible sources for potential invaders: (1) an external species pool; that is, species in nearby habitats; and (2) an internal species pool; that is, the soil seed bank or an "egg bank" of aquatic invertebrates (analogous to diapausing eggs in lake sediments; Marcus et al. 1994, Hairston et al. 1995). The main distinction between the two pools for additional species is that species of the external pool must somehow be transported from the surroundings by drift, dispersal, clonal growth, or some other mechanism, whereas species of the internal pool are already present but in a dormant state. For example, in temperate and tropical forests, many nonpioneer or late-successional species form a bank of "suppressed seedlings" in the forest understory, waiting for enhanced resource availability (especially light) to complete their life cycle (George and Bazzaz 1999). The presence of potential invaders from the internal species pool thus results from the community structure at the site at some point of time in the past.

Whether a new species can persist in the community depends on two conditions: (1) the abiotic conditions must allow survival of individuals until at least one successor is produced or has invaded, and (2) the competitive ability of the new species must be sufficient to allow establishment. In our model (see Figure 6.1), these two constraints are represented as filters; that is, mechanisms and conditions that reduce the size of the potential species pool. In a recent review of sowing experiments, Turnbull et al. (2000) found that for many plants, conditions for establishment are more stringent than conditions for germination. Therefore, in our model, a new species is considered part of the pool of established species if its population includes reproducing individuals as well as other life stages. The passage through the filters requires a certain amount of time, which varies from species to species even within the same functional group.

The model sees filtering as an ordered two-step process (see Figure 6.1). First, new species must pass an abiotic filter, which means that the chemical and physical conditions must allow colonization. Hence, the abiotic filter is a description of the edges of the fundamental niche for successful invaders. Only if species pass the first filter do they have an opportunity to compete with the established species and perhaps become part of the community. The biotic filter refers to the edges of the realized niche for successful invaders.

Dynamic Aspects of the Model

Because ecosystems and communities are open and dynamic (Rykiel 1985, De Leo and Levin 1997, Parker and Pickett 1997), the filters and their effects are by no means static but change over time. To a great extent, the filters are self-adjusting, because of feedback loops from the pool of established species. However, there are cues for external tuning as well. Environmental stress, as induced by pollutants, for example, will primarily affect the abiotic filter, in that only some specialist species may be able to tolerate extreme conditions. Disturbances as discrete events in time will also affect the abiotic filter; more important, they will change the biotic filter by resetting succession sequences or by creating gaps as competition-free hot spots for propagules of invading species. By definition, disturbances disrupt the ecosystem, community, or population structure and change the resources, substrate availability, or physical environment (White and Pickett 1985). Each particular disturbance generates new filtering effects that depend on (1) the type of disturbance, its intensity, the duration of the event, its frequency, and the spatial scale affected; (2) the interaction of disturbances; (3) the disturbance history; (4) the season at which the disturbance occurs; (5) the habitat characteristics; and (6) the regional and local species pool. Experimental studies clearly indicate, for example, that the plant species colonizing a disturbed site vary depending on the time of year when the disturbance occurred (Gross 1987).

In the model, this variation is implemented by an effect of external cues on the pool of established species. If stress and disturbances are severe enough, they will lead to local extinction, which in turn will affect the abiotic and particularly the biotic filter permeability via feedback loops. As a consequence, some species may disappear and others may appear for the first time. Henry et al. (1996) reported that among the 11 macrophyte species observed in a former channel of the Rhône River (France), four disappeared after a major flood, whereas 27 of 34 species observed after the flood were absent before the flood. When a disturbed ecosystem has entered the regeneration phase, the environment and the filters—in addition to extrinsic agents such as water, frost, wind, and inputs from the atmosphere—are steadily modified by the organisms' activity. Hence, even if we are considering a very long period of time, a return of the filters to exactly the same predisturbance state is very unlikely.

Finally, the abiotic and biotic filters are dependent on each other because they are connected by the pool of established species (see Figure 6.1). If, for example, the abiotic filter opens because environmental conditions are becoming tolerable for more and more species as the ecosystem recovers, the biotic filter will close because more and more sectors of possible niche dimensions for new species become occupied. Hence, increasing interspecific competition will make it incrementally more difficult for additional species to establish. In contrast, when the abiotic filter is almost closed due to extreme environmental conditions, the biotic filter widens because the few specialists still surviving reveal competitive dominance only because of the peculiar circumstances.

Filter Components

Because species differ in environmental requirements and behavior, two species may be excluded from the same community of interest for different reasons. Weiher and Keddy (1995) showed experimentally that different environmental conditions gave rise to different communities in sites with identical species pools. The proposed mechanism is that the environmental filters sort out species lacking traits required for that habitat, and competitive interactions subsequently modify community composition. The dynamic filter model is very similar to this idea, only our biotic filter is not restricted to competition. In this section we consider the two filters and their components separately.

The Abiotic Filter

Moisture and salinity may represent the most selective abiotic factors for one hypothetical plant species at one site, whereas another species may be excluded from the same site because of a combination of other unfavorable factors—such as soil structure and light conditions, for example. Many restoration studies have shown that the abiotic filter often represents the major barrier to the reappearance of certain species. For example, physical conditions, lack of nutrients, and toxicity usually prevent plant establishment and growth in mined lands (Bradshaw 1997; also see Chapter 16). According to various authors (for example, Clark 1997; also see Chapters 5 and 7), it is impossible to achieve self-sustaining restoration on a site that itself does not offer nonbiological functional equivalence. According to this view, merely restoring the physical habitat of streams could be enough if no toxic residuals impair the environment and nearby sources of colonizing species are adequate (Cairns 1995).

Examples from ecological restoration after disturbance or degradation suggest a general applicability of the first step in the two-step process. Some species, however, may need biotic interactions from the very beginning in order to recolonize a degraded site (see the next section), so this representation may not be valid for all species.

The Biotic Filter

Competition has been a major theme in ecology for nearly a century, and many authors have looked at the role of competition in assembling communities (see Keddy 2001). In the absence of competition, most plants grow best on fertile soils (Marschner 1986), and even halophytic species grow best under nonsaline conditions (Grace 2001b). It has been shown experimentally (for example, Leps 2001), however, that many plant species that are able to reach a site and grow well under the given abiotic conditions are eliminated from the community by competition.

In terms of the dynamic filter model, competitive exclusion is an important—but not the only—component of the biotic filter. Especially after disturbances that leave resources easily available, competition initially will be of lesser or even marginal importance compared to other biotic interactions. That established species may favorably change the abiotic constraints of a habitat for other species is well recognized in basic ecology and has been incorporated in succession concepts; for example, the facilitation pathway in Connell and Slatyer (1977).

The role of facilitation is perceived most easily in extreme environments. Mosses, for example, serve multiple purposes in the functioning of headwater streams; in particular, they are accumulators of fine detritus, a food supply for collector-gathering invertebrates (Laasonen et al. 1998). A positive relationship between moss biomass and macroinvertebrate diversity was documented in many studies (for example, Englund 1991, Suren 1991, Vuori and Joensuu 1996). In terrestrial primary successions, facilitation by nitrogen-fixing species has been found in many regenerating ecosystems, including strandlines (Davy and Figueroa 1993), volcanoes (Del Moral 1993), glacial moraines (Viereck 1966), and china clay wastes (Roberts et al. 1981). However, the interactions between nitrogen fixers and other species are also characterized by the full range of competitive interactions (Walker 1993). Finally, so-called nurse plants create micropatches for colonization by other species. Nurse plants include members of several groups: algae (Booth 1941), lichens (Cooper and Rudolph 1953), mosses (Clarkson and Clarkson 1983), liverworts (Griggs 1933), species that encourage free-living nitrogen-fixing bacteria (Hirose and Tateno 1984), and a wide range of other plants (Heath 1967, Tsuyusaki 1987, Suzán et al. 1996, Wied and Galan 1998; also see Chapter 14). So, apart from competitive exclusion, the degree of facilitation seems to be a crucial component of the biotic filter.

In terms of interactions between organisms of different functional groups, the presence of some plants may depend on certain pollinators, mycorrhizae (see Chapter 10), and other microorganisms. Also, trophic interactions are relevant, since herbivores may be excluded from a site because of a lack of food plants (see Chapter 9), and colonization and establishment of predators are possible only with adequate prey resources. Similar arguments apply to

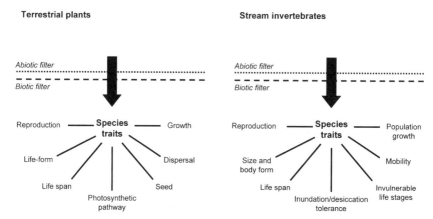

FIGURE 6.2 Examples of morphological, physiological, and life history traits important in terrestrial plants (*left*) and stream invertebrates (*right*). Species with specific traits at a given site are the result of filter discrimination for or against them over scales ranging from intrageneration to evolutionary time.

parasites that depend on particular hosts and to mutualists that depend on specific partners. Interactions between grassland plants and other groups of organisms suggest that the presence or absence of such functional groups as pathogens (Ryser 1993) and mycorrhizal fungi (Francis and Read 1994, Moora and Zobel 1996)—or such herbivores as arthropods, mollusks, and rodents (Hulme 1994, 1996)—may be more important than interspecific competition in determining the colonization success of invading seeds (Zobel et al. 2000).

Selection of Species Traits

So far, we have presented the dynamic filter model with an emphasis on the reasons that lead to the exclusion of species or that facilitate invasions. As illustrated in Figure 6.2, however, the environmental filters presumably act on traits rather than on species (Keddy 1992, Poff 1997, Booth and Larson 1999, Keddy and Weiher 1999). In searching for general patterns, therefore, it is important to collect information by screening many attributes, from relative growth rate to shoot thrust, in a large number of species (see Grime and Hunt 1975; Grime 1979; Grime et al. 1981, 1988). Application of the dynamic environmental filter model will be facilitated by a more careful consideration of general species traits and by further identification of the relative importance of specific traits in particular systems.

Recently, in both terrestrial and aquatic ecosystems, several attempts have been made to explain the presence or absence of taxa in particular communities on the basis of specific morphological, physiological, or life history traits (see the overview in Table 6.1). Weiher et al. (1999) proposed a common core list of plant traits to predict species responses to such disturbances as fire, grazing, and land-use change. To allow comparisons across regions, the list addresses the fundamentals that are common to nearly all plant species; that is, dispersal, establishment, and persistence. At present, a group of experts is working on an electronic database seen as a species-by-traits matrix, which considers vegetative and generative traits for a large part of the flora of central Europe (Poschlod et al. 2000). In stream ecology, a similar approach to enable biomonitoring of invertebrate communities at large spatial scales also has been developed in recent years using the genus taxonomic level (Dolédec et al. 1999, Charvet et al. 2000, Statzner et al. 2001).

Benefits and Open Questions

This book focuses on assembly rules in regenerating ecosystems; that is, the effects that established species and their interactions (as well as the abiotic environment) have on the order of establishment of additional species in a given degraded habitat. Ideas very similar to our model have been suggested repeatedly with respect to ecological restoration, and so this conceptual approach seems to be appropriate to address such systems. However, related concepts are also found in the literature on species community structure, disturbance, and succession. Although our model was derived from the study of two degraded ecosystems and aims to represent the process of regeneration, it would also adequately describe natural successions in nondegraded systems, because a degraded state, if such a state can be defined properly at all, is not explicitly incorporated in the model structure. In terms of the dynamic filter model, a degraded ecosystem would be distinguished by a somewhat odd current state of the established species pool and a particularly narrow mesh size of the abiotic filter due to severe environmental stress. The overlap of our model with existing concepts in both restoration ecology and succession theory is by no means disappointing, however. In fact, the overlap is rather promising because it may indicate that such a conceptual approach provides a possible link between basic ecological theory and problem-oriented restoration practice.

At the Dornburg workshop (described in the preface), we discussed several aspects of the model structure intensively without coming to a final agreement, which indicates that some clarification of the concept is still

TABLE 6.1

Examples of Taxa-Related Traits Studied in Different Communities

Communities	No. of taxa	Traits considered	Reference
Vascular plants of the British flora	502	Seed bank, seed weight, seed shape, dispersal mechanism, seedling characteristics, germination requirements, life-form, ecological strategy (CSR model), canopy height and structure, lateral spread, leaf phenology, mycorrhiza relations, flowering time and duration	Grime et al. (1988)
Wetland plants along the Ottawa River, Canada	88	Aboveground biomass, height, crown cover, stem diameter, capacity of lateral spread, "tussockness," number of rhizomes, form index of leaf shape, leaf unit area, height plasticity, crown plasticity	Weiher et al. (1998)
Alpine plants in NW Caucasus, Russia	42	Aboveground biomass, seed weight, seed production, seed yield, seed bank, reproduction effort, expenditure per offspring, relative growth rate, morphology index	Onipchenko et al. (1998)
Terrestrial plants in west–central Argentina	100	Photosynthetic pathway, ratio of leaf area to leaf weight, leaf succulence, C storage, vegetative spread, canopy height, life span, investment into support tissue, ramification at ground level, drought avoidance, thorniness, shoot phenology, seed size, seed shape, seeds per plant, dispersal mode, pollination mode, reproductive phenology	Díaz et al. (1998)
Riverine macrophytes of the Rhône River, France	34	Vegetative dissemination, flowering frequency, species distribution, relationship to overflow frequency, relationship to flow velocity, relationship to substrate grain size	Henry et al. (1996)
Benthic insects in tributaries of the Taieri River, New Zealand	35	Body length, adult mobility, habitat specificity, clinger/nonclinger, streamlined/flattened form, developmental stages outside the streams	Townsend et al. (1997)
Invertebrates in European running waters	179	Maximal size, longevity, reproductive cycles per year, aquatic stages, reproductive strategy, dispersal form, durable form, respiration organs, food, feeding habitats, locomotion-to-substratum relationships	Statzner et al. (2001)

needed. For example, in the original version of the model, we thought of disturbances as sudden changes of the filter mesh size; if the disturbances were severe enough, they would lead to local extinction of one or several species from the established species pool. Alternatively, it was suggested that the disturbance regime itself might be a filter (see Chapter 17). We also considered whether the socioeconomic aspects of restoration projects should be treated as external constraints of possible actions or should represent another filter (see Chapter 5), leading to a multilayer filter model. Most of the workshop participants agreed with the abiotic filter concept, but views deviated with respect to the biotic filter and the related feedback loops. The definition of this filter, therefore, seems to be somewhat difficult, and future discussions should concentrate on this aspect of the model.

Although two filters make up the basic model structure, we must be aware that a clear distinction between them may be difficult in some types of ecosystems. In a highly diverse and structured community, for example, the two interconnected and interdependent filters are more likely to have a simultaneous effect than to be temporally separated in a two-step process. Assessment of the abiotic and biotic filters may also depend on the spatial scales of interest. In marine soft-sediment habitats with disturbances at scales of up to 10 m², for example, both environmental conditions and biotic interactions control the successional dynamics of faunal communities (Zajac et al. 1998). It seems necessary to distinguish the effects of the abiotic filter from those of the biotic filter on relatively small scales in particular, because the relative importance of biotic interactions is much less at larger scales. On the other hand, if the spatial scale considered is too small, sampling units could exclude significant numbers of interacting individuals (Freckleton and Watkinson 2001).

Another important aspect of the model concerns the species pools. Several workshop participants felt uncomfortable with unstructured species pools, because in this view all species are identical. Particularly for the pool of established species, it was suggested that it may be essential to differentiate between core and satellite species, or—with even better reasoning—between ecological drivers and passengers (in the sense of Walker 1992). At present, however, it is impossible to determine whether such distinctions are really needed in the model. Conceptual models are conceptual because they brutally simplify complexity and ignore all the features that make systems particular. Hence, it is their natural fate to be blamed for being too simplistic. As a rule with all modeling, however, one should keep a model structure as simple as possible unless there are compelling reasons for increased complexity. Only if the model is applied to different systems and ecological situations and if relevant cases cannot fit the conceptual framework at all should changes be made.

The next important issue of discussion was the concept of multiple stable states of systems and the transition thresholds in between that must be passed to achieve progress in a restoration process (Hobbs and Norton 1996, Hobbs and Harris 2001; also see Chapter 5). The dynamic filter model seems to fit best to a more or less gradual change in filter permeability and pool composition, so probably the interactions between the compartments must be assumed to be nonlinear to allow for stepwise responses such as transition thresholds. If the conditions for thresholds could be identified structurally for theoretical reasons, they would also be of great importance for the restoration practitioner, because they suggest sensitive phases for "kicking" a system in the preferred direction. Restoration action in other phases may be inefficient or, even worse, may have detrimental effects.

We also were not able to reach agreement about how assembly rules are structurally related to the dynamic filter model. In our original conception, the biotic filter is a proxy of the established species with all their competitive, facilitative, inhibitory, trophic, and so forth, properties. Assembly rules of regenerating communities manifest in the pattern of species turnover that results from species invasions and subsequent readjustments of the permeability of the biotic (and partly also the abiotic) filter. Another view was that assembly rules are directly represented by the filters, because they determine the sequence of species turnover. A third idea was that the internal processes, which are largely ignored by the model, are equivalent to the assembly rules. Finally, an integrating concept tried to combine the different arguments: assembly rules determine what species are allowed to pass the filters at what time and by what mechanism. This divergent discussion was rather enlightening. On the one hand, it demonstrated that *assembly rule* as a scientific term is vaguely defined—or understood—and has multiple uses (see Chapter 3). On the other hand, it was obvious that restoration ecology is not yet scientifically mature enough to allow a clear distinction between pattern and process. This is a serious deficiency that needs more careful consideration in the future.

Apart from the many open questions, however, there was agreement that the dynamic filter model also has promising benefits. It not only provides an encouraging new approach in the search for a conceptual framework for the field of restoration ecology but also allows consideration of such practical restoration management actions as manipulating pools and flows within the same framework. Seeding, planting, and facilitation of species establishment with nurse plants all aim to increase the pool of established species by active addition of new species, which has consequences for both the biotic and abiotic filters to help the target species to establish. Establishing dispersal corridors may be interpreted as enlarging the external species pool, whereas planned disturbances such as mowing or burning can be represented as manipulations of the biotic filter by decreasing the established species pool. Unaided recovery of degraded systems can be formalized as a gradual opening of the abiotic filter, accompanied by a gradual narrowing of the biotic filter.

Conclusions

The dynamic filter model rests on the idea that new species can enter the pool of established species only if they are able to pass the abiotic and biotic filters, and that the permeability of the two filters changes dynamically due to feedback loops. For a given species, it would be important to identify the strongest filtering effect, because one or more components of the abiotic or biotic filter may prevent the species from successfully colonizing a disturbed site and persisting there. Environmental filters may exclude species by discriminating for or against particular traits over scales ranging from intrageneration to evolutionary time, so the strongest filtering effects will often differ between species. To make predictions about the establishment of a given species at relatively small spatial scales, much information is needed about the potential invader and the local pool of established species. Since it is very time-consuming and almost impossible to collect all this information for all species of interest, priority should be given to general functional traits that will allow comparisons among communities and across biogeographic regions that differ in their taxonomic composition (for example, Grime 1979, Grime et al. 1988).

As a concept, the dynamic filter model is not substantially new. Rather, it binds together the often-invoked metaphor of filter processes with ideas of linked species pools at different spatial scales (for example, Keddy 1992; Palmer et al. 1997; Poff 1997; Zobel 1997; Díaz et al. 1998, 1999). What is new in our model, however, is its dynamic nature, because none of the components or flows are thought of as being fixed. At its present state, the dynamic filter model is a suggestion, not a fully tested, ready-to-go solution, and using it may best reveal its strengths and deficiencies. As the next essential step, restoration practitioners are invited to see if they can place their specific system and management options into this scheme. We expect that this is possible, because the model is formulated generally enough to be widely applicable. Indeed, some practitioners will feel forced into narrow pathways and will argue that they have a different description in mind that fits their particular system much better. However, that is exactly what a framework should do. It provides only a very limited set of available description options, so "forcing" a particular case into this frame means that it becomes comparable to other cases on the basis of a least common denominator (see Chapter 2).

One more crucial question turned up in our discussion and urgently remains to be answered: If the dynamic filter model is in fact able to describe regeneration and restoration as well as natural succession, what then is special about restoration ecology? Are the problems in fact so specific that it is justified to treat restoration ecology as a discipline on its own, or are all the relevant issues covered by succession and disturbance theory? Maybe the only difference, after all, is just that the endpoint of succession is not defined by species interaction alone (that is, in the sense of some ecologically determined climax state) but by some agreed-upon restoration goal (that is, in the sense of a desired system state). If so, the search for an original conceptual framework for restoration ecology would be meaningless because we can already find such frameworks in ecology textbooks. In this case, efforts instead should concentrate on adapting existing concepts to the particular problems in ecological restoration. We think the question is indeed open, and the answer to it can become clearer only if restorationists are much more aware of the conceptual dimensions of their everyday work and consider which conceptual approaches are useful for them—and why.

Acknowledgments

We thank all the participants at the Dornburg assembly rules workshop for their crucial contributions to the discussion, which helped us to understand our model much better. We are very grateful to Richard Hobbs, Paul Keddy, Tim Nuttle, and Martin Zobel for their thoughtful comments on earlier drafts of the chapter and to the *Deutsche Forschungsgemeinschaft* (DFG, German Research Council) for financial support of our work over the years.

REFERENCES

Booth, B. D.; and Larson, D. W. 1999. Impact of language, history, and choice of system on the study of assembly rules. In *Ecological Assembly Rules: Perspectives, Advances, Retreats*, ed. E. Weiher and P. A. Keddy, 206–229. Cambridge: Cambridge University Press.

Booth, W. E. 1941. Algae as pioneers in plant succession and their importance in erosion control. *Ecology* 22:38–46.

Bradshaw, A. D. 1997. Restoration of mined lands: using natural processes. *Ecological Engineering* 8:255–269.

Cairns, J. 1995. Restoration ecology: protecting our national and global life support system. In *Rehabilitating Damaged Ecosystems*, ed. J. Cairns, 1–12. Boca Raton, Fla.: Lewis Publishers.

Charvet, S.; Statzner, B.; Usseglio-Polatera, P.; and Dumont, B. 2000. Traits of benthic macroinvertebrates in semi-natural French streams: an initial application to biomonitoring in Europe. *Freshwater Biology* 43:277–296.

Clark, M. J. 1997. The magnitude of the challenge: an outsider's view. In *Restoration Ecology and Sustainable Development*, ed. K. M. Urbanska, N. R. Webb, and P. J. Edwards, 353–377. Cambridge: Cambridge University Press.

Clarkson, B. R.; and Clarkson, B. D. 1983. Mt. Tarawera. 2. Rates of change in the vegetation and flora of the high domes. *New Zealand Journal of Ecology* 6:107–119.

Clements, F. E. 1916. *Plant Succession*. Publication No. 242. Washington, D.C.: Carnegie Institution.

Clements, F. E. 1936. The nature and structure of the climax. *Journal of Ecology* 22:9–68.

Connell, J. H.; and Slatyer, R. O. 1977. Mechanisms of succession in natural communities and their role in community stability and organization. *American Naturalist* 111:1119–1144.

Cooper, R.; and Rudolph, E. D. 1953. The role of lichens in soil formation and plant succession. *Ecology* 34:805–807.

Davy, A. J.; and Figueroa, M. E. 1993. The colonization of strandlines. In *Primary Succession on Land*, ed. J. Miles and D. W. H. Walton, 113–145. Oxford: Blackwell.

De Leo, G. A.; and Levin, S. 1997. The multifaceted aspects of ecosystem integrity. *Conservation Ecology* 1:3.

Del Moral, R. 1993. Mechanisms of primary succession on volcanoes: a view from Mt. St. Helens. In *Primary Succession on Land*, ed. J. Miles and D. W. H. Walton, 79–100. Oxford: Blackwell.

Díaz, S.; Cabido, M.; and Casanoves, F. 1998. Plant functional traits and environmental filters at a regional scale. *Journal of Vegetation Science* 9:113–122.

Díaz, S.; Cabido, M.; and Casanoves, F. 1999. Functional implications of trait-environment linkages in plant communities. In *Ecological Assembly Rules: Perspectives, Advances, Retreats*, ed. E. Weiher and P. A. Keddy, 338–362. Cambridge: Cambridge University Press.

Dolédec, S.; Statzner, B.; and Bournard, M. 1999. Species traits for future biomonitoring across ecoregions: patterns along a human-impacted river. *Freshwater Biology* 42:737–758.

Drake, J. A. 1990. The mechanics of community assembly and succession. *Journal of Theoretical Biology* 147:213–233.

Englund, G. 1991. Effects of disturbance on stream moss and invertebrate community structure. *Journal of the North American Benthological Society* 10:143–153.

Francis, R.; and Read, D. J. 1994. The contribution of mycorrhizal fungi to the determination of plant community structure. *Plant and Soil* 159:2593–2602.

Freckleton, R. P.; and Watkinson, A. R. 2001. Nonmanipulative determination of plant community dynamics. *Trends in Ecology and Evolution* 16:301–307.

George, L. O.; and Bazzaz, F. A. 1999. The fern understory as an ecological filter: growth and survival of canopy tree-seedlings. *Ecology* 80:846–856.

Gleason, H. A. 1917. The structure and development of the plant association. *Bulletin of the Torrey Botany Club* 44:463–481.

Gleason, H. A. 1926. The individualistic concept of the plant association. *Bulletin of the Torrey Botany Club* 53:7–26.

Grace, J. B. 2001a. The roles of community biomass and species pools in the regulation of plant diversity. *Oikos* 92:193–207.

Grace, J. B. 2001b. Difficulties with estimating and interpreting species pools and the implications for understanding patterns of diversity. *Folia Geobotanica* 36:71–83.

Griggs, R. F. 1933. The colonization of the Katmai ash, a new and inorganic soil. *American Journal of Botany* 20:92–111.

Grime, J. P. 1979. *Plant Strategies and Vegetation Processes.* Chichester, England: John Wiley and Sons.

Grime, J. P.; Hodgson, J. G.; and Hunt, R. 1988. *Comparative Plant Ecology: A Functional Approach to Common British Species.* London: Unwin Hyman.

Grime, J. P.; and Hunt, R. 1975. Relative growth-rate: its range and adaptive significance in a local flora. *Journal of Ecology* 63:393–422.

Grime, J. P.; Mason, G.; Curtis, A. V.; Rodman, J.; Band, S. R.; Mowforth, M. A. G.; Neal, A. M.; and Shaw, S. 1981. A comparative study of germination characteristics in a local flora. *Journal of Ecology* 69:1017–1059.

Gross, K. L. 1987. Mechanisms of colonization and species persistence in plant communities. In *Restoration Ecology: A Synthetic Approach to Ecological Research*, ed. W. R. Jordan; M. E. Gilpin, and J. D. Aber, 173–188. Cambridge: Cambridge University Press.

Hairston, N. G., Jr.; van Brunt, R. A.; and Kearns, C. M. 1995. Age and survivorship of diapausing eggs in a sediment egg bank. *Ecology* 76:1706–1711.

Heath, J. P. 1967. Primary conifer succession, Lassen Volcanic National Park. *Ecology* 48:270–275.

Henry, C. P.; Amoros, C.; and Bornette, G. 1996. Species traits and recolonization processes after flood disturbances in riverine macrophytes. *Vegetatio* 122:13–27.

Hirose, T.; and Tateno, M. 1984. Soil nitrogen patterns induced by colonization of *Polygonum cuspidatum* on Mt. Fuji. *Oecologia* 61:218–223.

Hobbs, R. J.; and Harris, J. A. 2001. Restoration ecology: repairing the earth's ecosystems in the new millennium. *Restoration Ecology* 9:239–246.

Hobbs, R. J.; and Norton, D. A. 1996. Towards a conceptual framework for restoration ecology. *Restoration Ecology* 4:93–110.

Hulme, P. E. 1994. Seedling herbivory in grassland: relative impact of vertebrate and invertebrate herbivores. *Journal of Ecology* 82:873–880.

Hulme, P. E. 1996. Herbivores and the performance of grassland plants: a comparison of arthropod, mollusc and rodent herbivory. *Journal of Ecology* 84:43–51.

Jones, C. G.; Lawton, J. H.; and Shachak, M. 1994. Organisms as ecosystem engineers. *Oikos* 69:73–86.

Jones, C. G.; Lawton, J. H.; and Shachak, M. 1997. Positive and negative effects of organisms as physical ecosystem engineers. *Ecology* 78:1946–1957.

Keddy, P. A. 1992. Assembly and response rules: two goals for predictive community ecology. *Journal of Vegetation Science* 3:157–164.

Keddy, P. A. 2001. *Competition.* 2nd ed. Dordrecht, The Netherlands: Kluwer.

Keddy, P. A.; and Weiher, E. 1999. Goals, obstacles, and opportunities for community ecology. In *Ecological Assembly Rules: Perspectives, Advances, Retreats*, ed. E. Weiher and P. A. Keddy, 1–22. Cambridge: Cambridge University Press.

Laasonen, P.; Muotka, T.; and Kivijärvi, I. 1998. Recovery of macroinvertebrate communities from stream habitat restoration. *Aquatic Conservation: Marine and Freshwater Ecosystems* 8:101–113.

Leps, J. 2001. Species-pool hypothesis: limits to its testing. *Folia Geobotanica* 36:45–52.

Marcus, N. M.; Lutz, R.; Burnett, W.; and Cable, P. 1994. Age, viability, and vertical distribution of zooplankton resting eggs from an anoxic basin: evidence of an egg bank. *Limnology and Oceanography* 39:154–158.

Marschner, H. 1986. *Mineral Nutrition of Higher Plants.* New York: Academic Press.

Moora, M.; and Zobel, M. 1996. Effect of arbuscular mycorrhiza on inter- and intraspecific competition of two grassland species. *Oecologia* 108:79–84.

Mueller-Dombois, D.; and Ellenberg, H. 1974. *Aims and Methods of Vegetation Ecology.* New York: John Wiley and Sons.

Onipchenko, V. G.; Semenova, G. V.; and van der Maarel, E. 1998. Population strategies in severe environments: alpine plants in the northwestern Caucasus. *Journal of Vegetation Science* 9:21–40.

Palmer, M. A.; Ambrose, R. F.; and Poff, N. L. 1997. Ecological theory and community restoration ecology. *Restoration Ecology* 5:291–300.

Parker, V. T.; and Pickett, S. T. A. 1997. Restoration as an ecosystem process: implications for the modern ecological paradigm. In *Restoration Ecology and Sustainable Development,* ed. K. M. Urbanska, N. R. Webb, and P. J. Edwards, 17–32. Cambridge: Cambridge University Press.

Poff, N. L. 1997. Landscape filters and species traits: towards mechanistic understanding and prediction in stream ecology. *Journal of the North American Benthological Society* 16:391–409.

Poff, N. L.; and Ward, J. V. 1990. Physical habitat template of lotic systems: recovery in the context of spatiotemporal heterogeneity. *Environmental Management* 14:629–645.

Poschlod, P.; Kleyer, M.; and Tackenberg, O. 2000. Databases on life history traits as a tool for risk assessment in plant species. *Zeitschrift für Ökologie und Naturschutz* 9:3–18.

Roberts, R. D.; Marrs, R. H.; Skeffington, R. A.; and Bradshaw, A. D. 1981. Ecosystem development on naturally colonized china clay wastes. I. Vegetation changes and overall accumulation of organic matter and nutrients. *Journal of Ecology* 69:153–162.

Rykiel, E. J. 1985. Towards a definition of ecological disturbance. *Australian Journal of Ecology* 10:361–365.

Ryser, P. 1993. Influences of neighbouring plants on seedling establishment in limestone grassland. *Journal of Vegetation Science* 4:195–202.

Statzner, B.; Bis, B.; Dolédec, S.; and Usseglio-Polatera, P. 2001. Perspectives for biomonitoring at large spatial scales: a unified measure for the functional composition of invertebrate communities in European running waters. *Basic and Applied Ecology* 2:73–85.

Suren, A. M. 1991. Bryophytes as invertebrate habitat in two New Zealand streams. *Freshwater Biology* 26:399–418.

Suzán, H.; Nabhan, G. P.; and Patten, D. T. 1996. The importance of *Olneya tesota* as a nurse plant in the Sonoran Desert. *Journal of Vegetation Science* 7:635–644.

Townsend, C. R.; Dolédec, S.; and Scarsbrook, M. R. 1997. Species traits in relation to temporal and spatial heterogeneity in streams: A test of habitat templet theory. *Freshwater Biology* 37:367–387.

Tsuyusaki, S. 1987. Origin of plants recovering on the Volcano Usu, Northern Japan, since the eruption of 1977 and 1978. *Vegetatio* 73:53–58.

Tumbull, L. A.; Crawley, M. J.; and Rees, M. 2000. Are plant populations seed-limited? A review of seed sowing experiments. *Oikos* 88:225–238.

van Breemen, N. 1995. How Sphagnum bogs down other plants. *Trends in Ecology and Evolution* 10:270–275.

van der Walk, A. G. 1981. Succession in wetlands: a Gleasonian approach. *Ecology* 62:688–696.

Viereck, L. A. 1966. Plant succession and soil development on gravel outwash of the Muldrow Glacier, Alaska. *Ecological Monographs* 36:181–199.

Vuori, K.-M.; and Joensuu, I. 1996. Impact of forest drainage on the macroinvertebrates of a small boreal headwater stream: do buffer zones protect lotic biodiversity? *Biological Conservation* 77:87–95.

Walker, B. H. 1992. Biodiversity and ecological redundancy. *Conservation Biology* 6:18–23.

Walker, L. R. 1993. Nitrogen fixers and species replacements in primary succession. In *Primary Succession on Land*, ed. J. Miles and D. W. H. Walton, 249–272. Oxford: Blackwell.

Weiher, E.; Clarke, G. D. P.; and Keddy, P. A. 1998. Community assembly rules, morphological dispersion, and the coexistence of plant species. *Oikos* 81:309–322.

Weiher, E.; and Keddy, P. A. 1995. The assembly of experimental wetland plant communities. *Oikos* 73:323–335.

Weiher, E.; van der Werf, A.; Thompson, K.; Roderick, M.; Garnier, E.; and Eriksson, O. 1999. Challenging Theophrastus: a common core list of plant traits for functional ecology. *Journal of Vegetation Science* 10:609–620.

White, P. S.; and Pickett, S. T. A. 1985. Natural disturbance and patch dynamics, an introduction. In *The Ecology of Natural Disturbance and Patch Dynamics*, ed. S. T. A. Pickett and P. S. White, 3–13. New York: Academic Press.

Wied, A.; and Galen, C. 1998. Plant parental care: conspecific nurse effects in *Frasera specios* and *Cirsium scopulorum*. *Ecology* 79:1657–1668.

Zajac, R. N.; Whitlatch, R. B.; and Thrush, S. F. 1998. Recolonization and succession in soft-sediment infaunal communities: the spatial scale of controlling factors. *Hydrobiologia* 375/376:227–240.

Zobel, M. 1997. The relative role of species pool in determining plant species richness: an alternative explanation of species coexistence? *Trends in Ecology and Evolution* 12:266–269.

Zobel, M.; Otsus, M.; Liira, J.; Moora, M.; and Möls, T. 2000. Is small-scale species richness limited by seed availability or microsite availability? *Ecology* 81:3274–3282.

Chapter 7

Beyond Ecological Filters: Feedback Networks in the Assembly and Restoration of Community Structure

Lisa R. Belyea

Restoration of degraded systems is perhaps the most challenging of all applications of ecological knowledge. Successful restoration demands a thorough understanding of the degraded and target systems, the particular site characteristics, and the autecology of target species—hard-won empirical knowledge directly applicable to particular sites and situations. But can ecology contribute any theoretical guidance that will increase the success rate and reduce the information demands of a purely empirical, case-by-case approach?

Ecological assembly is perhaps the branch of ecological theory most relevant to ecosystem restoration, but there are important differences in goals and approaches. Ecological assembly attempts to explain how the structure and function of a community and ecosystem develop, whereas restoration attempts to direct community development to a particular state. In restoration projects, humans impose value judgments to define targets for remedial action, so that a preferred direction for further development of the community is prescribed. Either the degraded system or the pathway leading from it to the target, or both, may have no analog in the assembly of natural systems. In this chapter, I consider how restoration—pushing the degraded system toward the target—fits within two types of conceptual models of ecological assembly: filter models and systems models.

Filter Models of Ecological Assembly and Restoration

Both the assembly of natural systems and the restoration of degraded systems occur within constraints imposed by the availability of potential colonists, by the demands of the abiotic environment, and by the interactions among

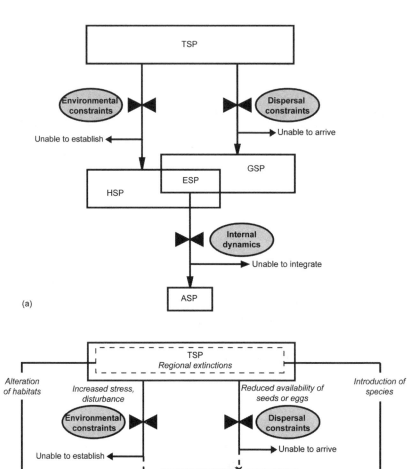

FIGURE 7.1 Filter model of changes to species pools during three stages of colonization in (a) a nondegraded and (b) a degraded ecosystem. Boxes represent species pools; solid-headed arrows with valves represent the determinants (dispersal constraints, environmental constraints, and internal dynamics) of the membership of each pool. Of the total species pool (TSP), only a subset (geographical species pool, GSP) will arrive at the site, either by dispersal or by emergence from a seed or egg bank. Species in another subset (habitat species pool, HSP) are capable of establishing under the environmental conditions at

constituent species (Belyea and Lancaster 1999). In much of the literature on community assembly (for example, Keddy 1992; also see Chapter 3), dispersal, abiotic conditions, and biotic interactions are portrayed as filters that exclude all but a select group of species or traits. In a selective process, the total pool of potential colonists is gradually whittled down to leave a subset of the original.

Terminology for the filter and species pool approach to community assembly differs among authors. I adopt the scheme of Kelt et al. (1995) because it is one of the few to link species pools to particular mechanisms of constraint or filtering (Figure 7.1a). The *total species pool* (TSP) is determined by biogeography and evolution; its membership includes all the species in a region. A subset of the TSP, the *geographical species pool* (GSP), is limited by dispersal constraints; membership includes only those species that can arrive at the target site by migration, dispersal, or release from a seed or egg bank. Another subset of the TSP, the *habitat species pool* (HSP), is defined by environmental constraints; membership includes only those species that can establish under the abiotic conditions at the target site. Overlap between the GSP and the HSP defines the *ecological species pool* (ESP); membership is limited to those species that overcome both dispersal and environmental constraints. An even smaller subset, the *actual species pool* (ASP), is limited by internal dynamics of the organisms already at the target site; membership includes only those species that can integrate into the existing community. Hence, constraints on the ASP operate at all phases (Moyle and Light 1996) of colonization: arrival of colonists, establishment of populations, and integration into the community (Figure 7.1a).

Ecosystem degradation may affect any or all of these phases, with consequent effects on species pools (Figure 7.1b). In geographical regions heavily influenced by human activity, the TSP may be depleted by regional extinctions. The GSP may be reduced if seed and egg banks at the degraded site are destroyed by loss of soil or sediment, or if dispersal from surrounding

---

the site. The intersection (ecological species pool, ESP) of these two subsets includes species that arrive at the site and establish viable populations. A subset (actual species pool, ASP) of these species integrates with existing members of the community. In (b), note that species pools in the degraded system (boxes with dashed outlines) are reduced relative to those in the nondegraded system (boxes with solid outlines). Dashed parts of arrows show links that are weakened. Open-headed arrows show human interventions to bypass particular constraints. Terminology follows Kelt et al. 1995 and Belyea and Lancaster 1999. Layout was inspired by Moyle and Light 1996.

areas is delayed or precluded by low rates of emigration from small and fragmented populations or by human-imposed barriers such as roads and inten-
sive agriculture. The HSP may be reduced if extremely stressful or highly disturbed abiotic conditions prevent establishment at the degraded site. The
combination of dispersal and environmental constraints will determine
effects on the ESP. Integration into the existing community may be precluded either by the presence of species having a negative effect on the
colonist (for example, predators, competitors) or by the absence of species
having a positive effect (for example, mutualistic symbionts). Consequently,
the ASP at a degraded site may be reduced relative to, or may differ substantially from, that at a nondegraded site. The aim inherent in many restoration activities is to bypass dispersal and environmental constraints, allowing
desired species to arrive and to establish (see Figure 7.1b). More rarely,
restoration activities aim to modify internal dynamics, allowing desired
species to integrate into the existing community.

As a hypothetical example of a practical application of the filter model,
consider a grassland degraded by acidic waste. Constraints on the plant
species pool might be identified by examining the seed bank (dispersal constraints?), by screening the ability of seeds to germinate and grow in the
degraded soil (environmental constraints?), or by testing the impact of invertebrate herbivores known to be present (internal dynamics?). Once the
mechanism of constraint is identified, remedial action can be taken to bypass
it. Acid-tolerant grass species may be planted directly, to overcome the lack
of a seed bank and so assist in the arrival phase. Lime may be added to the
soil, to raise soil pH and so assist in the establishment phase. Or seedlings
may be covered with mesh, to exclude herbivores and so assist in the integration phase.

Although filter models can provide insight into what is limiting membership of the species pool and so point to specific interventions, they tell
us little about how the community is structured or how the ecosystem functions. Even if we are interested only in restoring a particular list of species
to a site, is it sufficient simply to have all the species present? Or do we
need to know something about community structure (relative abundance
of species, trophic relationships, spatial arrangements) and dominant processes (competitive exclusion, biogeochemical cycling, soil development)?
If we treat the degraded or restored community simply as a black box containing a list of species, we forfeit insight into how constraints on its membership might change over time or with particular interventions. We need
a model for assembly and restoration that is more sophisticated than the filter approach.

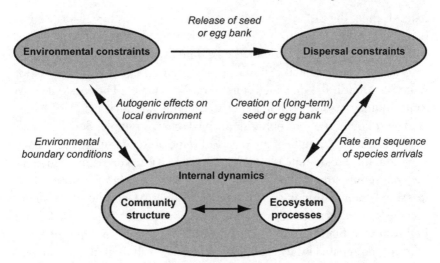

Figure 7.2 Systems model showing interactions among environmental constraints, dispersal constraints, and internal dynamics. Note that internal dynamics include both community structure and ecosystem processes.

Systems Models of Ecological Assembly and Restoration

Instead of focusing on species pools and ecological filters (see Figure 7.1), an alternative approach is to focus on the relationships among determinants of species pools (Figure 7.2). I will refer to this as a systems model approach. One of its distinguishing features is that processes occurring within the community or ecosystem (its internal dynamics) both respond to and alter the external constraints of dispersal and abiotic environment. For example, lake microcrustacea may produce an egg bank that becomes buried in the sediment, allowing recolonization after a disturbance that eliminates the original population. Or bottom-feeding fish may disturb the lake sediments, thereby increasing turbidity in the water column and shifting the competitive balance between algae and macrophytes. Hence, systems models can reveal feedback networks in which the biotic community changes the constraints within which it operates, leading to further changes in the community (see Figure 7.2).

Systems models help to answer two practical questions that summarize the challenge of restoration: Where is the degraded state in relation to the target? And how can ecosystem processes and community structure be manipulated to reach the target? As with multivariate analyses (Weiher and Keddy 1995), the systems approach allows us to map the degraded state, the target, and natural assembly trajectories onto a common system of coordinates. Hence, the distance between degraded state and target—as well as the progress of restoration—can be measured quantitatively (Keddy 1999). If the coordinate system is defined operationally by variables that summarize many of the essential features of the system, the plot also can suggest strategies to push the degraded state toward the target. The most effective strategies may involve reinforcing or weakening particular feedback loops.

The restoration of eutrophic shallow lakes (for example, Scheffer et al. 1993) is a classic example of the practical value of systems models. In many regions of the world, increased nutrient loading has led to major changes in the community structure of shallow lakes. The system can be summarized in a simple plot of turbidity versus nutrient loading. At low phosphorus loading, the water is clear and the community is dominated by macrophytes and herbivores. At high phosphorus loading, the water is turbid and the community is dominated by algae and bottom-feeding fish. Both the turbid- and clear-water states can exist at intermediate phosphorus concentrations. These "alternative stable states" are problematic because, in the absence of other restoration activities, phosphorus loading must be reduced to very low levels to ensure transition from the turbid- to the clear-water state. Successful restoration depends on forcing turbidity below a critical threshold long enough for macrophytes to become established. One of the most effective ways of decreasing turbidity enough to flip the system back to a clear-water state is to remove the fish (for example, Moss et al. 1996): direct resuspension of bottom sediments during fish feeding is reduced, and herbivore populations that regulate algal concentrations recover in the absence of predation. This simple, powerful model (Scheffer et al. 1993) of system feedbacks demands a deep understanding of the system under study, but the effort is rewarded by immense insight into how the system responds to interventions.

Systems models of ecological assembly complement, rather than replace, filter models. At the early stages of a restoration project, filter models may help to identify key drivers and suggest strategies to bypass constraints that prevent the arrival or establishment of particular species. Systems models may be more helpful for issues that involve more than one species: assessing how the community is changing in relation to the target, identifying the key drivers, and suggesting strategies for modifying community structure and ecosystem processes in ways that will transform the whole system.

A Case Study: The Assembly of Spatial Patterns on Peatlands

In this section, I will review how a specific community structure—surface patterning on peatlands—assembles naturally. Throughout, I will emphasize the feedback networks linking community structure to ecosystem processes and external constraints. After plotting the natural assembly trajectory in a functionally meaningful way, I will use both filter and systems models to assess the effects of peatland degradation and strategies for restoration.

Peatland Development and the Spatial Structure of Peatland Communities

A peatland, or mire, is an area of land covered by waterlogged and semi-decomposed organic matter. Mires are broadly separated into fens, which receive inputs of minerogenous water, and bogs, which do not (Økland et al. 2001). In the northern temperate region, peat deposits have been accruing since the most recent deglaciation and may range from less than a meter to more than 10 meters in thickness. Peat does not accumulate in bogs because of high rates of litter production but because of low rates of anaerobic decay in the anoxic layer below the water table (Clymo 1984). Even as the bog grows in thickness, the water table remains near the surface because losses of water are exceeded by inputs (Ingram 1982). Averaged over periods of decades, the net rate of water storage must equal the net rate of peat accumulation (Belyea and Clymo 2001). On shorter time scales, the amount of peat lying in the oxic layer above the water table—the surface microtopography—fluctuates with climate from year to year (Belyea and Clymo 2001).

The thickness of the oxic surface layer may vary spatially as well as temporally. The surfaces of many bogs are characterized by distinct microforms, ranging from dry hummocks, in which the water table rises only to within 50–100 cm of the surface, to pools that are permanently water-filled (Figure 7.3). Together, these microforms make a gradient of water table depth that is occupied by distinct associations of plants and animals, ranging from terrestrial to aquatic. The spatial arrangement of different types of microforms also gives a horizontal structure to the community. Aerial views of surface patterns formed by the co-occurrence of densely vegetated areas and open water are spectacular. The shape, orientation, and spacing of the pools seem to be related to landscape topography; for example, pools are often elongated parallel to topographic contours. The formation of these distinct, nonrandom surface patterns is a highly visual example of the spatial assembly of community structure.

Assembly of Natural Spatial Patterns

The assembly of surface patterns depends on the peatland surface diverging into at least two different types of microforms. It also depends on an initially random arrangement of microforms becoming organized into a nonrandom

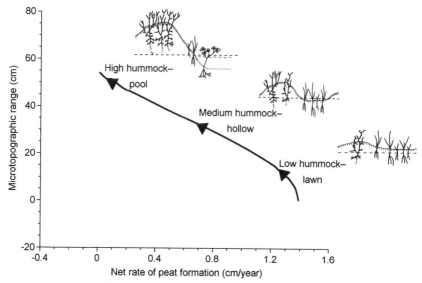

FIGURE 7.3 Assembly of surface microtopography on peatlands. Arrows show the inferred direction of microtopographic development, with values derived from Belyea and Clymo 2001. Microtopographic range is the height difference separating microforms that are stable at the corresponding net rate of peat formation. Line drawings show representative cross sections of the surface layer at three stages of development. In each cross section, the horizontal line shows the position of the water table; note that diversity of plant species increases with microtopographic range.

pattern. The precise mechanisms involved are still under debate, but we now have at least a partial understanding of the process.

The peatland surface diverges into different types of microforms in response to long-term changes in the rate of water storage (Belyea and Clymo 2001). Over the course of a bog's development, the water table always remains close to the surface of the growing peat deposit. Hence, the rate at which water is stored in the peat matches, on average, the rate at which the peat accumulates. As the peat deposit grows, changes in physical hydrology are likely to decrease the rate of water storage (Hilbert et al. 2000, Belyea and Clymo 2001), necessitating a decrease also in the net rate of peat accumulation. This matching change in peat accumulation can be only partially accounted for by increasing losses of material from the thickening layer of anoxic peat (Clymo 1984). Inputs of material to the anoxic layer must also decrease. This decrease in input, or rate of peat formation, is achieved by microform development (Belyea and Clymo 2001; also see Figure 7.3). Once past a minimum size (unstable point), a lawn or small hummock grows taller until decay limits further growth and the rate of peat accumulation matches the rate of water storage. At this stable point, year-to-year fluctuations in water table depth are compensated by minor adjustments in hummock height. In contrast, a lawn that falls below the unstable point will lose height, and the hollow, then pool, formed in its place will continue to deepen. If the rate of water storage is large and positive, the peatland surface is dominated by lawns that accumulate peat very quickly. As the rate of water storage decreases, the surface differentiates into hummocks and pools that accumulate peat very slowly or lose more peat by decay than they gain by new production. Hence, if water storage change is low enough, the relationship between rate of peat accumulation and water table position drives a divergent succession (Sjörs 1990) of the surface into hummocks and pools (see Figure 7.3).

Microforms can become organized into nonrandom patterns through a positive feedback loop: microforms alter local environmental constraints, which in turn affect the distribution of the microforms. The proposed mechanism is that the physical properties of different microforms change local pathways of water flow, so that some parts of the bog surface receive more water than others (Ivanov 1981, Swanson and Grigal 1988, Sjörs 1990). Pools tend to initiate on the wetter parts, whereas hummocks tend to initiate on the drier parts (Ivanov 1981, Swanson and Grigal 1988, Sjörs 1990). Swanson and Grigal (1988) simulated this network of processes to verify that it could drive pattern formation. Their simulations demonstrate that under certain combinations of environmental constraints (for example, slope and water input rate), striking patterns can indeed emerge from local interactions among microforms. Most notably, a banded pattern reminiscent of the string-and-flark systems of northern fens (Sjörs 1961) emerges for certain combinations of parameters. The mechanism involves a feedback loop between peat growth and water flow: water tends to pond upslope of hummocks, and hummocks tend to initiate on drier sites adjacent to existing hummocks. The surface patterns change with alteration of the model parameters. For example, banded patterns fail to emerge if the slope is reduced below a critical value. Hence, the same set of microforms (and plant species), interacting in the same way, can assemble into very different community structures under different environmental constraints.

Peatland Degradation and Restoration

Human disturbance of peatlands in northern temperate regions takes two main forms, each causing a distinct alteration of surface structure. Many

raised bogs have been drained and the peat mechanically harvested for horticulture or for power generation. The remnant cutover peatlands are flat
expanses of bare peat, devoid of vegetation and spatial structure, other than
drainage ditches orientated downslope. On shallower peat, such as the extensive, treeless blanket bogs of northern Scotland, the main human intervention is drainage of the peat and the planting of exotic tree species (Warren
2001). In these plantations, the peatlands retain most of their ground-layer
vegetation, but the surface microtopography is exaggerated and reoriented
downslope into a parallel series of ditches and spoil heaps. In both cases,
drainage has a major impact on the ability of the system to store water and
accumulate peat.

On cutover sites, removal of surface vegetation and upper peat layers
presents a further challenge to restoration. The ability of the system to accumulate peat is eliminated. Harvesting may expose deep peat that is in contact with mineral-rich groundwater, thereby reducing nutrient limitation at
the peatland surface. Removal of surface peat also removes the seed bank,
placing additional constraints on the arrival of peat-forming plant species.
Even if the water table has been raised to its former level, the surface of a
cutover peatland may experience a severely drying microclimate that prevents the establishment of native hydrophilous species (Campeau and
Rochefort 1996).

On planted peatlands, the challenge of restoration is to eradicate nonnative conifers from artificially dry microsites. The creation of spoil heaps
extends the range of niches available, so that regeneration of trees from the
seed bank occurs long after the adults have been felled (P. Baker, Royal Society for the Protection of Birds, personal communication). Exaggeration of the
microtopography also has an indirect, negative effect on peat accumulation,
but it is one that might reverse if capacity for water storage is restored.

Regardless of the cause or nature of peatland degradation, the goal of
restoration is often to return the degraded site to as near its original state as
possible, in terms of both ecological function and habitat for native flora and
fauna (Wheeler and Shaw 1995). Restoration of the vertical structure of the
peatland surface is an integral, if implicit, part of this goal because the microtopography provides the niches required for a diverse range of species. The
most effective strategy may be to allow the peatland surface to redevelop
from its most primitive state; that is, by rejoining the natural assembly trajectory where microtopography is subdued (lawns to low hummocks) and
peat accumulation rates are fastest (Figure 7.4). Microforms can then
develop naturally to sizes that match rates of peat accumulation and water
storage. For planted peatlands, an exaggerated vertical structure must be

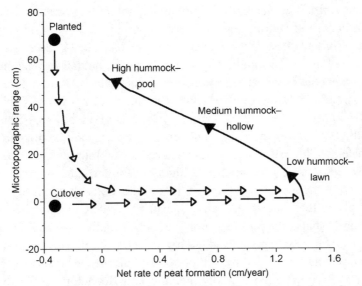

FIGURE 7.4 Possible trajectories for restoration of cutover and planted peatlands. Solid line and arrows show the natural assembly trajectory, as in Figure 7.3. Solid circles show the approximate coordinates of degraded sites. Open arrows denote possible trajectories for restoration.

reduced. For cutover peatlands, the main barrier is reestablishment of a peat-forming vegetation. In both cases, restoration of the ability to store water is the first priority.

Filter (Figure 7.5) and systems (Figure 7.6) approaches can be combined to assess how species pools and ecological processes might be restored using direct or indirect intervention (see Figures 9.5 and 9.6 in Chapter 9). In both types of degraded site, water losses are reduced directly by blocking drains or, more rarely, by installing waterproof membranes along the perimeter. On cutover peatlands, drains are sometimes filled with peat and the bare surface reshaped to create dams or bunds that reduce overland flow (Figure 7.6a). Opportunities for such large-scale engineering work are more limited on planted peatlands, because the surface vegetation is still intact. Recent restoration of planted sites in northern Scotland included felling exotic trees and placing them into the ditches intact. The expectation is that branches of the felled trees will act as a climbing frame, enhancing growth of the peat-forming moss, *Sphagnum*, in the ditches (N. Russell, Royal Society for the Protection of Birds, personal communication). If the peat formed in the ditches is sufficiently dense, it may impede water loss by blocking the drains

(Figure 7.6b). The decrease in microtopographic range would have the added benefit of wetting the spoil heaps, perhaps enough to discourage reestablishment of exotic conifers from the seed bank; that is, decreasing the HSP (see Figures 7.5b and 7.6b). In cutover sites, natural regeneration of *Sphagnum* onto bare peat occurs very slowly, due to loss of the diaspore bank and to inhospitably dry conditions during summer (Campeau and Rochefort 1996). Constraints on arrival (Figures 7.5a and 7.6a) can be bypassed by the introduction of regenerating fragments of *Sphagnum* (Campeau and Rochefort 1996, Ferland and Rochefort 1997). Establishment and growth of *Sphagnum* can be promoted (see Figures 7.5a and 7.6a) by increasing humidity at the bog surface through the introduction and establishment of nurse plants, the installation of physical barriers (mulch, shade cloth), or the creation of shallow depressions amid low ridges (Salonen 1992, Campeau and Rochefort 1996, Ferland and Rochefort 1997). Creation of gentle micro-topography has the added benefit of opening up niches for species with a range of moisture requirements (Ferland and Rochefort 1997); that is, increasing the HSP (see Figure 7.5a).

Although the problems of restoring cutover and planted peatlands differ markedly, filter and systems models provide a common context. The filter model (see Figure 7.5) helps to identify the stage of colonization at which bottlenecks (or bulges) occur and hence helps to identify the type of inter-vention necessary to overcome (or enhance) constraints on arrival, estab-lishment, and integration. The systems model allows quantitative assessment of the progress of restoration (see Figure 7.4) and identifies particular feed-back loops that could be reinforced by intervention (see Figure 7.6). Together, the two types of models provide a useful conceptual framework for practical restoration.

Conclusions

Let us return to the question of whether ecological theory can be of practi-cal benefit in restoration. I think the answer is yes, with the proviso that the theory applied must go beyond ecological filters on species colonization. If we view the filters as dynamic (see Chapter 6), then we must have some basis for predicting how the filters will change as a community assembles or reassembles. The only basis for such predictions is an understanding of how community structure and ecosystem processes interact with each other and with external constraints. The feedback loops inherent to systems models are the mechanisms by which ecological filters change over time.

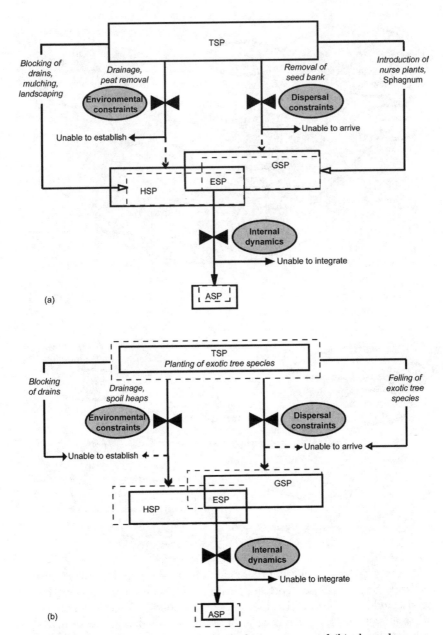

FIGURE 7.5 Filter models for restoration of (a) cutover and (b) planted peatlands. Refer to Figure 7.1 for an explanation of symbols and abbreviations. Note that species pools in cutover peatlands are reduced, and those of planted peatlands are expanded, relative to the nondegraded system.

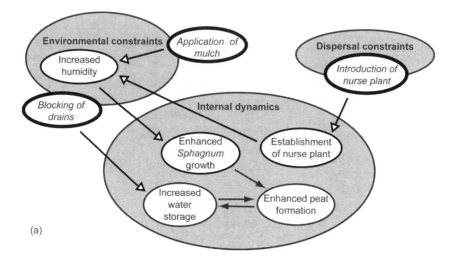

(a)

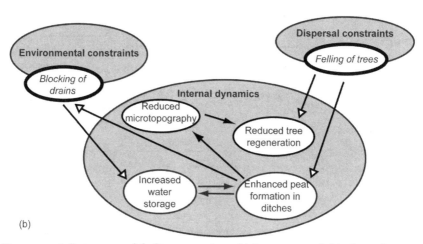

(b)

FIGURE 7.6 Systems models for restoration of (a) cutover and (b) planted peatlands. Shaded ellipses represent environmental constraints, dispersal constraints, and internal dynamics, as in Figure 7.2. Solid-headed arrows show links among processes. Interventions for restoration are denoted by heavily outlined ellipses and open-headed arrows.

A related question asks if we can identify and apply assembly rules in restoration. Definitions differ widely, but I consider assembly rules to be general principles, applicable to a broad range of ecosystem and community types (Belyea and Lancaster 1999). A common complaint is that such a general definition is so vague as to be useless. A specific example, to which I have alluded already, may answer this complaint. Spatial patterns may emerge from a few general principles that govern local interactions in vastly different systems: spiral colonies of bacteria (Ben-Jacob et al. 1994); radially symmetric *Hydra* (Turing 1952); banded communities of bush and bare ground (Lejeune and Tlidi 1999); and, perhaps, pool and hummock complexes on peatlands. The rules can be summarized in two sentences:

1. Positive feedback, or cooperativity, acts on a small scale to amplify initially random fluctuations.
2. These self-reinforcing changes are kept in check by negative feedback, or inhibition, operating on a larger spatial scale.

These two simple rules underlie a general model of system feedback that can explain the dynamics of very different systems or of the same system under different environmental constraints (Turing 1952). Of course, the model's parameters will take different meanings and values according to the system under study, and its boundary conditions will vary with the particular site under consideration. The identification of general assembly rules does not eliminate the need either for a deep understanding of species' characteristics, requirements, and interactions or for a thorough knowledge of particular sites. Nonetheless, they can and do provide powerful insight into how ecosystem and community structure develop (for example, Lejeune and Tlidi 1999).

At present, I think it is more useful to talk of guidelines for restoration rather than of rules. The first priority in any restoration program should be to reestablish key ecosystem processes. For degraded peatlands, water storage and peat accumulation are the key processes. In some systems, particular species may be required to restore particular processes: *Sphagnum* for peat formation, for instance. Establishment of those particular species should become a priority, however, only after major environmental constraints have been redressed. It would be pointless to carry out *Sphagnum* introductions on a cutover peatland, for example, until the water table could be maintained near the peatland surface. Once the major ecosystem processes have been restored, interventions should be geared toward establishing species that occur early in the natural assembly trajectory. On peatlands, for example, this strategy suggests the reestablishment of a gently undulating microtopography. It is highly unlikely that all steps leading from the primitive state to the target could be reconstructed by human interventions. The most effective strategy, therefore, may be to allow the degraded system to redevelop of it own accord, within the new constraints acting on it.

Acknowledgments

I am grateful to the organizers for inviting me to the Dornburg workshop and to all the participants for lively and thought-provoking debate. Comments and suggestions from Martin Zobel and Daniel Campbell improved the manuscript. This work was supported by the Natural Environment Research Council, UK (Grant GR3/11369, awarded to Jill Lancaster).

REFERENCES

Belyea, L. R.; and Clymo, R. S. 2001. Feedback control of the rate of peat formation. *Proceedings of the Royal Society: Biological Sciences* 268:1315–1321.

Belyea, L. R.; and Lancaster, J. 1999. Assembly rules within a contingent ecology. *Oikos* 86:402–416.

Ben-Jacob, E.; Schochet, O.; Tenenbaum, A.; Cohen, I.; Czirok, A.; and Vicsek, T. 1994. Generic modelling of cooperative growth patterns in bacterial colonies. *Nature* 368:46–49.

Campeau, S.; and Rochefort, L. 1996. *Sphagnum* regeneration on bare peat surfaces: field and greenhouse experiments. *Journal of Applied Ecology* 33:599–608.

Clymo, R. S. 1984. The limits to peat bog growth. *Philosophical Transactions of the Royal Society of London: Biological Sciences* 303:605–654.

Ferland, C.; and Rochefort, L. 1997. Restoration techniques for *Sphagnum*-dominated peatlands. *Canadian Journal of Botany* 75:1110–1118.

Hilbert, D. W.; Roulet, N.; and Moore, T. 2000. Modelling and analysis of peatlands as dynamical systems. *Journal of Ecology* 88:230–242.

Ingram, H. A. P. 1982. Size and shape in raised mire ecosystems: a geophysical model. *Nature* 297:300–303.

Ivanov, K. E. 1981. *Water Movement in Mirelands*. London: Academic Press.

Keddy, P. A. 1992. Assembly and response rules: two goals for predictive community ecology. *Journal of Vegetation Science* 3:157–164.

Keddy, P. A. 1999. Wetland restoration: the potential for assembly rules in the service of conservation. *Wetlands* 19:716–732.

Kelt, D. A.; Taper, M. L.; and Meserve, P. L. 1995. Assessing the impact of competition on community assembly: a case study using small mammals. *Ecology* 76:1283–1296.

Lejeune, O.; and Tlidi, M. 1999. A model for the explanation of vegetation stripes (tiger bush). *Journal of Vegetation Science* 10:201–208.

Moss, B.; Stansfield, J.; Irvine, K.; Perrow, M.; and Phillips, G. 1996. Progressive restoration of a shallow lake: a 12 year experiment in isolation, sediment removal and bio-manipulation. *Journal of Applied Ecology* 33:71–86.

Moyle, P. B.; and Light, T. 1996. Biological invasions of fresh water: empirical rules and assembly theory. *Biological Conservation* 78:149–161.

Økland, R. H.; Økland, T.; and Rydgren, K. 2001. A Scandinavian perspective on ecological gradients in north-west European mires: reply to Wheeler and Proctor. *Journal of Ecology* 89:481–486.

Salonen, V. 1992. Effects of artificial plant cover on plant colonization of a bare peat surface. *Journal of Vegetation Science* 3:109–112.

Scheffer, M.; Hosper, S. H.; Meijer, M.-L.; Moss, B.; and Jeppesen, E. 1993. Alternative equilibria in shallow lakes. *Trends in Ecology and Evolution* 8:275–279.

Sjörs, H. 1961. Surface patterns in Boreal peatland. *Endeavour* 20:217–224.

Sjörs, H. 1990. Divergent successions in mires, a comparative study. *Aquilo Seria Botanica* 28:67–77.

Swanson, D. K.; and Grigal, D. F. 1988. A simulation model of mire patterning. *Oikos* 53:309–314.

Turing, A. M. 1952. The chemical basis of morphogenesis. *Philosophical Transactions of the Royal Society of London* 237:37–72.

Warren, C. 2001. "Birds, Bogs and Forestry" revisited: the significance of the Flow Country controversy. *Scottish Geographical Journal* 116:315–337.

Weiher, E.; and Keddy, P. A. 1995. The assembly of experimental wetland plant communities. *Oikos* 73:323–335.

Wheeler, B. D.; and Shaw, S. C. 1995. *Restoration of Damaged Peatlands.* London: HMSO.

PART THREE

Assembly Rules and Community Structure

Here we explore how focusing on different levels of organization can provide important insights about how communities assemble during restoration and regeneration. The following chapters demonstrate how practitioners can use aspects of community structure to indicate where a degraded ecosystem is along its recovery trajectory. Chapter 8 examines the relative merits of a species-based versus a morphologically based assessment of community structure in plankton communities of freshwater ecosystems. Chapter 9 continues this theme, focusing on functional groups across several trophic levels in terrestrial grasslands. Chapter 10 reviews the significance of arbuscular mycorrhizal fungi for community development and restoration of various kinds of terrestrial ecosystems. Chapter 11 abstracts community membership to the classification of vegetation types and explores how assembly rules influence the transition from one vegetation type to another. Chapter 12 applies stable isotope analysis of different trophic levels to investigate how tightly linked the food web is along a restoration trajectory in a degraded grassland.

Chapter 8

Self-Organization of Plankton Communities: A Test of Freshwater Restoration

CARMEN ROJO

Restoring ecosystems is not an easy task. Restoring ecosystem structure (that is, specific species) is proving much harder than restoring ecosystem functions (see Lockwood and Pimm 1999). It seems that restoration at the community level may require more focus on both community function (see Chapter 16) and dynamics (see Chapter 17). This proposal is in accordance with recent ideas about ecological assembly rules and dynamics (Drake 1990, 1991). Weiher and Keddy (1999) suggested some implications of the difficulty of restoring ecosystem function for restoration goals; for example, because a long-term community assembly trajectory occurred before perturbation, it is almost impossible to reach the same predisturbed state following a restoration protocol. Moreover, restoration can be considered a large disturbance in itself; consequently, a new assembly trajectory may be initiated with an unpredictable end.

What is predictable, however, is that in a long-term, postrestoration community study, one could observe a persistent community (Powell and Steele 1995, Lockwood and Pimm 1999). When a community achieves a new dynamic equilibrium and regulated state (Pahl-Wostl 1995), the end of the restoration disturbance seems to be clear. Therefore, the restoration effort would be considered successful when certain broad goals are achieved and the system dynamics show that the community is in a new equilibrium state or progressing along an acceptable trajectory.

In many cases, the main goal in restoration of aquatic systems is an endpoint stable state with a significantly lower concentration of phytoplankton. Therefore, to monitor the restoration, it may be appropriate to use variables related to phytoplankton dynamics. Two common approaches to freshwater restoration management are the control of nutrient loading and food-web manipulation. The first approach links a positive relationship between primary production (for example, by phytoplankton) and nutrient concentrations (Vollenweider and Kerekes 1980). Food-web manipulations are supported by the demonstrated relationship between primary producers and consumers (Brylinsky and Mann 1973), and by the trophic cascade hypothesis (Carpenter and Kitchell 1993). That is, modifying some trophic level of consumers modifies the concentration of primary producers, which decreases as the concentration of direct consumers increases. These two approaches are applied in specific restoration management strategies: (1) restoring only the physical environment (that is, nutrient loading; Gaedke and Schweizer 1993); (2) biomanipulating the food web (Goldyn et al. 1997); and (3) a combination of both (Søndergaard et al. 1997).

Measuring the success of freshwater restoration achieved by manipulating abiotic conditions is very difficult. For example, after nutrient reduction, a phytoplankton community may differ significantly from the predisturbance state (Ruggiu et al. 1998). Thus, we cannot predict the new community of a restored freshwater system, even when the desirable environmental conditions are reachieved. So how can we evaluate restoration success or failure at the community level?

Evaluating successful restoration achieved by biomanipulation is also difficult. It appears that many changes in the ecosystem after restoration (such as establishment of macrophytes, decrease in nutrient concentrations, changes in phytoplankton composition, and some alternative clear-water states) are insufficient indices to allow us to conclude that a lower amount of phytoplankton biomass has reached a stable state (Perrow et al. 1997). This difficulty results among other causes, from the fact that effects of management are often observed over too short a period. Thus, analysis of longer-term (greater than 10 years) case histories would be helpful (Perrow et al. 1997).

It is important to stress that when the restoration goal is to reduce phytoplankton production, the success of restoration implies that both the variables used to describe the phytoplankton effectively indicate an improved condition and that this community state is stable. The new challenge in restoration ecology is to quantify the success of restoration in different ecosystems appropriately. So, first of all, what variable related to primary producers should be used? It seems that variables that represent traits or functions of a community are most predictable (Keddy and Weiher 1999, Lockwood and Pimm 1999). Therefore, their use as a tool in restoration ecology is being widely recommended (Lockwood and Pimm 1999), instead of, for example, restoration focused on particular species (Palmer et al. 1997). In phytoplankton, for example, for a given selective abiotic constraint, functional traits (functional groups) could be predictors of a set of possible dominant species (PROTHEC model in Reynolds et al. 2000, Elliott et al. 2001). Similarly, the distribution of phytoplankton biomass in different body-size classes seems to be sensitive to the ecosystem trophic state and can signal that a community is in a self-organized steady state (Rojo and Rodríguez 1994, Jørgensen et al. 1998, Cottingham 1999). Finally, another possible variable is the total biomass of primary producers, which is well related to the freshwater trophic state (Vollenweider and Kerekes 1980).

Another important question is the variation of state variables in the system (for an example in terrestrial systems, see Chapter 17). The ecosystem is dynamic, and after every restoration, at least three components of its dynamics are expected: the self-variation (autogenic, or variation caused by interspecific interactions), the response to the environment (allogenic; for example, seasonality), and the response to the restoration as a disturbance (allogenic). For example, the trajectory of a planktonic variable will always fluctuate with seasonal periodicity (Rojo et al. 2000b). Over a longer time scale, the dynamics can show alternative states between clear and turbid periods in wetlands with macrophytes (Scheffer et al. 1993). The task, therefore, is to decide when the fluctuations have a stressor component caused by the restoration-disturbance and when they are the result of a self-organized process or autogenic dynamics (Pahl-Wostl 1995). Unfortunately, to analyze the community trajectory, long-term data must be available (Perrow et al. 1997, Powell and Steele 1995), and few restored ecosystems are monitored over the long term after restoration management ceases.

To detect whether the stress period of disturbance and restoration has stopped affecting the plankton community, it is necessary to study the community's structure or its dynamics. The structural study must show that the community is in a stable state. This can be done, for example, by analyzing the community size spectrum (or normalized size spectra) of biomass, because the spectrum is sensitive to environmental stress and also reflects when such a community is in a self-organized critical state (Platt 1985, Jørgensen et al. 1998, Cottingham 1999). Studying community dynamics involves selecting a variable, result, or expression of plankton interactions (for example, phytoplankton biomass), measured over the long term. Then this time series must be analyzed to find evidence of self-organizing processes.

These two approaches (studying structure or studying dynamics) need a little introduction. First, the description of size distributions is an important aspect of research in many disciplines that deal with size-structured systems

(Vidondo et al. 1997). Normalized biomass size spectra (hereafter NB-SS) are used to examine and quantify the distribution of plankton biomass among multiple size classes of organisms (Sprules et al. 1983, Rojo and Rodríguez 1994). NB-SS is a linear model of community structure that relates the logarithm of normalized biomass (similar to the density) within each size class to the logarithm of average body size within the class. The slope of this linear model should be –1 when a system is in equilibrium, as it would be for seston in the oligotrophic ocean (Sheldon et al. 1972, Rodríguez and Mullin 1986). A simple example can help to explain the transformation from a biovolume distribution in body size classes to an NB-SS distribution (Figure 8.1). The slope is related to the trophic state: it is less steep when the system is more eutrophic and larger organisms are predicted (Sprules and Munawar 1986). Also, the NB-SS linear model should show more unexplained variance when the system is under stress or perturbed (Sprules and Munawar 1986, Cottingham 1999). So the coefficient of determination of the size-abundance relationship could be a measure of the disturbance regime in a given system, with lower values expected in systems that are strongly perturbed (far from steady state) and higher values in systems closer to a steady state (Choi et al. 1999). Moreover, when NB-SS is a linear model, the distribution of normalized biomass in size classes is a power law, which suggests that the community would behave as a self-organized critical system not ordered by external influences (Bak 1996, Jørgensen et al. 1998, Choi et al. 1999), as we discuss below.

The second approach is based on the concept of self-organization (SO). As defined by Camazine et al. (2001), self-organization is a process in which a pattern at the global level of a system emerges solely from numerous interactions among the lower-level components of the system. Moreover, the rules specifying interactions among the system's components are executed using local information, without reference to the global pattern. Self-organization occurs exclusively in complex systems (Cambell 1993), and such systems derive much of their structure from this process (Kauffman 1993). In self-organization, the system's pattern and structure are generated spontaneously through dynamic feedback among its components, without an external ordering influence (Kauffman 1993, Camazine et al. 2001). Ecological systems can be studied using this complex systems approach: the assembly trajectory must be seen as a self-organizing process built from a variety of interacting components (for example, individuals, populations, guilds, and so forth). In this sense, communities exist on a continuum from essentially random assemblages to highly integrated self-organized structures (Drake et al. 1999). This continuum will be related to some assembly rules and an assem-

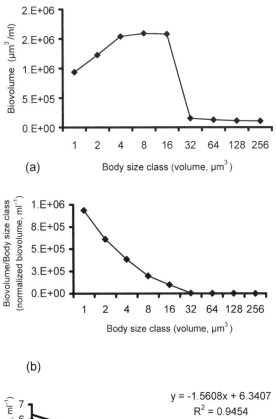

(a)

(b)

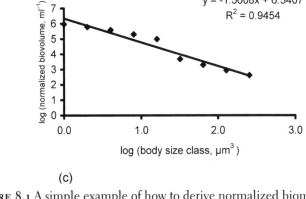

$y = -1.5608x + 6.3407$
$R^2 = 0.9454$

(c)

FIGURE 8.1 A simple example of how to derive normalized biomass size spectra from a hypothetical ecological community. (a) Total biovolume (or biomass) of organisms of each body size class. (b) Normalized biovolume as total biovolume of each class divided by the body size of this class. (c) Logarithm of normalized biovolume versus log of body size class; that is, the NB-SS of the community in question.

bly trajectory; therefore, community assembly and restoration should be strongly linked (see Chapter 3).

A self-organization process tends to go through a critical state in which the open, dynamic system does not vary despite changes, such as additions and losses of elements. This critical state, or self-organized criticality (SOC), was illustrated by Bak et al. (1988) with a sandpile analogy: when sand is continually poured onto a plate, the pile eventually reaches a critical state (slope of cone). At this critical state, the rate of sand flowing over the plate (sand avalanches) equals the rate of sand input into the system, and the shape of the sand cone remains the same. Systems at this critical state display a frequency-magnitude distribution of events (the sand avalanches) that is best described by a simple inverse power law of the form $D(s) = s^{-w}$ where s is the size or magnitude of the event, $D(s)$ is the frequency of occurrence of event size s, and w is the spectral coefficient, ranging from 0 to 2. SOC models can be analyzed relatively easily by testing if the frequency distribution of changes of a given variable exhibits a power law (Bak et al. 1988, Solé et al. 1999; for a critical point of view, see Fukami et al. 1999). Variable changes (descriptors of a system) that conform to a power-law distribution imply that, at least, the assemblage is a complex system with a self-organized equilibrium dynamic (Pahl-Wostl 1995). Most research into self-organization has been conducted via mathematical simulation modeling (Murray 1989, Kauffman 1993), but SOCs have empirical examples in most systems. For instance, the magnitude and frequency of occurrence of earthquakes fits a power law (Guttenberg-Richter law in Bak 1996). The application of mathematical modeling to community ecology has just begun. Some studies that have embraced this approach deal with avifauna of the Hawaiian Islands (Keitt and Marquet, 1996), rain forests (Solé and Manrubia, 1995), grasslands (Wilson et al. 1996), and planktonic primary producer biomass changes over time (Jørgensen et al. 1998).

In summary, to establish when a restoration project should end (for example, when oligotrophication occurs), I propose that attainment of a self-organized critical state should be a target endpoint of restoration. When reduced primary producers, after the restoration-disturbance process, achieve a self-organized critical state, then the plankton community is no longer being affected by stressors (nutrients or food-web disturbances) and restoration is finished.

This chapter aims to illustrate the difficulty of predicting plankton response to environmental changes when species composition or even trophic relationships are used as descriptors of community structure. Phytoplankton communities inhabiting very similar water bodies are compared using different structural variables (species composition or morphological traits). I emphasize the idea that when the restoration goal (for example, lower concentration of nutrients in a lake) is achieved, the restoration can be deemed a success at the community level if the system seems self-organized and thus is in a stable, unstressed condition.

Study Sites

Some of the data sets used for this research come from ecosystems I have studied intensively; for example, the Las Tablas de Daimiel National Park wetland in central Spain (Sánchez-Carrillo and Álvarez-Cobelas 2001; Rojo et al. 2000b; Ortega et al. 2000, 2002a) and the La Safor Regional Park marsh on the Mediterranean coast of Spain (Rodrigo et al. 2001). Other data sets result from calculations and modifications based on bibliographic information about Lake Trummen, a small, shallow Swedish lake (Björk and Digerfeldt 1965), and Lake Constance, a large, deep alpine lake at a site in Germany (Elster 1982). Reducing the input of nutrients restored Lake Trummen and Lake Constance, and their oligotrophication histories are well described (Cronberg 1982, Gaedke and Schweizer 1993, Bäuerle and Gaedke 1998). I used series of phytoplankton biomass from both lakes, from some years before their restoration until many years later. From Lake Trummen, the time series (fresh weight of phytoplankton in water volume, mg/L) was from 1968 to 1978; the end of the eutrophic period was in 1969, and primary producer biomass decreased after 1970 (Cronberg 1982). From Lake Constance, the time series (biovolume of phytoplankton in integrated water column, cm^3/m^2) was 1979 to 1989; the restoration effort started in 1984 (Bäuerle and Gaedke 1998). Unfortunately, the time series is incomplete, so the years used were from 1979 to 1982 (as period before restoration) and from 1985 to 1989 (as period after restoration).

Cluster Analysis

There are many multivariate analytical methods (Podani 2000); I used a method common in phytoplankton studies. I carried out Q-mode analyses (clustering samples) to look for similarities between samples of phytoplankton and their ordination (Legendre and Legendre 1979). I calculated similarity using Pearson's coefficient of correlation and clustered samples using the UPGMA linkage method (Podani 2000). When groups of samples are clustered based on their phytoplankton composition, using these methods, at least some assemblages of species are repeated.

The variables used were the biomass of phytoplankton species in the first analysis and the biomass of six phytoplankton morphotypes in the second analysis. In both analyses, data were collected from a site in the Las Tablas de Daimiel wetland, sampled monthly over a 6-year period (1996–2001) and observed by inverted microscope (Rojo et al. 2000b). The six morphotypes were colonies, such as *Dictyosphaerium ehernbergii*; small single cells with greatest axial linear dimension (GALD) less than 20 µm (for example, *Monoraphidium minutum*); large single cells with GALD greater than 20 µm (for example, *Nitzschia acicularis*); small flagellates with GALD less than 20 µm (for example, *Rhodomonas minuta*); large flagellates with GALD greater than 20 µm (for example, *Euglena acus*); and filaments (for example, *Planktothrix agardhii*). The third cluster analysis was done using the biomass of morphotypes inhabiting two different sites within the wetland over 2 years (sampled monthly).

First, I will illustrate the lack of pattern in plankton community structure when comparing their species assemblages. When comparing communities at higher levels of classification (taxonomic classes, morphological groups, and so forth), however, some similar community structures appear. For example, in the Las Tablas de Daimiel eutrophic wetland, information about plankton communities was very diverse, depending on the level of organization analyzed (Figure 8.2). Samples from the Molemocho (MM) site (1 site sampled monthly over 6 years) were not clustered when their phytoplankton species assemblages were compared and revealed no similarity between sample compositions (Figure 8.2a). When the same samples were analyzed for similarity based on the morphotype matrix (colonies, large and small single cells, large and small flagellates, and filaments), the community structures clustered in at least two periods: winter-spring and summer-autumn (Figure 8.2b). There were only a few samples in a third miscellaneous group.

Using the matrix of morphotypes from two water-connected sites (MM and Puente Navarro, PN) in the same wetland over 2 years (Figure 8.2c), some clusters appeared. Therefore, different sites (MM, PN) may have similar assemblages of morphotypes. However, in this analysis, one cluster grouped winter-spring MM assemblages with summer-autumn PN communities, another cluster included winter-spring PN communities with summer-autumn MM assemblages, and yet another comprised a mixture of communities from both sites and all time periods.

Linear Regressions

A relationship may be expected between primary and secondary plankton producers (Brylinsky and Mann 1973). To test this possible relationship, a

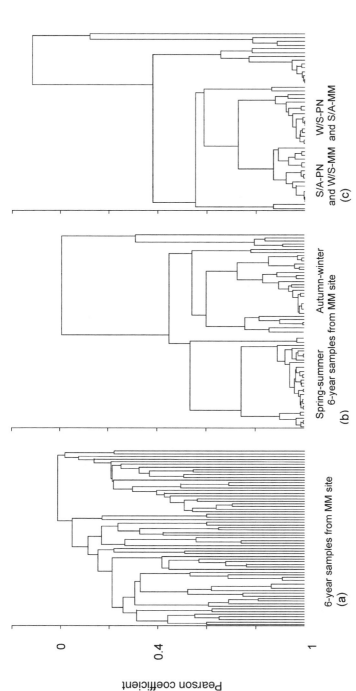

FIGURE 8.2 Comparison of effect of clustering phytoplankton based on species biomass (a) and biomass of morphotypes — colonies, large single cells, small single cells, large flagellates, small flagellates, or filaments; (b) and (c). (a) and (b) cluster the samples from a single central site (MM; see text) in the Las Tablas de Daimiel National Park wetland in Spain, sampled monthly over 6 years. They demonstrate the different information on possible assemblages that can be found using different groupings. (c) shows samples from two connected sites (MM and PN, see text) in the same wetland, sampled monthly over 2 years, where samples were clustered on the basis of the biomass. S/A: samples from summer and autumn; W/S: samples from winter and spring.

linear regression on log-transformed data was carried out between biomass of phytoplankton and biomass of zooplankton that inhabited four different areas of the Las Tablas de Daimiel wetland—Molemocho (MM), Puente Navarro (PN), Filtro Verde (FV), and Patagallina (PG)—over a 3-year period sampled monthly (data from Ortega et al. 2000, Rojo et al. 2000b).

A well-known relationship exists between phytoplankton abundance and zooplankton abundance, but there are many cases in which the relationship is not a simple linear function. For example, no relationship between phytoplankton biomass and total zooplankton was found (Figure 8.3) in any of the four different areas in the Las Tablas de Daimiel National Park wetland over a period of 3 years. Pearson's correlation coefficients of phytoplankton and zooplankton log-transformed data relationships were, in FV, EN, MM, and PN sites: 0.02, 0.02, 0.28, and 0.12, respectively. These results mean that regression coefficients of these linear functions are not significantly different from 0, being $p < 0.01$.

Self-Organized Criticality in Plankton Communities

I calculated the normalized biomass size spectra (NB-SS) of plankton from 31 sites sampled at La Safor, where a remarkably broad nutrient gradient occurs (total phosphorus concentrations, TP, were between 0.004 and 7.700 mg/L). Sites were sampled on the same spring day. The samples were analyzed using a Coulter counter. Counted particles were grouped in size classes (octave scale), using a scale starting with size class 1–2 μm^3 and continuing (2–4 μm^3, 4–8 μm^3, and so forth) up to 65,536–131,072 μm^3 (this range corresponds to 1–63 μm of equivalent spherical diameter). I calculated the normalized biomass (β_i) of each size class (i) by dividing the total biovolume (B_i) in that class by the class's range amplitude (Δw_i. Note that when classes are in octave scale, the range of each class is the same value as the class's lowest limit: w_i). These distributions were fitted to linear regressions in log-log plots; log (β_i) = a + b log (w_i) (Rodríguez and Mullin 1986). The regression coefficients (b) are the same spectral coefficients found in the power law (Vidondo et al. 1997, Jørgensen et al. 1998). Looking for a relationship between NB-SS and the trophic state of a freshwater ecosystem, the regression coefficients of this NB-SS were related to TP concentration in each site.

Community Trajectory

Time series of phytoplankton biovolume (B) show a trajectory of irregular temporal courses with increases ($B_t - B_{t+1} < 0$) and decreases ($B_t - B_{t+1} > 0$)

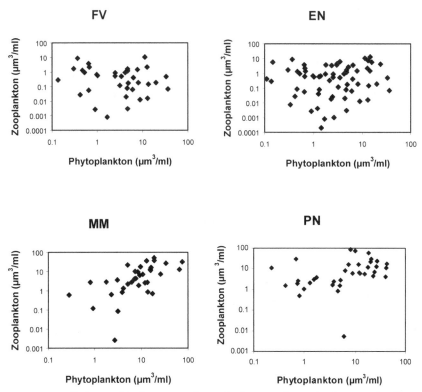

FIGURE 8.3 Relationship between phytoplankton and zooplankton in the four areas (called FV, EN, MM, and PN; see text) of the Las Tablas de Daimiel National Park wetland (sampled monthly for 3 years, 1996–1998), where an omnivorous copepod, rather than the herbivorous *Daphnia*, is the main phytoplankton consumer.

of biovolume (the decreases are also called avalanches, Jørgensen et al. 1998). A plot of avalanche magnitudes (m) versus the number of decreases with magnitude larger than a given magnitude m (a Pareto distribution, in fact) can be fitted to a power law. To illustrate the fact that plankton can exhibit an SOC and can be a useful tool to test the regulated endpoint of restoration, I used two long time series of phytoplankton biovolume (time scale lower than a fortnight) from Lake Constance, Germany (Gaedke and Schweizer 1993), and Lake Trummen, Sweden (Cronberg 1982). Both lakes underwent eutrophication and were restored to a less nutrient-rich state.

Results

Self-organization in plankton and its relationship to ecosystem restoration can be illustrated with some examples. Changes in mean phytoplankton biomass have been observed in two very different lakes, Lake Trummen and Lake Constance (Figures 8.4 and 8.5), during eutrophication and after restoration. In Lake Trummen, mean phytoplankton biomass was 30 mg/L (standard deviation, SD, 34 mg/L) before restoration and 7 mg/L (SD, 7 mg/L) after restoration. Expressed in terms of biovolume: in Lake Constance, mean phytoplankton biovolume was 15 cm^3/m^2 before restoration (SD, 17 cm^3/m^2) and 10 cm^3/m^2 (SD, 9 cm^3/m^2) after restoration. Biomass became stationary after a restoration period around a mean value lower than the value during the trophic-stressed period and similar to values existing before eutrophication (Cronberg 1982, Gaedke and Schweizer 1993, Sommer et al. 1993).

Evaluating whether an SOC is an attractor in phytoplankton dynamics, I find (see Figures 8.4 and 8.5) that the slope of the relationship between cumulative frequency of biomass avalanches versus avalanche size was statistically significant and close to -1 in the restored period, but not before restoration. In Lake Trummen before the restoration (1968–1969), the average soluble reactive phosphorus (SRP) was 185 µg/L and its slope was -0.63 (standard error = 0.079, $t_{0.05}$ = 2.57, degrees of freedom = 5), a slope significantly different from -1. However, studying the phase of reoligotrophication in Lake Trummen (1970–1978; average SRP was 70 µg/L) after manipulation, the slope was not statistically different from -1 (slope = -1.15, standard error = 0.098, $t_{0.05}$ = 2.26, g.l. = 8). The other example, Lake Constance, was eutrophic until the early 1980s (104 µg SRP/L) when it showed an avalanche distribution slope of -0.72, significantly different from -1 (standard error = 0.044, $t_{0.05}$ = 2.02, g.l. = 41). But when the period of reoligotrophication took place (45 µg SRP/L), the slope was -1.12 (not significantly different from -1, standard error = 0.074, $t_{0.05}$ = 1.62, g.l. =15).

The study of size-normalized biomass spectra in 31 sites (on the same day) at La Safor (Figure 8.6) showed a clear relationship between the size structure of a community and the trophic state of the water (as judged by total phosphorus, Figure 8.6a). These spectra were the distribution of normalized biovolume in size classes. For example, at the most oligotrophic point in the marsh (total phosphorus 0.004 mg/L; Figure 8.6b), the slope was steeper (more negative) than in one of the hypertrophic sites (total phosphorus 2367 mg/L; Figure 8.6c). This change in the slope was due mainly to the appearance of populations with larger body size (compare x-axes between Figures 8.6b and 8.6c).

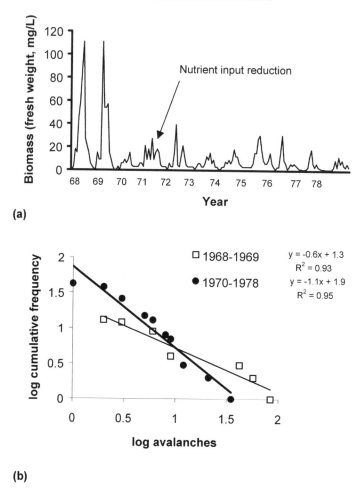

(a)

(b)

FIGURE 8.4 (a) Phytoplankton biomass in Lake Trummen (after Cronberg 1982). (b) Log-log plot of the relationship between the magnitude of phytoplankton biomass decrease in Lake Trummen (log "avalanches") versus the frequency with which the indicated magnitude of decrease is exceeded (log "cumulative frequency"), in the before and after restoration periods. Regression coefficients are significantly different to 0, $p < 0.01$ from both periods.

Discussion

In the following discussion, I explore weak patterns of plankton and their implications for restoration. To illustrate the weakness of some plankton patterns, I clustered some very similar sites from the same wetland. This single example has two important implications: (1) the degree of convergence between community structures varies with the analyzed structural variable (Samuels and Drake 1997), and (2) similar planktonic communities can exist

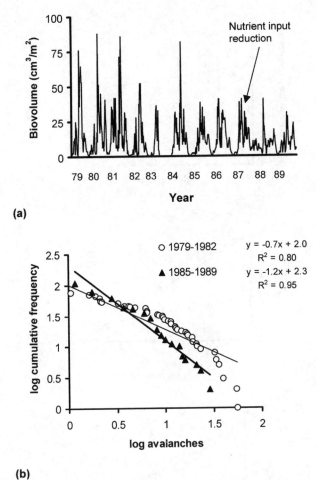

FIGURE 8.5 (a) Phytoplankton biomass in Lake Constance (after Gaedke and Schweizer 1993). (b) Log-log plot of the relationship between the magnitude of phytoplankton biomass decrease in Lake Constance (log "avalanches") versus the frequency with which the indicated magnitude of decrease is exceeded (log "cumulative frequency"), in the before and after restoration periods. Regression coefficients are significantly different to 0, $p < 0.01$ from both periods.

under different freshwater conditions (Reynolds et al. 2000, Rojo et al. 2000a). Both issues have a clear relevance to restoration efforts. In Lake Balaton (Hungary), measures to reverse eutrophication (reducing phosphorus) began in the mid-1980s. Total phytoplankton biomass and composition of eukaryotic plankton have not changed much, but cyanoprokariota composition appears to have been sensitive to trophic interannual changes (Padisák and Reynolds 1998). In Bighorn Lake (a fishless alpine lake in the United States), biomanipulation was carried out by stocking and removing a nonnative salmonid.

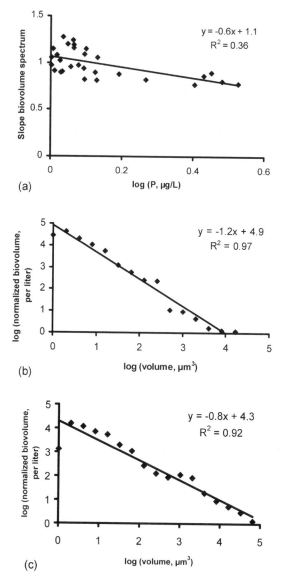

Figure 8.6 Size spectra of phytoplankton community across different trophic states. Data are from 31 simultaneously sampled sites in La Safor (Mediterranean marsh, Spain). (a) Regression coefficients of size spectra versus the total phosphorus concentration in each site. (b) Normalized size spectrum in an oligotrophic site of La Safor (0.004 mg P/L). (c) Normalized size spectrum in a hypertrophic onsite of La Safor (2.367 mg P/L). Regression coefficients are significantly different to 0, $p < 0.01$.

There, the proportion of large zooplankton increased after treatment, but the overall biomass of primary producers did not change (Parker et al. 2001).

To determine the degree to which community composition can be predicted, we must answer an important question: If an association of populations can indicate a specific environment, then, conversely, can we expect a defined community composition for each environmental condition? The answer is no, and restoration ecology must adapt to this reality (Samuels and Drake 1997, Jenkins and Buikema 1998, Rojo et al. 2000a). For example, in the process of nutrient reduction in Lago Maggiore (Italy) over the last 10 years, when preeutrophication nutrients concentrations were achieved, very different phytoplankton structures were observed interannually, and these are dissimilar to those of the preeutrophication community (Ruggiu et al. 1998). In an alpine lake (Snowflake Lake, United States), the zooplankton structure was studied over a period of 35 years after trout stocking (McNaught et al. 1999). This long-term work revealed that zooplankton structure (composition and abundance distribution) did not recur, even under similar conditions.

Despite the fact that the species composition of a community is unpredictable, a set of traits can be expected (Elliott et al. 2001, Reynolds 2001), and restoration of those traits should be explored in greater depth. Similarly, a revision of the old well-known freshwater restoration studies, analyzing the pattern of traits instead of the species composition, should be done. For example, at another level of scale, I studied a case in which no relationship between grazers and phytoplankton dynamics was apparent. This unexpected occurrence was observed in Las Tablas de Daimiel and was explained first by the alternating composition of large and small grazers, provoked by the planktivorous activity of mosquito fish; second by the presence of omnivorous copepods instead of big grazers such as *Daphnia*; and third by the relationship between ciliates and autotrophic picoplankton (Ortega-Mayagoitia et al. 2000, Ortega-Mayagoitia et al. 2002a, Ortega-Mayagoitia et al. 2002b). Many other cases could be of interest; for example, McNaught et al. (1999) showed that an omnivorous presence prevented a simple model of a trophic cascade from being useful; Parker et al. (2001) did not achieve any variation in phytoplankton biomass after a biomanipulation effort in Bighorn Lake; the same circumstances were observed in the Czech reservoirs, where the introduction of piscivorous fish affected planktivorous fish, but cascade effects did not reach primary producers (Seda et al. 2000). In many other cases, restoration based on the cascade hypothesis was a success; for example, the well-known biomanipulation project in Lake Mendota (Lathrop et al. 2002). Because many restoration works focus on reducing the biomass of phytoplankton through grazer manipulation, the weakness of this relationship should be taken into account.

Looking for Self-Organization in Plankton as a Test of Successful Restoration

The response of a pelagic ecosystem to restoration can be slow. The new conditions, along with new producer concentrations, will be stable only after several years (Perrow et al. 1997). This fact can be related to the beginning of a new approach to studying plankton dynamics: Is there any empirical evidence of self-organization at the plankton community level? Is there any relationship between the signal of SOC and system stability? The answers are probably yes (Jørgensen et al. 1998). Like any set of interacting populations, a plankton community is a complex system (Putman 1994), but not all species assemblages can be self-organizing. For example, species-poor systems or systems experiencing very frequent, intense, or highly variable disturbances cannot self-organize (Drake et al. 1999). A plankton community that is not a species-poor or highly disturbed system, however, can be self-organizing. A self-organized community is a regulated system (Samuels and Drake 1997), and only one or two single rules may guide its trajectory (Drake et al. 1999)—a trajectory without stressors. There are some studies on self-organization in plankton: for example, phytoplankton group dynamics in the Bay of Marseille (Gregori et al. 2001), phytoplankton variability over wide scale ranges (Lovejoy et al. 2001), plankton ecological exergy in lakes (Alvarez-Cobelas and Rojo 2000), and plankton community response to an artificially induced feedback (Petersen 2001). At the moment, however, only one study has looked for SOC using the power-law distribution of different plankton descriptors (Jørgensen et al. 1998), and a short review of this topic in limnological communities prompts further exploration of it (Choi et al. 1999).

The results presented in this chapter show SOC in phytoplankton: organisms distributed in size classes did fit a power law in accordance with Jørgensen's (1998) hypothesis. Moreover, some interesting relationships were found between the fit to a power law and environmental conditions. Two patterns were observed. First, during long, unperturbed periods; normalized biomass spectra changed toward less steep slopes (indicating a greater proportion of large organisms than in a distribution with a slope equal to −1). Second, the same trend was observed in a trophic gradient: an inverse relationship between the trophic state and the slope of the size spectrum (Sprules and Munawar 1986, Rodríguez et al. 1990, Echevarría and Rodríguez 1994, Rojo and Rodríguez 1994). These patterns are consistent with the idea that autogenic succession toward larger sizes occurs during long, undisturbed periods (Reynolds 1984) and that community size (for example, associations of large algae such as P. agardhii, inedible algae increasing) is directly related to trophic state (Harris 1986, Schindler 1990, Watson et al. 1992).

In the same way, temporal variation of phytoplankton size spectra in a gravel pit sampled weekly throughout a year (Rojo and Rodríguez 1994) showed its sensitivity to change from autogenic to allogenic periods (from unperturbed water column stability to the mixing period). Through a long unperturbed period (stratification of the water column), normalized biomass spectra changed toward less negative slopes (an increase of larger cells). In addition, results presented here indicate a significant relationship between the biomass spectra coefficient of regression and the nutrient concentration in the system. Recently, Cottingham (1999) published results of normalized size spectra after the manipulation of nutrients and food webs in three lakes, showing a range of slopes similar to those reported here. These results mirror those discussed in this chapter; that is, eutrophication promotes shallower slopes of normalized size spectra, illustrating that the size structure of a community is more predictable as a result of manipulation than, for example, the taxonomic structure, particularly species composition (Cottingham 1996).

The results presented here are another demonstration of self-organized critical phenomena in primary producer dynamics related to system perturbations. Of course, these results are not conclusive proof of an inverse relationship between disturbance and SOC measured as a power law. Rather, the results show only that frequencies in plankton producer variations are similar to those in other self-organizing critical systems, as pointed out by Jørgensen et al. (1998), and that the slope is sensitive to frequent or intense disturbances in the system (Drake et al. 1999). More studies of this type would be interesting, which raises another question: Which mechanisms and processes are necessary to generate a self-organized community state? Experiments are needed, and careful monitoring during a long-term period after restoration management could be akin to a natural experiment, as pointed out by Lockwood and Samuels (see Chapter 4).

A Final Consideration

According to Keddy (1999), we now have enough knowledge about ecosystems to know what broad, multiple steps are necessary for restoration (even though we may not be able to restore a given ecosystem). Success is not a specific, expected change in a short time, but rather the achievement of a better general state, in the sense of coming close to the preperturbed state of the system, which remains independent of the manipulations used in restoration. Therefore, in aquatic ecosystems, the reduction of primary producers, as the goal, will lead to success if dynamics of primary producers remain regulated as a self-organized process within the plankton system.

Acknowledgments

I thank Marie Lionard and Dolores Ferrer for their help in obtaining data about phytoplankton composition in Las Tablas de Daimiel and, in addition, for performing counts with the cell counter. This research has been supported by grants from the Spanish Ministry of the Environment (BOS2002-2003). I thank James Drake, Jill Lancaster, Tim Nuttle, Vicky Temperton, and Richard Hobbs for constructive comments on the manuscript. Discussions of this chapter's ideas with Miguel Álvarez-Cobelas contributed to its organization and expression.

REFERENCES

Alvarez-Cobelas, M.; and Rojo, C. 2000. Ecological goal functions and plankton communities in lakes. *Journal of Plankton Research* 22:729–748.

Bak, P. 1996. *How Nature Works: The Science of Self-Organized Criticality.* New York: Springer-Verlag.

Bak, P.; Tang, C.; and Wiesenfeld, K. 1988. Self-organized criticality. *Physical Review* 38:364–374.

Bäuerle, E.; and Gaedke, U. 1998. *Lake Constance. Characterization of an Ecosystem in Transition.* Advances in Limnology 53. Stuttgart, Germany: Schwezerbart.

Björk, S.; and Digerfeldt, G. 1965. Notes on the limnology and post-glacial development of Lake Trummen. *Botaniska Notiser* 118:305–325.

Brylinsky, M.; and Mann, K. H. 1973 An analysis of factors governing productivity in lakes and reservoirs. *Limnology and Oceanography* 18:1–14.

Camazine, S.; Deneubourg, J. L.; Franks, N. R.; Sneyd, J.; Theraulaz, G.; and Bonabeau, E. 2001. *Self-Organization in Biological Systems.* Princeton: Princeton University Press.

Carpenter, S. R.; and Kitchell, J. F. 1993. *The Trophic Cascade in Lakes.* Cambridge: Cambridge University Press.

Choi, J. S.; Mazumder, A.; and Hansell, R. I. C. 1999. Measuring perturbation in a complicated, thermodynamic world. *Ecological Modelling* 117:143–158.

Cottingham, K. L. 1999. Nutrients and zooplankton as multiple stressors of phytoplankton communities: evidence from size structure. *Limnology and Oceanography* 44:810–827.

Cronberg, G. 1982. Phytoplankton changes in Lake Trummen induced by restoration. *Folia Limnologica Scandinavica* 18:5–119.

Drake, J. A. 1990. Communities as assembled structures: do rules govern pattern? *Trends in Ecology and Evolution* 5:159–164.

Drake, J. A. 1991. Community-assembly mechanics and the structure of an experimental species ensemble. *American Naturalist* 137:1–26.

Drake, J. A.; Zimmerman, G. R.; Purucker, T.; and Rojo, C. 1999. On the nature of the assembly trajectory. In *Ecological Assembly Rules. Perspectives, Advances, Retreats,* ed. E. Weiher and P. Keddy, 233–250. Cambridge: Cambridge University Press.

Echevarría, F.; and Rodríguez, J. 1994. The size distribution of planktonic biomass during a deep bloom in a stratified reservoir. *Hydrobiologia* 284:113–124.

Elliott, L. A.; Reynolds, C. S.; and Irish, A. E. 2001. An investigation of dominance in phytoplankton using the PROTECH model. *Freshwater Biology* 46:99–108.

Elster, H. J. 1982. Neuere Untersuchungen über die Eutrophierung und Sanierung des Bodensees. *GWF-Wasser Abwasser* 123:277–287.

Fukami, T.; Zimmermann, C. R.; Russell, G. J.; and Drake, J. A. 1999. Self-organized criticality in ecology and evolution. *Trends in Ecology and Evolution* 14:321.

Gaedke, U.; and Schweizer A. 1993. The first decade of oligotrophication in Lake Constance. I. The response of phytoplankton biomass and cell size. *Oecologia* 93:268–275.

Goldyn, R.; Kozak, A.; and Romanowicz, A. 1997. Food-web manipulation in the Maltanski Reservoir. *Hydrobiologia* 342/343:327–333.

Gregori, G.; Colosimo, A.; and Denis, M. 2001. Phytoplankton group dynamics in the Bay of Marseille during a 2-year survey based on analytical flow cytometry. *Cytometry* 44:247–256.

Harris, G. 1986. *Phytoplankton Ecology*. London: Chapman and Hall.

Jenkins, D. G.; and Buikema, A. L. 1998. Do similar communities develop in similar sites? A test with zooplankton structure and function. *Ecological Monographs* 68:421–443.

Jørgensen, S. E.; Mejer, H.; and Nielsen, S. N. 1998. Ecosystem as self-organizing critical systems. *Ecological Modelling* 111:261–268.

Kauffman, S. A. 1993. *The Origins of Order*. Oxford: Oxford University Press.

Keddy, P. 1999. Wetland restoration: the potential for assembly rules in the service of conservation. *Wetlands* 19:716–732.

Keddy, P.; and Weiher E. 1999. The scope and goals of research on assembly rules. In *Ecological Assembly Rules: Perspectives, Advances, Retreats*, ed. E. Weiher and P. Keddy, 1–23. Cambridge: Cambridge University Press.

Keitt, T. M.; and Marquet, P. A. 1996. The introduced Hawaiian avifauna reconsidered: evidence for self-organized criticality? *Journal of Theoretical Biology* 182:161–167.

Lathrop, R. C.; Johnson, B. M.; Johnson, T. B.; Vogelsang, M. T.; Carpenter, S. R.; Hrabik, T. R.; Kitchell, J. F.; Magnuson, J. J.; Rudstam, L. G.; and Stewart, R. S. 2002. Stocking piscivores to improve fishing and water clarity: a synthesis of the Lake Mendota biomanipulation project. *Freshwater Biology* 47:2410–2424.

Legendre, L.; and Legendre, P. 1979. *Ecologie Numerique* 2. Paris: Masson.

Lockwood, J. L.; and Pimm, S. 1999. When does restoration succeed? In *Ecological Assembly Rules: Perspectives, Advances, Retreats*, ed. E. Weiher and P. Keddy, 363–393. Cambridge: Cambridge University Press.

Lovejoy, S.; Currie, W. J. S.; Tessier, Y.; Claereboudt, M. R.; Bourget, E.; Roff, J. C.; and Schertzer, D. 2001. Universal multifractal and ocean patchiness: phytoplankton, physical fields and coastal heterogeneity. *Journal of Plankton Research* 23:117–141.

McNaught, A. S.; Schindler, D. W.; Parker, B. R.; Paul, A. J.; Anderson, R. S.; Donald, D. B.; and Agbeti, M. 1999. Restoration of the food web of an alpine lake following fish stocking. *Limnology and Oceanography* 33:127–136.

Murray, J. D. 1989. *Mathematical Biology*. Berlin: Springer-Verlag.

Ortega-Mayagoitia, E.; Rodrigo, M. A.; Rojo, C.; and Alvarez-Cobelas, M. 2002a. Picoplankton dynamics in a hypertrophic semiarid wetland. *Wetlands* 22:575–587.

Ortega-Mayagoitia, E.; Rojo, C.; and Armengol, J. 2000. Structure and dynamics of zooplankton in a semi-arid wetland, the National Park Las Tablas de Daimiel Spain. *Wetlands* 20:629–638.

Ortega-Mayagoitia, E.; Rojo, C.; Rodrigo, M. A. 2002b. Factors masking the trophic cascade in shallow eutrophic wetlands: evidence from a microcosm study. *Archiv für Hydrobiologie* 155:43–63.

Padisák, J.; and Reynolds, C. S. 1998. Selection of phytoplankton association in Lake Balaton, Hungary, in response to eutrophication and restoration measures, with special references to the cyanoprokaryotes. *Hydrobiologia* 284:41–53.

Pahl-Wostl, C. 1995. *The Dynamic Nature of Ecosystems*. Chichester, England: John Wiley and Sons.

Palmer, M. A.; Ambrose, R. F.; and Poff, N. L. 1997. Ecological theory and community restoration ecology. *Restoration Ecology* 5:291–300.

Parker, B. R.; Schindler, D. W.; Donald, D. B.; and Anderson, R. S. 2001. The effects of stocking and removal of a nonnative salmonid on the plankton of an alpine lake. *Ecosystems* 4:334–345.

Perrow, M. R.; Meijer, M. L.; Dawidowicz, P.; and Coops, H. 1997. Biomanipulation in shallow lakes: state of the art. *Hydrobiologia* 342/343:355–365.

Petersen, J. E. 2001. Adding artificial feedback to a simple aquatic ecosystem: the cybernetic nature of ecosystems revisited. *Oikos* 94:533–547.

Platt, T. 1985. Structure of the marine ecosystem: its allometric basis. *Canadian Bulletin of Fisheries and Aquatic Science* 213:55–64.

Podani, J. 2000. *Introduction to the Exploration of Multivariate Biological Data.* Leiden, The Netherlands: Backhuys Publishers.

Powell, T. M.; and Steele, J. H. 1995. *Ecological Time Series.* New York: Chapman and Hall.

Putman, R. J. 1994. *Community Ecology.* London: Chapman and Hall.

Reynolds, C. S. 1984. *The Ecology of Freshwater Phytoplankton.* Cambridge: Cambridge University Press.

Reynolds, C. S. 1994. The ecological basis for the successful bio-manipulation of aquatic communities. *Archiv für Hydrobiologie* 130:1–33.

Reynolds, C. S. 2001. Emergence in pelagic communities. *Scientia Marina* 65:5–30.

Reynolds, C. S.; Dokulil, M.; and Padisák, J. 2000. Understanding the assembly of phytoplankton in relation to the trophic spectrum; where are we now? *Hydrobiologia* 424:147–152.

Rodrigo, M. A.; Rojo, C.; Armengol, X.; and Mañá, M. 2001. Heterogeneidad espacio-temporal de la calidad del agua en un humedal costero: el marjal de La Safor Valencia. *Limnetica* 20:329–339.

Rodríguez, J.; and Mullin, M. M. 1986. Relation between biomass and body weight of plankton in a steady state oceanic ecosystem. *Limnology and Oceanography* 31:361–370.

Rojo, C.; Ortega-Mayagoitia, E.; and Álvarez-Cobelas, M. 2000a. Lack of pattern among phytoplankton assemblages. Or, what does the exception to the rule mean? *Hydrobiologia* 424:133–139.

Rojo, C.; Ortega-Mayagoitia, E.; Rodrigo, M. A.; and Álvarez-Cobelas, M. 2000b. Phytoplankton structure and dynamics in a semiarid wetland, the National Park Las Tablas de Daimiel Spain. *Archiv für Hydrobiologie* 148:397–419.

Rojo, C.; and Rodríguez, J. 1994. Seasonal variability of phytoplankton size structure in a hypertrophic lake. *Journal of Plankton Research* 6:317–335.

Ruggiu, D.; Morabito, G.; Panzani, P.; and Pugnetti, A. 1998. Trends and relations among basic phytoplankton characteristics in the course of the long-term oligotrophication of Lago Maggiore, Italy. *Hydrobiologia* 369/370:243–257.

Samuels, C. L.; and Drake, J. A. 1997. Divergent perspectives on community convergence. *Trends in Ecology and Evolution* 12:427–432.

Sánchez-Carrillo, S.; and Álvarez-Cobelas, M. 2001. Nutrient dynamics and eutrophication patterns in a semi-arid wetland: the effects of fluctuating hydrology. *Water, Air and Soil Pollution* 127:12–27.

Scheffer, M.; Hosper, S. H.; Meijer, M. L.; Moss, B.; and Jeppesen, E. 1993. Alternative equilibria in shallow lakes. *Trends in Ecology and Evolution* 8:275–279.

Schindler, D. W. 1990. Experimental perturbations of whole lakes as tests of hypotheses concerning ecosystem structure and function. *Oikos* 57:25–41.

Seda, J.; Hejzlar, J.; and Kubecka, J. 2000. Trophic structure of nine Czech reservoirs regularly stocked with piscivorous fish. *Hydrobiologia* 429:141–149.

Sheldon, R. W.; Prakash, A.; and Sutcliffe, W. H. 1972. The size distribution of particles in the ocean. *Limnology and Oceanography* 17:327–340.

Solé, R. V.; and Manrubia, S. C. 1995. Are the rainforests self-organized in a critical state? *Journal of Theoretical Biology* 173:31–40.

Solé, R. V.; Manrubia, S.; Benton, M.; Kauffman, S.; and Bak, P. 1999. Criticality and scaling in evolutionary ecology. *Trends in Ecology and Evolution* 14:156–160.

Sommer, U.; Gaedke, U.; and Schweizer, A. 1993. The first decade of oligotrophication of Lake Constance. II. The response of phytoplankton taxonomic composition. *Oecologia* 93:276–284.

Søndergaard, M.; Jeppesen, E.; and Berg, S. 1997. Pike *Esox lucius* L. stocking as a biomanipulation tool 2. Effects on lower trophic levels in Lake Lyng, Denmark. *Hydrobiologia* 342/343:319–325.

Sprules, W. G.; Casselman, J. M.; and Shuter, B. J. 1983. Size distribution of pelagic particles in lakes. *Canadian Journal of Fisheries and Aquatic Sciences* 40:1761–1769.

Sprules, W. G.; and Munawar, M. 1986. Plankton size spectra in relation to ecosystem productivity, size and perturbation. *Canadian Journal of Fisheries and Aquatic Sciences* 43:1789–1794.

Vidondo, B.; Prairie, Y. T.; Blanco, J. M.; and Duarte, C. M. 1997. Some aspects of the analysis of size spectra in aquatic ecology. *Limnology and Oceanography* 42:184–192.

Vollenweider, R. A.; and Kerekes, J. 1980. The loading concept as basis for controlling eutrophication philosophy and preliminary results of the OECD programme on eutrophication. *Progress in Water Technology* 12:5–38.

Watson, S.; McCauley, E.; and Downing, J. A. 1992. Sigmoid relationships between phosphorus, algal biomass, and algal community structure. *Canadian Journal of Fisheries and Aquatic Sciences* 49:2605–2610.

Weiher, E.; and Keddy, P. 1999. Assembly rules as general constraints on community composition. In *Ecological Assembly Rules. Perspectives, Advances, Retreats*, ed. E. Weiher and P. Keddy, 251–271. Cambridge: Cambridge University Press.

Wilson, J. B.; Wells, T. C.; Trueman, I. C.; Jones, G.; Atkinson, M. D.; Crawley, M. J.; Dodd, M. E.; and Silvertown, J. 1996. Are there assembly rules for plant species abundance? An investigation in relation to soil resources and successional trends. *Journal of Ecology* 84:527–538.

Chapter 9

Functional Group Interaction Patterns Across Trophic Levels in a Regenerating and a Seminatural Grassland

Winfried Voigt and Jörg Perner

There is hardly any question that natural ecosystems in our world are endangered; they are being destroyed at an alarming rate. As a consequence, the restoration of disturbed or devastated ecosystems is an important social concern of our time, and putting this task into practice is a major challenge for ecologists, in particular for restoration ecologists. Much progress has been made in recent years (Packard and Mutel 1997, Urbanska et al. 1997, Gilbert and Anderson 1998, Middleton 1999, Perrow and Davy 2002), but in many restoration projects, sustained progress has not yet been completely secured because of the many different factors discussed in other chapters of this book.

The primary necessary condition allowing the survival and establishment of desired species and preventing colonization by undesired species in habitats under restoration is undoubtedly the return of abiotic conditions to a state similar to the state that existed before the impact of any disturbances; that is, to install appropriate environmental filters (Keddy 1989, 1992; Díaz et al. 1999; also see Part II, "Ecological Filters as a Form of Assembly"). The restoration of these conditions is far from easy because they interact in many ways. Furthermore, we still know little about the small-scale environmental heterogeneity that characterizes natural systems—heterogeneity that is likely to be very difficult to reproduce for many habitats.

Another issue is how extensive our approach to restoration projects should be. Any restoration process must affect community and ecosystem scales, and so it is necessary to consider all trophic levels—carnivores, herbivores, and decomposers—and not just desired elements of the vegetation or individual species. We believe, therefore, that it is vital not to focus merely on the restitution of plant species composition, as is usually the case in terrestrial ecosystem restoration, but as far as possible to include other trophic levels as well. It is a widespread but unjustified assumption that once the vegetation has been successfully reassembled, the consumer levels will inevitably develop unassisted to complete the restored ecosystem. This view implies that consumers play only a minor, or even a negligible, role in restoration processes; but it is clear that herbivores alone dramatically affect both the direction and rate of succession (Brown and Southwood 1987; Lawton 1987; Brown et al. 1988; Brown and Gange 1989, 1992) and so may well have the same effect in restoration.

Of course, it is difficult enough to restore vegetation successfully, and so the deliberate inclusion of consumers raises plenty of new and more complicated problems. However, given a certain knowledge about the structural and functional associations of ecosystems, and perhaps even rules for assembling the elements of ecosystems, are we able to use this knowledge purposefully in restoration practice? It would be ideal if, as an initial step, we could use our knowledge of assembly to track and measure restoration success at the community level, including not only vegetation development but also the development of other trophic levels.

We apply innovative statistical techniques to compare two grassland ecosystems—with similar vegetation, topography, and geology—to reveal patterns of species composition and to search for the assembly rules that may generate them. To examine the role of consumers, as well as of vegetation, we include in our comparison species groups that represent all trophic levels. We designate the system at Leutratal, which has a "desirable" community structure, as our reference. We designate the system at Steudnitz, after the end of several decades of industrial pollution, as regenerating. We believe that the data and results from Steudnitz apply directly to restoration, even though it is self-regenerating (unassisted restoration; Bradshaw 1992, 2000) and has undergone no manipulation (true restoration, van Diggelen et al. 2001). The processes, however, are essentially the same (Davis et al. 2001), which allows information derived from regenerating ecosystems to be applied in restoration ecology (Pickett et al. 2001).

We used a very large database for our analysis, including more than 1000 species, but we are aware of the problems in deriving general patterns from only two systems—especially because general community process rules are considered contingent rather than fixed (Lawton 1999). Generalization requires a greater number of replications than we were able to exploit; nevertheless, we hope that our approach will encourage further studies and meta-analyses to scrutinize our findings.

A Functional Approach

To reveal general structures better and to increase the likelihood of finding the underlying community pattern rules, we did not use species composition directly in our study but scaled up to a higher, more abstract level of integration. This approach reduces the number of dimensions and clarifies the outcome of the analysis. This higher level need not be a larger spatial scale (Lawton 1999) and, in any case, the spatial scale is usually predetermined by the restoration design, as is the time span of the project. To achieve greater integration, with more comprehensible community units, and yet retain a useful link between structure and function, we allocated species to functional groups or guilds (Root 1967, 2001; Simberloff and Dayan 1991; Steneck 2001). Functional groups are composed of species that share ecological characteristics and play equivalent roles in communities. They also often contain different taxa and therefore cross phylogenetic borders. Functional groups provide a low-resolution, but nevertheless accurate, tool for predicting ecosystem alterations (Körner 1994, Steneck 2001) if they contain quantitative information derived from an extensive range of species; that is, from a community study. The focus on functional groups is more advanced in plant ecology than in animal ecology (Smith et al. 1997), and there are some well-developed, albeit disputed, concepts for plant functional groups (Körner 1994, Tilman 2001). For this reason, it might have been thought prudent to restrict our analysis to plants (Pickett 2001), but we believe that we cannot neglect the effects of consumers on restoration. There are many interactions (not necessarily trophic) between functional groups within and between different trophic levels. Producers can control one or more consumer levels or, conversely, consumers can control producers or control producers and other consumer levels, in every possible combination (Fretwell 1977, 1987; Oksanen et al. 1981). Estimating and controlling consumer functions during restoration might improve the outcome considerably—although, so far, little attempt has been made to do so in practice. Only a few studies include both restoration aspects and assembly rules across all trophic levels, and most of these are restricted to aquatic ecosystems (Tatrai et al. 1997a, 1997b; Jeppesen et al. 1998; Nicholls 1999; Tallberg et al. 1999; Burgi and Stadelmann 2002; Degans and de Meester 2002; Gulati and van Donk 2002; Kalchev et al. 2002; Mehner et al. 2002; Svensson and Stenson 2002). Even rarer is the application of this approach to terrestrial communities (Korthals et al. 2001); in addition, theoretical studies deal only with single taxa or with small numbers of taxa or functional groups (Fox 1987, 1999; Wilson 1989, 1999a, 1999b; Fox and Brown 1993; Handel 1997; Majer 1997; Kelt and Brown 1999).

Grassland Studies

Our comparative analysis uses the results of a field study of regeneration in Steudnitz, which was affected over several decades by emissions of a phosphate fertilizer factory. This factory closed in 1990; relieved of the anthropogenic stress, unassisted regeneration of the grassland system has been taking place ever since (Heinrich et al. 2001; also see Chapter 13 for a more detailed historical and topographical description). The question we address is whether there are specific patterns, and underlying assembly rules, that explain and predict the ongoing process of regeneration. To answer this question, we compare the data and patterns of the disturbed grassland with those derived from a methodologically similar succession study in Leutratal, an undisturbed grassland of the same type. Leutratal has a plant community that is valued by society, as reflected in its status as a nature reserve. This community is the result of centuries of management as hayfields but is now undergoing changes of its own after the cessation of traditional land management.

There are advantages in studying assembly rules for species establishment in grasslands. Grasslands are very abundant and species-rich ecosystems that have been widely studied (Barnard 1964, Spedding 1971, Coupland 1979, French 1979, Breymeyer and Van Dyne 1980, Estes et al. 1982, Van Andel et al. 1987, Hillier et al. 1990, Curry 1994, Murray et al. 2000). Furthermore, the responses to any disturbances, and the subsequent regeneration processes, are usually rapid and most impressive: dramatic structural changes can be observed within a few years (Van Andel et al. 1987). The size and distribution of plants and invertebrate consumers permit study areas of manageable size.

The Disturbed Grassland

The location, management, and pollution history of the regenerating grassland at Steudnitz are described in Chapter 13. In 1990, three permanent plots (on the lower (L), middle (M), and upper (U) slopes) were set up along a hill behind the emissions source, at successively greater distances from the former source of pollution (a fertilizer factory). Site L is thus the most, and U the least, disturbed site. All three sites were subdivided into 24 plots (squares) of 5×5 m. The vegetation on these sites was assessed annually until 1999 (Heinrich et al. 2001) using the Braun-Blanquet combined cover-abundance estimation method (Kent and Coker 1994). Within these permanent plots, or in their immediate vicinity, and at equivalent distances to the original pollution source, the arthropod community was sampled from

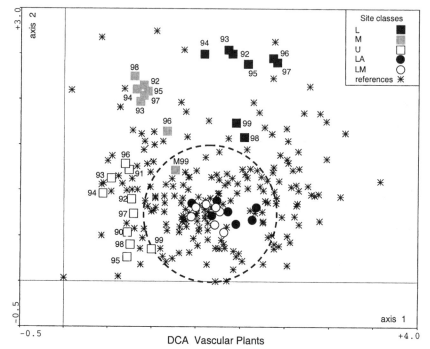

FIGURE 9.1 Joint ordination (DCA) of 200 xerothermic grassland sites (Roscher 2000) in eastern Thuringia (stars), the Leutratal series (circles), and the Steudnitz series (squares). The numbers next to the site symbols indicate the year of data sampling. Typical undisturbed grasslands are empirically defined to be within the circle (1.5 standard deviation units in diameter).

1992 to 1996 using standardized sweep-netting techniques (10 by 10 randomly distributed double swings per site; 300-mm-diameter linen net; Witsack 1975) at midday every 2 weeks from April to October. The number of individuals per species was determined and summed for each year.

The Reference Grassland

The reference data were gathered in exactly the same way in a grassland nature reserve in the Leutratal Valley 22 km south of Steudnitz. This seminatural grassland, hay fields for many centuries, was abandoned in 1970, and since then natural succession to scrub and woodland has been taking place. At this site, vegetation was assessed in two different plots, A and M, every 4 years from 1972 to 1996, and the arthropods were sampled in 1972–1974, 1983–1985, and 1987–1989 (Müller et al. 1978, Heinrich et al. 1998). Plot A, located at the base of the slope, represented a fresh *Bromus* grassland;

whereas plot M, at the middle slope, is a typical meso-xerophile *Bromus* grassland. We established that Leutratal (despite the changes in land use since 1970) was a suitable reference for calcareous grasslands (a socially valued ecosystem type) in eastern Thuringia, by comparing its vegetation to that of 200 unpolluted grassland sites. The data from these sites, which cover a large geographic area and represent a large variety of histories and management methods (Roscher 2000), were submitted to a detrended correspondence analysis (DCA). This basic multivariate technique locates sites on the basis of their vegetation in the phase space determined by the DCA axes. The more similar the quantitative species composition of the vegetation, the closer the sites in the plot. The centroid of the scatter (the point of highest density of points) then represents a point with the typical, or average, vegetation composition. We a priori defined sites as representative if they lay within 1.5 standard deviates of the centroid. Standard deviates are convenient because the DCA axes scale in units of standard deviation (SD) (Figure 9.1). The site at Leutratal lies well within this region, demonstrating that it is an appropriate reference for undisturbed seminatural grassland and represents a regeneration (restoration) goal that is possible to achieve.

Definition of Functional Groups

We classified plants (autotrophs) as mosses, grasses, annual forbs, perennial forbs, and woody plants and the invertebrates according to their foraging characteristics (feeding guilds) within trophic levels (Table 9.1).

Use of Relationships Within and Between Different Trophic Levels to Differentiate Between Succession and Regeneration

Wilson and Roxburgh (1994) introduced the concept of guild proportionality: the idea that the proportions of plant species of different guilds are almost constant across differently developed sites of a comparable community type. Following this general idea, and extending it to whole communities, we ask whether or not there is a general hierarchical pattern of functional groups across all trophic levels in grasslands and if there is a difference in proportions between disturbed (Steudnitz) and undisturbed (Leutratal) grasslands. To answer these questions, we studied the interrelationships among all functional groups for which we had data across trophic levels. To compare whole functional groups with different quantitative species compositions, we applied simple and partial Mantel tests (Mantel 1967, Sokal 1979, Manly 1986, Legendre and Troussellier 1988, Legendre and Fortin 1989, Legendre

TABLE 9.1

The Classification of Functional Groups Within the Different Trophic Levels

Trophic level	Abbreviation	Guild	Taxon
Primary producers	PAF	Annual forbs	Spermatophyta
	PGR	Grasses	Spermatophyta
	PMO	Mosses	Bryophyta
	PPF	Perennial forbs	Spermatophyta
	PWP	Woody plants	Spermatophyta
Herbivores	HCW	Stage-specific chewers	Coleoptera
	HLC	Lifetime chewers	Caelifera
	HMI	Miners	Diptera (Brachycera)
	HTS	Cell-tissue suckers	Heteroptera
	HVS	Vascular suckers	Auchenorrhyncha
Carnivores	CAC	Attacking chewers	Ensifera
	CBS	Biting suckers	Diptera (Brachycera)
	CCH	Chewing hunters	Coleoptera
	CPA	Parasitoids	Diptera
	CSH	Sucking hunters	Heteroptera
	CWS	Web-spinners	Araneae
Detritivores	DBS	Biting suckers	Diptera (Brachycera)
	DCW	Chewers	Coleoptera

and Legendre 1998, Fortin and Gurevitch 2001). We used the SMT (version 1.2) and PMT (version 1.1) programs written by Eric Bonnet (2001) (eb_ce@yahoo.fr). The Mantel test is a multivariate symmetrical method that analyzes the relationship between two matrices (functional groups) based on previously calculated distance (or similarity) matrices. This calculated relationship is a special form of matrix correlation. Unlike the simple correlation of two data matrices, it measures the extent to which the variations in the distances (similarity) of one matrix correspond to the variations in a second distance (similarity) matrix (Legendre 1993). The calculated standardized Mantel statistic r_M or the correlation between any two functional groups is interpreted in the same way as Pearson's product-moment correlation coefficient, but a permutation test procedure must be used to test its statistical significance. This is necessary because the distances in the matrices compared are not independent values; moreover, in our study, even the original variables (species abundances) are both temporally and spatially autocorrelated. Mantel tests account for such autocorrelation patterns because they are based on resemblance matrices, the cells of which correspond exactly and relate to the same spatial or temporal samples. In theory, r_M ranges from –1 to +1, but because our hypothesis is that two resemblance matrices (functional groups) can be related only positively or not at all, the test is inevitably one-tailed. In the case of a significant relationship, the value of r_M obtained should be positive; otherwise, the values should scatter around zero. Negative r_M would be obtained, however, if one matrix was of distance measures and the other was the corresponding matrix of similarity measures, because similarity is generally negatively correlated with distance (Legendre and Legendre 1998). We therefore do not compare distance and similarity matrices, but only those of the same type.

A high positive correlation between two functional groups (for example, grasses and vascular suckers) should not necessarily be interpreted as a high degree of trophic association (for example, herbivory) but rather as a concordant temporal or spatial pattern. These patterns, however, may be the result of a similar degree of species turnover caused by the same abiotic processes, the same abundance dynamics of important species, or similar factors. Therefore, r_M measures an average association between the members of two functional groups rather than specific interactions.

We summarize and represent the associations between the functional groups of different trophic levels in terms of r_M as dendrograms of hierarchical classification and as ordination plots of principal component analyses (PCA) (Figures 9.2 and 9.3). In Leutratal, the Mantel correlation between functional groups within plants is very high (Table 9.2) and is always highly significant ($p \leq 0.001$). It is lower between herbivore functional groups and still lower among the carnivore groups; that is, the correlation between functional groups decreases systematically with increasing trophic level affiliation. The correlation between functional groups across different trophic levels, however, is relatively small (see Figure 9.2) and decreases with ascending trophic rank. There are three well-separated trophic level clusters (see Figure 9.2a) that are arranged pretty much according their trophic rank from right to left (see Figure 9.2b). Exceptions to this pattern are the group of lifetime chewers (HCL; caeliforous grasshoppers), which is included in the carnivore cluster, and the group of biting sucking decomposers (DBS; Diptera), which is included in the herbivore cluster. The variance within trophic levels increases with trophic rank (see Figure 9.2b); this is possibly a general phenomenon, since it was also found for climatic sensitivity (Voigt et al. 2003).

In contrast to Leutratal, the relationship between functional groups within and across trophic levels in Steudnitz appears to be rather irregular (see Figure 9.3). There are only two big clusters in the dendrogram, and the functional groups of different trophic levels are quite intermingled, so the PCA plot shows a different arrangement than that for Leutratal. Herbivores,

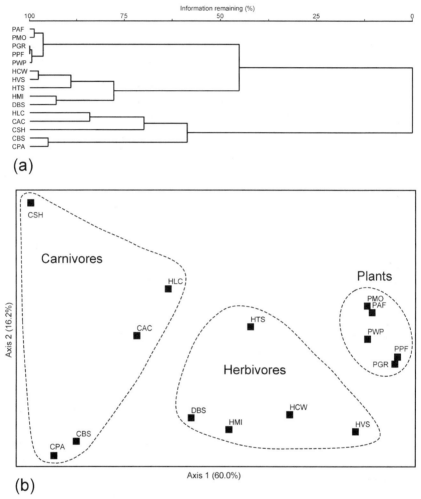

FIGURE 9.2 Two different ways of visualizing the relationships between functional groups of different trophic levels for the undisturbed grassland system Leutratal using the standardized Mantel statistic r_M. (a) Dendrogram of hierarchical classification (method: WARD) and (b) ordination plot of a principal component analysis (PCA). Plants, herbivores, and carnivores are enveloped separately by dashed lines. For abbreviations of functional groups, see Table 9.1.

carnivores, and detrivores form well-separated clusters, but the plants overlap the herbivores and carnivores considerably (see Figure 9.3b) and do so even along PCA axis 3 (not shown here). Some plant groups are more highly correlated with particular carnivores than with herbivores, which is less often the case in Leutratal.

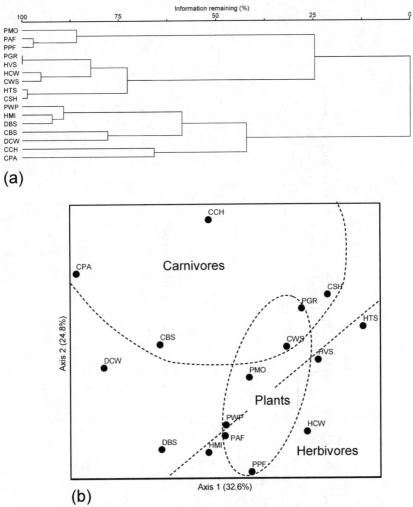

(a)

(b)

FIGURE 9.3 Two different ways of visualizing the relationships between functional groups of different trophic levels for the regenerating grassland system Steudnitz using the standardized Mantel statistic r_M. (a) Dendrogram of hierarchical classification (method: WARD) and (b) ordination plot of a principal component analysis (PCA). Plants, herbivores, and carnivores are enveloped separately by dashed lines. For abbreviations of functional groups, see Table 9.1.

There is empirical evidence that undisturbed ecosystems have a hierarchical structure of connectedness between trophic groups (O'Neill et al. 1986), and we suppose that in largely undisturbed grassland ecosystems, there is a regular hierarchical pattern of trophic levels based on the correlation

TABLE 9.2

Standardized Mantel Statistic r_M for the Correlations Between Functional Groups of Different Trophic Levels in the Undisturbed Grassland Leutratal

	PAF	PGR	PMO	PPF	PWP	HCW	HLC	HMI	HTS	HVS	CAC	CBS	CPA	CSH
PGR	0.80***	x												
PMO	0.80***	0.87***	x											
PPF	0.85***	0.98***	0.90***	x										
PWP	0.73***	0.85***	0.80***	0.86***	x									
HCW	0.54***	0.66***	0.52***	0.62***	0.59***	x								
HLC	0.27**	0.28***	0.32**	0.32***	0.41***	0.28**	x							
HMI	0.42***	0.47***	0.27***	0.44***	0.40***	0.65***	0.28*	x						
HTS	0.59***	0.50***	0.49***	0.54***	0.46***	0.56***	0.37***	0.39**	x					
HVS	0.69***	0.85***	0.62***	0.83***	0.79***	0.79***	0.29*	0.70***	0.56***	x				
CAC	0.08	0.31***	0.23**	0.31***	0.40***	0.41***	0.17*	0.21*	0.11	0.28*	x			
CBS	0.01	0.18**	0.08	0.20**	0.17*	0.34**	0.04	0.31*	0.43***	0.37***	0.23*	x		
CPA	-0.08	0.10	0.05	0.11*	0.10	0.39**	0.18	0.28	0.13	0.37***	0.15	0.63***	x	
CSH	0.04	-0.06	0.06	-0.02	0.10	0.17	0.15	0.05	0.32**	-0.04	0.12	0.10	-0.01	x
DBS	0.39***	0.39***	0.24**	0.40***	0.36***	0.40***	0.24***	0.52***	0.18*	0.44***	0.19*	0.25*	0.31*	-0.06

between functional groups. Even natural succession driven by changed general competitive conditions, as in the Leutratal grassland, may cause typical and stable patterns on a medium scale. All terrestrial ecosystems are more or less dynamic, in a continuous state of flux, and whether development is directional or not is basically a matter of scales. One could argue that, in both our sites, competitive conditions have changed and that restoration in Steudnitz is superimposed on natural succession (incipient invasion of scrub and trees). The difference between the sites, however, is that Leutratal grasslands are the result of centuries of mowing and grazing. This has allowed plenty of time for all possible invading species to pass the abiotic filters (see also Part II, "Ecological Filters as a Form of Assembly Rule") and eventually to establish in these grasslands while developing strong biotic interactions. Human management is an inherent feature of these grasslands; it is an incorporated disturbance (O'Neill et al. 1986). Thus, changes caused by reduced management or even disuse may interrupt interactions between functional groups and trophic levels gradually rather than suddenly. The high correlation values between the functional groups within trophic levels indicate a balanced temporal pattern in these groups. This concordance decreases with increasing trophic rank because fluctuation within different consumer levels is driven differently by abiotic conditions—for example, climate (Voigt et al. 2003)—which are caused, in part, by autogenic succession processes (such as the alteration of microclimate by vegetation height or density).

Unlike Leutratal, initial vegetation cover at Steudnitz was sparse, largely of the halophytic grass *Puccinellia distans*. Since 1990, however, regeneration has been dramatic (see Chapters 12, 13, and 14). The increasing canopy density and architectural complexity may well have provided anchorage points for web spinners and shelter for cursorial hunting beetles, leading to the high Mantel correlation between plants and carnivorous invertebrates.

To clarify the relationship between functional groups of different trophic levels and to check for spurious correlations, it is necessary to calculate r_M between two functional groups (herbivores and carnivores) while controlling for the effect of a third group (plants as the assumed autogenic driving forces of regeneration processes). This "partial" Mantel test is a step toward causal modeling using resemblance matrices (Legendre 1993, Legendre and Legendre 1998). In contrast to path analysis, we disregarded any possible unknown residual variables (for example, weather) but tried to elucidate the direct and indirect biotic effects of one trophic level on another. Following the causal models involving three matrices and the criteria given for model expectations (Legendre and Troussellier 1988, Legendre 1993), we evaluated all possible pairs of plant-herbivore-carnivore combinations.

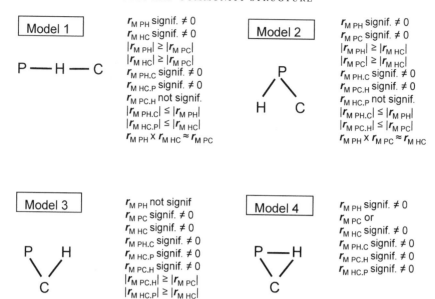

FIGURE 9.4 Criteria for the four possible models of correlative relationship between functional groups of the three trophic levels, in terms of the expected values for the simple and partial standardized Mantel statistic r_M (after Legendre and Legendre 1998). All conditions for each model must be satisfied to accept the model. However, model 4 is supported even if one, but only one, of both $r_{M\,PC}$ and $r_{M\,HC}$ are significantly different from zero. P: plants, H: herbivores, C: carnivores. By convention, partial Mantel correlation between two matrices (for example, H and C) when controlling for their relationships with a third matrix (for example, P) is written $r_{M\,HC.P}$.

There are four potential relationship patterns between the trophic levels (Figure 9.4). Detailed model characteristics and criteria to support any particular model type can be found elsewhere (Legendre 1993, Legendre and Legendre 1998), but the most important criteria for accepting a specified model are

1. Value, rank, and significance of $r_{M\,PH}$, $r_{M\,PC}$, and $r_{M\,HC}$ (r_M for combinations of any functional groups for plants (P), herbivores (H), and carnivores (C))

2. Decrease or increase of $r_{M\,HC.P}$ in comparison to $r_{M\,HC}$. The partial Mantel correlation between variables (matrices) H and C, when controlling for their relationships with P, is written $r_{M\,HC.P}$.

The ecological meaning of these models in detail (see Figure 9.4) is

- Model 1: No immediate effect of plants on carnivores.
- Model 2: Plants interact with both herbivores and carnivores; no relationship between herbivores and carnivores ("spurious" correlation).

- Model 3: Both plants and herbivores interact independently with carnivores; no correlation between plants and herbivores.
- Model 4: Strong interactions between all trophic levels.

All the simple Mantel correlations between Leutratal plant functional groups and herbivore groups ($r_{M\,PH}$) are highly significant (see Table 9.2). Partial Mantel correlations $r_{M\,HC.P}$ are in many cases higher than the simple Mantel correlations $r_{M\,HC}$, indicating that plant effects interfere with the relationship between herbivores and carnivores. This strongly supports model 4 (20.0 percent of all possible triplet combinations, hereafter abbreviated as TC, and 42.6 percent of all actually supported models of any type with significant associations between trophic level groups, hereafter abbreviated as MSA). However, because the values of $r_{M\,PC.H}$ are rather small, the relationship between plants and predators is weak, in contrast to the herbivore-carnivore relationship. When $r_{M\,HC.P} < r_{M\,HC}$ and $r_{M\,HC.P}$ significant $\neq 0$, then model 1 (TC: 12 percent, MSA: 25.5 percent) holds; otherwise, model 2 (TC: 15.0 percent, MSA: 31.9 percent) holds (see Figure 9.4). Model 2 means that there is no causality between a substantial part of herbivore and carnivore functional groups but only a spurious correlation caused by a common and similar response to vegetation or effects (for example, microclimate) caused by it. Causal relationships between herbivore and carnivore groups are not to be expected for two reasons. First, only a fraction of herbivores belongs to the prey spectrum of any particular carnivore, even of generalists. Second, distributions of potential trophic pairs may not be spatially and temporally congruent. It is interesting that 13 of 15 combinations supporting a significant model 2 include attacking chewers (CAC; ensiferous grasshoppers; not present in Steudnitz data), which cannot be classified unequivocally as pure predators. Most of these species are, at least temporarily, omnivores and eat flowers and buds, too.

Model 3 was not supported unequivocally in any case, either in Leutratal or in Steudnitz. Since it is not a plausible pattern, its absence is unsurprising.

We therefore conclude that ecological succession in Leutratal is characterized mainly by interactions between all trophic levels (model 4) and, less often, either by intervening, sequential relationships across trophic levels (model 1) or by no significant relationships between herbivores and carnivores (model 2). The results of the causal modeling of all significant functional combinations with Leutratal data using partial Mantel are summarized in an interaction web (Menge 1987, Price 2002) that shows connectedness as the number of connecting lines (Figure 9.5a).

Compared to Leutratal, the relationships between trophic functional groups in the regenerating grasslands at Steudnitz are not overly pronounced (Figure 9.5b). Only 15 of 85 possible combinations (17.7 percent compared

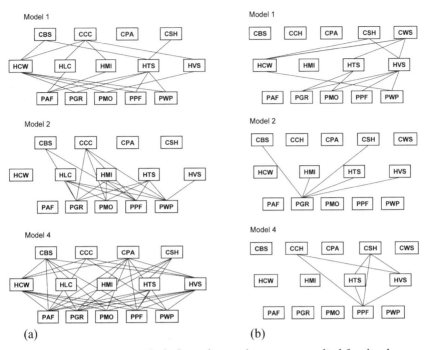

FIGURE 9.5 Interaction webs linking the trophic groups studied for the three supported models. (a) Leutratal; (b) Steudnitz. See Table 9.1 for abbreviations for individual functional groups.

to 47 percent in Leutratal) could be assigned unequivocally to any model type. Model 1 is dominant (TC: 10.6 percent, MSA: 60.0 percent), followed by models 2 and 4 (each with TC: 3.5 percent, MSA: 20.0 percent).

The dominance of model 1 means that there is no significant direct relationship between plants and invertebrate carnivores when the effects of herbivores are controlled for. In fact, this applies only to seven combinations of web spinners (CWS) with vascular suckers (leafhoppers; HVS) and with cell-tissue suckers (plant bugs; HTS), and two combinations of sucking hunters (true bugs; CSH) with vascular suckers (HVS) and with chewers (herbivore beetles; HCW). It is clear, though surprising given our earlier supposition, that both these predator groups are uninfluenced either by plant architecture or by microclimate caused by vegetation but are strongly associated with prey patterns.

The most important distinction between Leutratal and Steudnitz, however, is not the different distribution of model types between the two grassland systems. Rather, it is the different number of functional group combinations that support any model at all; that is, the connectedness within this system. In addition, the identity of the consumer functional groups included differs (compare Figure 9.5). It is noticeable that in Steudnitz, models 2 and 4 depend on a single but different producer group, grasses (PGR) and perennial forbs (PPF). Furthermore, there is no significant relationship between annual forbs (PAF) and any consumer group. The presence of many annual species appears to be rather episodic and, because of the rapid species turnover in this plant group, it is possible that no consumer group is able to keep up with the yearly changing patterns and thus no clear relationship develops. Likewise, and probably for the same reason, the parasitoids (CPA), as highly specialized carnivores, show no significant relationship to any herbivore group in Steudnitz. This effect is also likely to apply, though less strongly, to the relationships between all the other functional groups in Steudnitz.

Because, in contrast to Steudnitz, both annual forbs and parasitoids in Leutratal grasslands show many strong interactions at least partially described by model 4, these functional groups appear to be differential or key groups for conditions in semidry grasslands (see the next section in this chapter).

In the Leutratal grasslands, every functional group is significantly linked with one group of another trophic level in at least one of the three models (see Figure 9.5). In particular, the interaction web for the dominant model 4 shows the amply developed interactions—that is, the high connectedness—in such grasslands. Why are there so many more clear relationships in Leutratal than in Steudnitz?

The answer appears to lie in the higher species diversity in Leutratal grasslands, which causes a higher connectedness between plants, invertebrate herbivores, and carnivores. More species in all functional groups increases the chance of finding a suitable species (resource) in a functional group of the trophic level below. However, we detected no significant correlation between the summarized species richness of all functional group pairs versus the standardized Mantel statistic (r_M) for the plant-herbivore, herbivore-carnivore, or plant-carnivore comparisons.

A complementary explanation might be greater host plant or prey specialization, with consumers in largely undisturbed grasslands having become more adapted to their resources over the very long period these grasslands have existed. This still must be proved, but it is strongly supported by the weak interactions between annuals and any herbivorous groups, and between parasitoids and herbivores in Steudnitz. However, if true, it would imply lower consumer competition in Leutratal than is indicated by the r_M values between functional groups among herbivores and predators (see Table 9.2);

TABLE 9.3

Standardized Mantel Statistic r_M for the Correlations Between the Functional Groups of Different Trophic Levels in the Regenerating Grassland System Steudnitz[1]

	PAF	PGR	PMO	PPF	PWP	HCW	HMI	HTS	HVS	CBS	CCH	CPA	CSH	CWS	DBS
PGR	0.19**	x													
PMO	0.56****	0.24**	x												
PPF	0.70****	0.17*	0.64***	x											
PWP	0.22*	0.28**	−0.12	0.13	x										
HCW	0.31	0.39**	0.34	0.57*	0.63***	x									
HMI	0.35	0.44**	0.31	0.70**	0.62**	0.79**	x								
HTS	0.20	0.75*	0.75*	0.49*	0.29	0.84***	n.p.	x							
HVS	0.14	0.80***	0.34**	0.40***	0.51***	0.54***	0.52***	0.73***	x						
CBS	0.34	0.54***	0.76***	0.23	0.19	0.23	0.33	n.p.	0.28	x					
CCH	−0.19	0.40**	−0.03	−0.21	−0.04	0.07	0.13	0.40	0.35**	0.60	x				
CPA	−0.14	−0.04	0.28	−0.20	−0.26	−0.15	0.15	n.p.	−0.21	0.24	0.08	x			
CSH	0.44	0.72**	0.54*	0.06	0.25	0.57*	n.p.	0.71***	0.59***	n.p.	0.52**	n.p.	x		
CWS	0.16	0.38***	0.22	0.22	0.41**	0.61***	0.40	0.81***	0.51*	0.51*	0.31*	0.01	0.68***	x	
DBS	0.34	0.07	0.42	0.45*	0.55*	0.50	0.33	n.p.	0.28	0.34	−0.08	0.08	n.p.	0.43*	x
DCW	−0.02	0.12	−0.16	−0.16	0.34*	0.23	0.57*	0.01	0.20*	0.83**	0.01	0.15	0.14	0.23	0.65**

[1]n.p. = calculations not possible because corresponding years for the groups to compare were not available.

but, after all, r_M represents any kind of interaction and is not solely determined by competition. Disturbed grasslands should be characterized by more generalist herbivores and predators, with higher intratrophic level competition (Table 9.3) not waxing and waning as their prey species increase and decrease, but having patterns driven, either via plants or directly, by the same abiotic forces. These results inform our interpretation of the PCA plots (see Figures 9.2b and 9.3b). The position of the plant cluster between herbivores and carnivores in Steudnitz (see Figure 9.3b) indicates a tendency to model 2; that is, an independent effect of plants on both herbivores and carnivores. It is interesting that the detritivores, weakly represented in our study, lie between the plants and herbivores in Leutratal, probably because these species originate from both phytophagous and zoophagous clades. In the PCA graph for Steudnitz (see Figure 9.3b), the detrivores form an independent cluster.

The Mantel tests, in summary, indicate that in undisturbed grassland plots in Leutratal, statistically significant biotic effects dominate the relationships between functional groups of different trophic levels. In the regenerating grassland plots at Steudnitz, however, a substantial part of all relationships may be directly controlled by abiotic factors or indirectly mediated or caused by vegetation structure.

Different driving forces for development in the two grassland systems have obviously caused different patterns of hierarchical trophic structure, manifested in a different proportion of connectedness between the functional groups they consist of.

The Change of Proportions Between Plant Functional Groups

Severe disturbances have a lasting effect on ecosystems, causing structural changes and possibly a transition to alternative stable states (see Chapters 5 and 17) or even new systems. In the folded surface model (Allen 1998), adapted from the general community model, such shifts become apparent when we calculate an index that reflects the current developmental direction or describes the stage reached. An appropriate index to characterize a grassland community might be the ratio of the abundance (or species richness) of the grass group to the abundance (or species richness) of other producer and consumer functional groups. In grasslands, as the name tells us, grass species usually constitute the characteristic functional producer group. Hence, we should expect distinct changes of absolute or relative abundance (or species richness) primarily in the grass group. In particular, when studying regeneration processes following strong disturbances, we should be

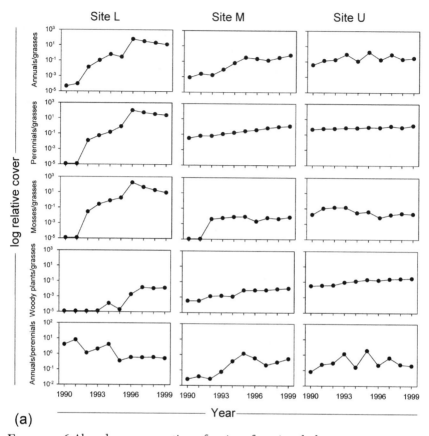

(a)

FIGURE 9.6 Abundance proportion of various functional plant groups to grasses and to perennial forbs, respectively, in the semidry grassland sites (a) L, M, and U at Steudnitz, and (b) A and M at Leutratal.

able to reveal altered proportions between grasses and most of the other functional groups of both producers and consumers. Therefore, we compared the temporal development of such proportions in the Steudnitz grasslands with that of the Leutratal grasslands. Plotting the changes in abundance proportions of plant functional groups to grasses for the three differently polluted sites at Steudnitz shows that, as expected, the most dramatic alterations have occurred in the vegetation closest to the former polluter (plot L) (Figure 9.6a). Starting from 1990, the relative abundances in this grassland plot changed within a period of 6 years, with a decline in grass species indicating a continuous development away from the prevailing grassland system. Since 1996, this process has slowed or has taken a different course, but the ratio values are still quite high. In the plots farther from the pollution source (M and U), the retrograde process is much weaker and the trajectory is more mono-

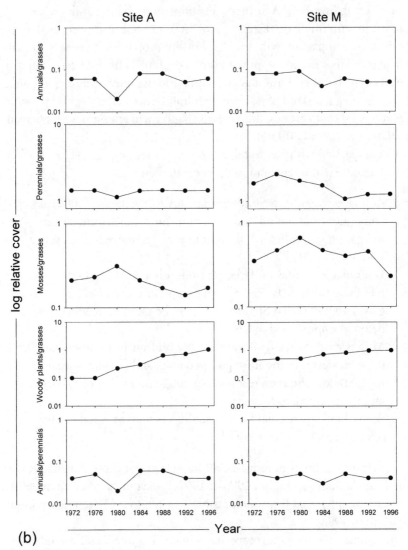

(b)

FIGURE 9.6 (continued)

tonic. Only the trajectory of the proportion of mosses to grasses declines again to its initial value after having peaked in the mid-1990s. The proportion of annual forbs to perennial forbs varies from plot to plot and, for the most part, is much higher than the same proportion in Leutratal.

By contrast, in Leutratal (Figure 9.6b), but over a different period, all proportions fluctuate around a more or less constant mean value, indicating an approximately constant proportionality of these functional groups. One exception is the increasing ratio of woody plants to grasses due to succession (scrub invasion). A further exception, for probably similar reasons, is that in the intermediate-distance plot (M), annual and perennial forbs decrease in comparison with grasses, but the proportion between annuals and perennials remains approximately constant. The highest similarity appears to be between Leutratal grasslands and the most distant, and hence least disturbed, grassland plot (U) at Steudnitz. However, except for the ratio of perennial forbs to grasses, the ratios of all groups to grasses have higher values than they do in Leutratal.

In comparison with Leutratal, reference values (average ±SD; $n = 14$), may characterize the regeneration process in Steudnitz:

1. Annual forbs/grasses: Starting from very low values (≤ 0.001) as a consequence of pollution disturbance, the ratio has since been increasing very quickly from year to year, far above the reference value of 0.1 ± 0.02.
2. Perennial forbs/grasses: Low, continuously increasing to the reference value of 1.5 ± 0.32. An exception is the most polluted grassland plot (U) at Steudnitz, where the quotient reached 100 before dropping to about 20 in 1999.
3. Mosses/grasses: Always lower than the undisturbed reference of 0.4 ± 0.15. However, in the most polluted grassland (L) after 5 years of regeneration, the ratio reached a much higher value of 170 in 1996 and then decreased to about 5.
4. Annual forbs/perennial forbs: Usually less than or equal to the reference value of 0.04 ± 0.01

Apart from annuals/perennials, all trajectories of these density (cover) ratios show a sharp decline in 1996 or 1997, indicating a different direction of development at that time and possibly a switch to a different community stage (Allen 1998).

In a similar way, the ratios of species richness (Figure 9.7) reflect the details of the regeneration process (Leutratal as reference: average ±SD; $n = 14$):

1. Annual forbs/grasses is always larger than the reference 0.60 ± 0.25.
2. Perennial forbs/grasses is always smaller than the reference 4.00 ± 0.77.
3. Mosses/grasses is always smaller than the reference 1.29 ± 0.32. However, in the most polluted grassland (plot L), the ratio shot from 0 to 7 for the first 4 years of regeneration due to the low number of grass species at that time.
4. Annual forbs/perennial forbs is always larger than the reference 0.16 ± 0.09.

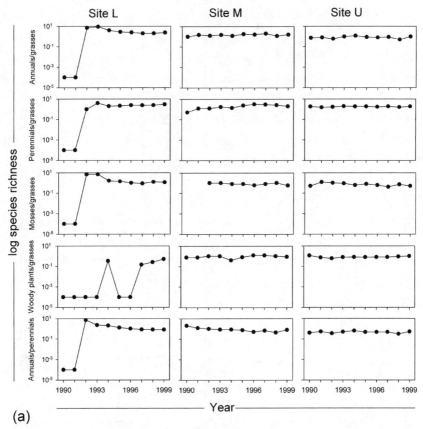

(a)

FIGURE 9.7 Richness proportion of several plant functional groups to grasses and to perennial forbs in the semidry grassland sites (a) L, M, and U at Steudnitz and (b) A and M at Leutratal.

Apart from the richness ratio of woody plants to grasses, all ratios for the reference grassland Leutratal show a slight decrease over time, reflecting the ongoing natural succession (scrub invasion).

Both the abundance ratio and the species richness ratio in the grasslands at Steudnitz show that the plot (U) farthest from the pollution is very close to the undisturbed grasslands of Leutratal.

To summarize the interdependence between various functional groups and the grasses group, we examined joint phase diagrams of their species richness in both Steudnitz and Leutratal (Figure 9.8). This enabled us to understand the general relationship between species richness of different functional groups, while showing the chronological development in the different grassland sites. As additional information, samples are included from

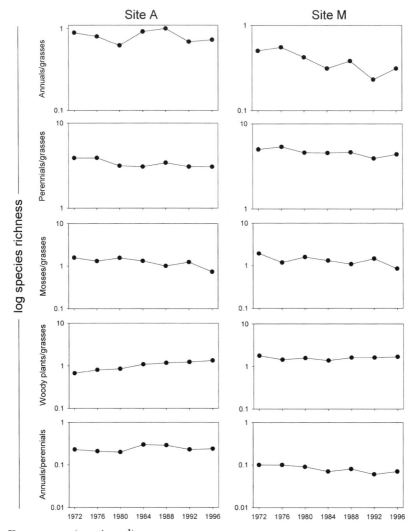

FIGURE 9.7 (continued).

a different grassland type, a very dry grassland—LS; indicated by crosses (+) in Figure 9.8)—with harsh environmental conditions that was sampled in exactly the same way as plots A and M and that is in the field adjacent to them. In contrast to sites A and M, which are man-made, LS has not been managed recently (if ever), and so it represents a near-natural grassland persisting over a long period under almost constant environmental conditions (that is, it shows only slight fluctuations caused by weather variations, but no succession).

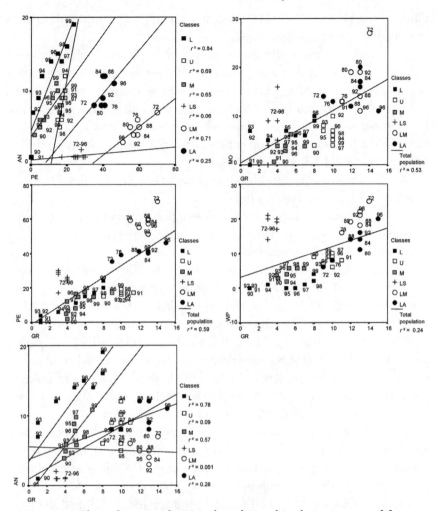

FIGURE 9.8 Phase diagrams showing the relationships between several functional groups and grasses (species richness). Straight lines represent linear trends in single sites; or, if the relationship is similar for all sites, a general linear trend (indicated as "total population" in legend). GR: grasses, AN: annual forbs, PE: perennial forbs, MO: mosses, WP: woody plants. LA and LM: sites A and M at Leutratal; L, M, U: sites L, M, and U at Steudnitz; LS: very dry grassland of a different type than A and M at Leutratal (additional comparison). Numbers next to site symbols indicate the year of data collection.

The position of each grassland type in the graph indicates that there are probably specific trajectories for each type. Because it is often subject to drought and heat stress, this dry grassland is characterized by a rather low diversity and by reduced interactions between functional groups (also see

Chesson and Huntly 1997), and so it is, in a general way, comparable to the disturbed Steudnitz grassland plots.

Perennials compared with grasses show a clear positive linear relationship, which indicates that the proportions of these two functional groups remain almost constant during the course of regeneration and succession in this type of semidry grassland. This result may also indicate a negative relationship between disturbance and diversity (richness), which is best indicated by the temporal sequence of plot L (the most polluted) at Steudnitz (the dry grassland LS is an outlier here).

Woody plants compared with grasses show a linear relationship. The dry grassland LS is a distant outlier here.

Mosses compared with grasses also show a linear relationship. In the dry grassland LS, the richness of grass species is almost constant, but the number of moss species fluctuates strongly.

Annual forbs compared with grasses give a very different picture. Across the various grassland plots of Steudnitz and Leutratal, a negative relationship is shown as the plots are arranged from top left (U) to bottom right (LM). However, within each plot (the time series), there is, at least for Steudnitz, a clear positive relationship. The higher the level of disturbance (the early years of plots L and M at Steudnitz) or of harsh environmental conditions (for example, the dry grassland plot LS), the lower the fraction of annual forbs. During the regeneration process, however, the number of annual forb species increases much faster than the number of grass species.

The ratio of annual forbs to perennial forbs gives a similar picture. All grassland plots, even from Leutratal, have a unique positive (temporal) relationship between the two functional plant groups. The intersection of the model lines with the x-axis indicates that the proportion annual forbs/perennial forbs is always < 1 for the Leutratal plots (excluding LS) and the less disturbed plot U at Steudnitz, but is > 1 for the Steudnitz plots L and M.

We conclude that the functional group annual forbs is the key group for regeneration processes in anthropogenically disturbed semidry grasslands (see also the previous section of this chapter). The proportion of this group to grasses and to perennial forbs might be an indicator for both the disturbance stage and the phase of regeneration in semidry grassland systems. If this relationship were confirmed for more grassland systems, this indicator of disturbance and of extent of regeneration could be used in restoration projects in the future.

Combining the information from the proportions between all functional plant groups in Steudnitz into a two-dimensional nonmetric multidimensional scaling (NMDS) plot (Figure 9.9) reveals that the three grassland plots

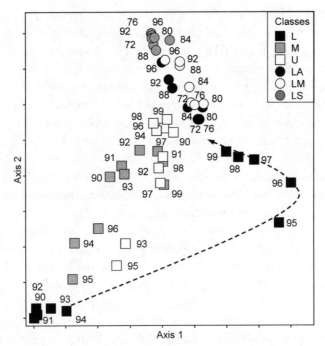

FIGURE 9.9 Nonmetric multidimensional scaling (NMDS) ordination diagram of proportions between functional plant groups in Leutratal and Steudnitz grasslands (variables: ratios annual forbs, perennial forbs, mosses, woody plants over grasses; and annual forbs/perennial forbs; distance measure: relative Euclid). LA and LM: sites A and M at Leutratal; L, M, U: sites L, M, and U at Steudnitz. Numbers next to site symbols indicate the year of data collection. The arrowed trajectory illustrates the dramatic jump in vegetation regeneration in site L within a few years, which takes a path toward the undisturbed grassland sites.

follow a different regeneration course depending on the degree to which they were previously disturbed. It is also likely that the nature of the disturbance affected the regeneration differently, too. In addition, Figure 9.9 shows a recent slowdown of changes (closer position of sequential year points) and, since 1997, a rapid transition of the formerly most disturbed plot L to the conditions represented by U.

Conclusions

Assembly rules are not recipes for building communities (Weiher and Keddy 1999). However, if they are not to be just an intellectual pleasure, they should be of some service to ecological theory and practice. For restoration ecology, we require enhanced predictability of the outcome of a particular restoration project or an aid for assessing the course and progress of reassembling a desired community. Such an assessment done based on single species, even ones highly characteristic of the target community, might be misleading because a species occurrence could be merely transient. Moreover, for various reasons, the whole desired restoration process—even if very promising at the beginning—may turn out not to be sustainable and consequently may fail in the end. Because ecosystems or communities rather than single species are restored, we must also consider multitrophic levels in both theoretical and applied restoration ecology. Individual trophic levels may be controlled from below or from above; consequently, their inherent processes cannot be explained adequately by considering only one level. Rather than using strictly defined species compositions, a community may be predicted best by its structure at functional group level (Steneck 2001), which gives only an indication about what is going on in any temporal ecosystem processes. Comparing the trajectories of single functional groups (Figure 9.8) or of whole communities (Figure 9.9), or looking at the proportions between particular functional groups (Figures 9.6 and 9.7), allows us to monitor the course and success of any restoration project provided that data of sufficient quantity and quality have been gathered. For many projects, this may be a crucial point, because such data collections are extremely time-consuming and expensive. Micro- and mesocosm studies might be a remedy (Lawton 1995, Naeem et al. 1995, Jones et al. 1998), but even in such studies time and money costs are very high and the application of findings from these very simple systems to real communities in the field remains questionable (Carpenter et al. 1995, Drake et al. 1996).

Because not all important invertebrate groups could be considered in our studies (for example, hymenoptera, aphids, mollusks, and some others), we have been not able to evaluate and test their interrelationships in the two grassland communities under study. Also, the strong and important links between trophically different aboveground and belowground components of ecosystems (recently identified by, for example, Scheu 2001, Gange and Brown 2002, Wardle 2002) had to be completely neglected.

Nevertheless, there are some theoretical and practical consequences of our findings. In our study of grassland ecosystems, we showed that the start of a regeneration (or restoration) history is characterized by the strongly fluctuating group of annual forbs, and this start is detectable independently of analysis of species richness or abundance data. There is no significant association, however, between that functional group and almost every other group of every single trophic level. Annual forbs show a high turnover rate in Steudnitz, and most of these species have a high food quality for herbivores but low chemical defenses against them (Strong et al. 1984, Masters and Brown 1997). As a result, generalists also dominate herbivorous groups, and their species richness is rather low in comparison with Leutratal. Consequently, the carnivores also appear not to be very specialized. This is probably best demonstrated by the weak role of parasitoids in Steudnitz (CPA; see Figure 9.9).

To generalize these findings cautiously, we conclude that any generalist producers (opportunists, ruderalists, r-strategists, and so forth; the meanings of these terms are not fully congruent) will dominate regenerating ecosystems and cause bottom-up control of life-history strategies and species richness on all consumer levels. For practical restoration, this producer fraction provides the key functional group or groups whose control (Brown and Gange 1989, 1992; Lawton and Brown 1994) or replacement by perennial producers should be advanced, if possible, by suitable manipulations. The inoculation, or at least the promotion, of diverse perennial producers should be the next complementary step, to enable a high diversity of consumer specialists for both herbivores and carnivores. In fact, to establish a natural dominance structure within the perennial plants as soon as possible, and thus to enable sustainable functioning of the restored ecosystem, attempts should be made to introduce a matching herbivorous fauna.

Future advances in ecology will undoubtedly enable us to control restoration and conservation of natural ecosystems more precisely than we can today.

Acknowledgments

We thank Marzio Fattorini, Vicky Temperton, and Stefan Halle for inviting us to participate in the Dornburg workshop, as well as Andrew Davis and Richard Hobbs, whose comments and suggestions helped to improve this chapter greatly. Particular thanks are due our colleagues Rudolf Bährmann, Bärbel Fabian, Wolfgang Heinrich, Günter Köhler, Dorit Lichter, Rolf Marstaller, and Friedrich W. Sander for contributing data and identifying species of some taxa. The data collections were carried out under several programs funded by the University of Jena, UFZ Leipzig-Halle GmbH, and by the *Deutsche Forschungsgemeinschaft* (DFG, German Research Council) as part of the Graduate Research Group.

REFERENCES

Allen, T. F. H. 1998. Community ecology: the issue at the center. In *Ecology*, ed. S. L. Dodson, T. F. H. Allen, S. R. Carpenter, A. R. Ives, R. L. Jeanne, J. F. Kitchell, N. E. Langston, and M. G. Turner, 315–382. New York: Oxford University Press.

Barnard, C., ed. 1964. *Grasses and Grasslands*. London: Macmillan.

Bradshaw, A. 1992. The biology of land restoration. In *Applied Population Biology*, ed. S. K. Jain and L. W. Botsford, 25–44. Dordrecht, The Netherlands: Kluwer.

Bradshaw, A. 2000. The use of natural processes in reclamation: advantages and difficulties. *Landscape and Urban Planning* 51:89–100.

Breymeyer, A. I.; and Van Dyne, G. M,, eds. 1980. *Grasslands, Systems Analysis and Man*. International Biological Programme 19. Cambridge: Cambridge University Press.

Brown, V. K.; and Gange, A. C. 1989. Differential effects of above- and below-ground insect herbivory during early plant succession. *Oikos* 54:67–76.

Brown, V. K.; and Gange, A. C. 1992. Secondary plant succession: how is it modified by insect herbivory? *Vegetatio* 101:3–13.

Brown, V. K.; Jepsen, M.; and Gibson, C. W. D. 1988. Insect herbivory: effects on early old field succession demonstrated by chemical exclusion methods. *Oikos* 52:293–302.

Brown, V. K.; and Southwood, T. R. E. 1987. Secondary succession: patterns and strategies. In *Colonization, Succession and Stability*, ed. A. J. Gray, M. J. Crawley, and P. J. Edwards, 315–337, Palo Alto, Calif.: Blackwell Science.

Burgi, H.; and Stadelmann, P. 2002. Change of phytoplankton composition and biodiversity in Lake Sempach before and during restoration. *Hydrobiologia* 469:33–48.

Carpenter, S. R.; Chisholm, S. W.; Krebs, C. J.; Schindler, D. W.; and Wright, R. F. 1995. Ecosystem experiments. *Science* 269:324–327.

Chesson, P.; and Huntly, N. 1997. The roles of harsh and fluctuating conditions in the dynamics of ecological communities. *American Naturalist* 150:519–553.

Coupland, R. T., ed. 1979. *Grassland Ecosystems of the World: Analysis of Grasslands and Their Uses*. International Biological Programme 18. Cambridge: Cambridge University Press.

Curry, J. P. 1994. *Grassland Invertebrates: Ecology, Influence on Soil Fertility and Effects on Plant Growth*. London: Chapman and Hall.

Davis, M. A.; Thompson, K.; and Grime, J. P. 2001. Charles S. Elton and the dissociation of invasion ecology from the rest of ecology. *Diversity and Distribution* 7:97–102.

Degans, H.; and De Meester L. 2002. Top-down control of natural phyto- and bacterioplankton prey communities by Daphnia magna and by the natural zooplankton community of the hypertrophic Lake Blankaart. *Hydrobiologia* 479:39–49.

Díaz, S.; Cabido, M.; and Casanoves, F. 1999. Functional implications of trait-environment linkages in plant communities. In *Ecological Assembly Rules: Perspectives, Advances, Retreats*, ed. E. Weiher and P. Keddy, 338–362. Cambridge: Cambridge University Press.

Drake, J. A.; Huxel, G. R.; and Hewitt, C. L. 1996. Microcosms as models for generating and testing community theory. *Ecology* 77:670–677.

Estes, J. R.; Tyrl, R. J.; and Brunken, J. N.; eds. 1982. *Grasses and Grasslands*. Norman: University of Oklahoma Press.

Fortin, M.-J., and Gurevitch, J. 2001. Mantel tests: spatial structure in field experiments. In *Design and Analysis of Ecological Experiments*, ed. S. M. Scheiner and J. Gurevitch, 308–326. New York: Oxford University Press.

Fox, B. J. 1987. Species assembly and the evolution of community structure. *Evolutionary Ecology* 1:201–213.

Fox, B. J. 1999. The genesis and development of guild assembly rules. In *Ecological Assembly Rules: Perspectives, Advances, Retreats*, ed. E. Weiher and P. Keddy, 23–57. Cambridge: Cambridge University Press.

Fox, B. J.; and Brown, J. H. 1993. Assembly rules for functional groups in North American desert rodent communities. *Oikos* 67:358–370.

French, N. R., ed. 1979. *Perspectives in Grassland Ecology: Results and Applications of the US-IBP Grassland Biome Study.* New York: Springer-Verlag.

Fretwell, S. D. 1977. The regulation of plant communities by food chains exploiting them. *Perspectives in Biology and Medicine* 20:169–185.

Fretwell, S. D. 1987. Food chain dynamics: the central theory of ecology? *Oikos* 50:291–301.

Gange, A. C.; and Brown, V. K. 2002. Soil food web components affect plant community structure during early succession. *Ecological Research* 17:217–227.

Gilbert, O. L.; and Anderson, P. 1998. *Habitat Creation and Repair.* New York: Oxford University Press.

Gulati, R. D.; and Van Donk, E. 2002. Lakes in the Netherlands, their origin, eutrophication and restoration: state-of-the-art review. *Hydrobiologia* 478:73–106.

Handel, S. N. 1997. The role of plant-animal mutualisms in the design and restoration of natural communities. In *Restoration Ecology and Sustainable Development*, ed. K. M. Urbanska, N. R. Webb, and P. J. Edwards, 111–132. Cambridge: Cambridge University Press.

Heinrich, W.; Marstaller, R.; Bährmann, R.; Perner, J.; and Schäller, G. 1998. Das Naturschutzgebiet "Leutratal" bei Jena-Struktur-und Sukzessionsforschung in Grasland-Ökosystemen. *Naturschutzreport* 14:1–424.

Heinrich, W.; Perner, J.; and Marstaller, R. 2001. Regeneration and secondary succession: a 10 year study of permanent plots in a polluted area near a former fertilizer factory. *Zeitschrift für Ökologie und Naturschutz* 9:237–253.

Hillier, S. H.; Walton, D. W. H.; and Wells, D. A., eds. 1990. *Calcareous Grasslands: Ecology and Management.* Huntingdon, UK: Bluntisham Books.

Jeppesen, E.; Sondergaard, M.; Jensen, J. P.; Mortensen, E.; Hansen, A. M.; and Jorgensen, T. 1998. Cascading trophic interactions from fish to bacteria and nutrients after reduced sewage loading: an 18-year study of a shallow hypertrophic lake. *Ecosystems* 1:250–267.

Jones, T. H.; Thompson, L. J.; Lawton, J. H.; Bezemer, T. M.; Bardgett, R. D.; Blackburn, T. M.; Bruce, K. D.; Canon, P. F; Hal., G. S.; Hartley, S. E.; Howson, G.; Jones, C. G.; Kampichler, C.; Kandeler, E.; and Ritchie, D. A. 1998. Impacts of rising atmospheric carbon dioxide on model terrestrial ecosystems. *Science* 280:441–443.

Kalchev, R. K.; Pehlivanov, L. Z.; and Beshkovaand, M. B. 2002. Trophic relations in two lakes from the Bulgarian Black Sea coast and possibilities for their restoration. *Water Science and Technology* 46:1–8.

Keddy, P. A. 1989. *Competition.* New York: Chapman and Hall.

Keddy, P. A. 1992. Assembly and response rules: two goals for predictive community ecology. *Journal of Vegetation Science* 3:157–164.

Kelt, D. A.; and Brown, J. H. 1999. Community structure and assembly rules: confronting conceptual and statistical issues with data on desert rodents. In *Ecological Assembly Rules: Perspectives, Advances, Retreats*, ed. E. Weiher and P. Keddy, 75–107. Cambridge: Cambridge University Press.

Kent, M.; and Coker, P. 1994. *Vegetation Description and Analysis: A Practical Approach.* Chichester: John Wiley and Sons.

Körner, C. 1994. Scaling from species to vegetation: the usefulness of functional groups. In *Biodiversity and Ecosystem Function*, ed. E. D. Schulze and H. A. Mooney, 117–140. Berlin, Heidelberg: Springer-Verlag.

Korthals, G. W.; Smilauer, P.; Van Dijk, C.; and Van der Putten, W. H. 2001. Linking above- and below-ground biodiversity: abundance and trophic complexity in soil as a response to experimental plant communities on abandoned arable land. *Functional Ecology* 15:506–514.

Lawton, J. H. 1987. Are there assembly rules for successional communities? In *Colonization, Succession and Stability* (26th Symposium of the British Ecological Society, Southampton, Hampshire, UK), ed. A. J. Gray, M. J. Crawley, and P. J. Edwards, 225–243. Palo Alto, Calif.: Blackwell Science.

Lawton, J. H. 1995. Ecological experiments with model systems. *Science* 269:328–331.

Lawton, J. H. 1999. Are there general laws in ecology? *Oikos* 84:177–192.

Lawton, J. H.; and Brown, V. K. 1994. Redundancy in ecosystems. In *Biodiversity and Ecosystem Function*, ed. E. D. Schulze and H. A. Mooney, 255–270. Berlin, Heidelberg: Springer-Verlag.

Legendre, P. 1993. Spatial autocorrelation: trouble or new paradigm? *Ecology* 74:1659–1673.

Legendre, P.; and Fortin, M.-J. 1989. Spatial pattern and ecological analysis. *Vegetatio* 80:107–138.

Legendre, P.; and Legendre, L. 1998. *Numerical Ecology*. 2nd English ed. Amsterdam, New York: Elsevier.

Legendre, P.; and Troussellier, M. 1988. Aquatic heterotrophic bacteria: modeling in the presence of spatial autocorrelation. *Limnology and Oceanography* 33:1055–1067.

Majer, J. D. 1997. Invertebrates assist the restoration process: an Australian perspective. In *Restoration Ecology and Sustainable Development*, ed. K. M. Urbanska, N. R. Webb, and P. J. Edwards, 212–237. Cambridge: Cambridge University Press.

Manly, B. F. J. 1986. *Multivariate Statistical Methods: A Primer*. London: Chapman and Hall. pp. 159.

Mantel, N. 1967. The detection of disease clustering and a generalized regression approach. *Cancer Research* 27:209–220.

Masters, G. J.; and Brown, V. K. 1997. Host plant mediated interactions between spatially separated herbivores: effects on community structure. In *Multitrophic Interactions in Terrestrial Systems*, ed. A. C. Gange and V. K. Brown, 217–237. London: Blackwell Science.

Mehner, T.; Benndorf, J.; Kasprzak, P.; and Koschel, R. 2002. Biomanipulation of lake ecosystems: successful applications and expanding complexity in the underlying science. *Freshwater Biology* 47:2453–2465.

Menge, B. A.; and Sutherland, J. P. 1987. Community regulation variation in disturbance competition and predation in relation to environmental stress and recruitment. *American Naturalist* 130:730–757.

Middleton, B. 1999. *Wetland Restoration, Flood Pulsing and Disturbance Dynamics*. Chichester, England: John Wiley and Sons.

Müller, H. J.; Bährmann, R.; Heinrich, W.; Marstaller, R.; Schäller, G.; and Witsack, W. 1978. Zur Strukturanalyse der epigäischen Arthropodenfauna einer Rasen-Katena durch Kescherfänge. *Zoologische Jahrbücher, Systematik* 105:131–184.

Murray, S.; White, R.; and Rohweder, M. 2000. *Pilot Analysis of Global Ecosystems: Grassland Ecosystems*. Washington, D.C.: World Resources Institute.

Naeem, S.; Thompson, L. J.; Lawler, S. P.; Lawton, J. H.; and Woodfin, R. M. 1994. Declining biodiversity can alter the performance of ecosystems. *Nature* 368:734–737.

Naeem, S.; Thompson, L. J.; Lawler, S. P.; Lawton, J. H.; and Woodfin, R. M. 1995. Empirical evidence that declining species-diversity may alter the performance of terrestrial ecosystems. *Philosophical Transactions of the Royal Society of London Series B, Biological Sciences* 347:249–262.

Nicholls, K. H. 1999. Evidence for a trophic cascade effect on north-shore western Lake Erie phytoplankton prior to the zebra mussel invasion. *Journal of Great Lakes Research* 25:942–949.

Oksanen, L.; Fretwell, S. D.; Arruda, J.; and Niemelä, P. 1981. Exploitation ecosystems in gradients of primary productivity. *American Naturalist* 118:240–261.

O'Neill, R. V.; DeAngelis, D. L.; Waide, J. B.; and Allen, T. F. H. 1986. A hierarchical concept of ecosystems. *Monographs in Population Biology* 23. Princeton: Princeton University Press.

Packard, S.; and Mutel, C. F. 1997. *The Tallgrass Restoration Handbook: For Prairies, Savannas, and Woodlands.* Washington, D.C.: Island Press.

Perrow, M. R.; and Davy, A. J. 2002. *Handbook of Ecological Restoration: Principles of Restoration,* vol. 1. Cambridge: Cambridge University Press.

Pickett, S. T. A.; Cadenasso, M. L.; and Bartha, S. 2001. Implications from the Buell-Small succession study for vegetation restoration. *Applied Vegetation Science* 4:41–52.

Price, P. W. 2002. Resource-driven terrestrial interaction webs. *Ecological Research* 17:241–247.

Root, R. B. 1967. The niche exploitation pattern of the blue-gray gnatcatcher. *Ecological Monographs* 37:317–350.

Root, R. B. 2001. Guilds. In *Encyclopedia of Biodiversity,* vol. 3, ed. S. A. Levin, 295–302. San Diego: Academic Press.

Roscher, C. 2000. Zur räumlichen und zeitlichen Heterogenität von Halbtrockenrasen (Mesobromion) im Mittleren Saaletal und in angrenzenden Gebieten. Doctoral thesis. Friedrich-Schiller-Universität, Jena.

Scheu, S. 2001. Plants and generalist predators as links between the below-ground and above-ground system. *Basic and Applied Ecology* 2:3–13.

Simberloff, D.; and Dayan, T. 1991. The guild concept and the structure of ecological communities. *Annual Review of Ecology and Systematics* 22:115–143.

Smith, T. M.; Shugart, H. H.; and Woodward, F. I. 1997. *Plant Functional Types: Their Relevance to Ecosystem Properties and Global Change.* Cambridge: Cambridge University Press.

Sokal, R. R. 1979. Testing statistical significance of geographic variation patterns. *Systematic Zoology* 28:27–231.

Spedding, C. R. W. 1971. *Grassland Ecology.* Oxford: Clarendon Press.

Steneck, R. S. 2001. Functional groups. In *Encyclopedia of Biodiversity,* vol. 3, ed. S. A. Levin, 121–139. San Diego: Academic Press.

Svensson, J. E.; and Stenson, J. A. E. 2002. Responses of planktonic rotifers to restoration measures: trophic cascades after liming in Lake Gardsjon. *Archiv für Hydrobiologie* 153:301–322.

Tallberg, P.; Horppila, J.; Vaisanen, A.; and Nurminen, L. 1999. Seasonal succession of phytoplankton and zooplankton along a trophic gradient in a eutrophic lake: implications for food web management. *Hydrobiologia* 412:81–94.

Tatrai, I.; Olah, J.; Paulovits, G.; Matyas, K.; Kawiecka, B. J.; Jozsa, V.; and Pekar, F. 1997a. Biomass dependent interactions in pond ecosystems: responses of lower trophic levels to fish manipulations. *Hydrobiologia* 345:117–129.

Tatrai, I.; Olah, J.; Paulovits, G.; Matyas, K.; Kawiecka, B. J.; Jozsa, V.; and Pekar, F. 1997b. Changes in the lower trophic levels as a consequence of the level of fish manipulation in the ponds. *Internationale Revue der gesamten Hydrobiologie* 82:213–224.

Tilman, D. 2001. Functional diversity. In *Encyclopedia of Biodiversity*, vol. 3, ed. S. A. Levin, 109–120. San Diego: Academic Press.

Urbanska, K. M.; Webb, N. R.; Edwards, P. J., eds. 1997. *Restoration Ecology and Sustainable Development*. Cambridge: Cambridge University Press.

Van Andel, J.; Bakker, J. P.; and Snaydon, R. W., eds. 1987. *Disturbance in Grasslands: Causes, Effects and Processes*. Dordrecht, The Netherlands: Dr W. Junk Publishers.

Van Diggelen, R.; Grootjans, A. P.; and Harris, J. A. 2001. Ecological restoration: state of the art or state of the science? *Restoration Ecology* 9:115–118.

Voigt, W.; Perner, J.; Davis, A. J.; Eggers, T.; Schumacher, J.; Bährmann, R.; Fabian, B.; Heinrich, W.; Köhler, G.; Lichter, D.; Marstaller, R.; and Sander, F. W. 2003. Trophic levels are differentially sensitive to climate. *Ecology* 84:2444–2453.

Wardle, D. A. 2002. *Communities and Ecosystems: Linking the Aboveground and Belowground Components*. Princeton: Princeton University Press.

Weiher, E.; and Keddy, P. A. 1999. Assembly rules as general constraints on community composition. In *Ecological Assembly Rules: Perspectives, Advances, Retreats*, ed. E. Weiher and P. Keddy, 251–271. Cambridge: Cambridge University Press.

Wilson, J. B. 1989. A null model of guild proportionality, applied to stratification of a New Zealand temperate rain forest. *Oecologia (Berl.)* 80:263–267.

Wilson, J. B. 1999a. Assembly rules in plant communities. In *Ecological Assembly Rules: Perspectives, Advances, Retreats*, ed. E. Weiher and P. Keddy, 130–164. Cambridge: Cambridge University Press.

Wilson, J. B. 1999b. Guilds, functional types and ecological groups. *Oikos* 86:507–522.

Wilson, J. B.; and Roxburgh, S. H. 1994. A demonstration of guild-based assembly rules for a plant community, and determination of intrinsic guilds. *Oikos* 69:267–276.

Witsack, W. 1975. Eine quantitative Keschermethode zur Erfassung der epigäischen Arthropodenfauna. *Entomologische Nachrichten* 8:123–128.

Chapter 10

Structure, Dynamics, and Restoration of Plant Communities: Do Arbuscular Mycorrhizae Matter?

CARSTEN RENKER, MARTIN ZOBEL, MAARJA ÖPIK,
MICHAEL F. ALLEN, EDITH B. ALLEN, MIROSLAV
VOSÁTKA, JANA RYDLOVÁ, AND FRANÇOIS BUSCOT

Restoration of plant communities inevitably requires an understanding of the functioning of natural communities, which are the main ecological forces that produce patterns. Following the historical line of Clements, the classical ecological theory explains the patterns of species composition and diversity of plant communities through the impact of the abiotic environment, differential dispersal, and the outcome of local biotic interactions. Since the 1950s and 1960s, the emphasis in theory has shifted toward biotic interactions within communities, especially resource partitioning and (avoidance of) competition in spatiotemporally heterogeneous environments (MacArthur and Connell 1966, MacArthur 1972, Whittaker and Levin 1977). Later, both equilibrium theory, explaining species coexistence in spatially heterogeneous environments (Tilman 1982, 1986), and nonequilibrium theory, explaining species coexistence in temporally heterogeneous environments (Connell 1978, Huston 1979, Chesson 1986), were thoroughly elaborated. Resource competition was still considered to be the driving force behind community patterns, though the role of herbivory was also considered quite often (reviewed by Brown and Gange 1990, Huntly 1991, Brown 1994). The role of symbiotic relationships, such as mycorrhiza, in the functioning of plant populations and communities was seldom taken into account, except in a few papers discussing the impact of mycorrhiza on competition among species representing different stages of plant community succession (for example, Janos 1982, E. Allen and M. Allen 1984).

In addition to the functioning of whole plant communities, attention has been paid to the dynamics of populations of single species. Why are certain species absent in most communities while others are frequent everywhere? (See Chapters 13 and 14.) Possible causes of plant species rarity have been discussed extensively (Rabinowitz 1981, Kunin and Gaston 1993). In certain cases, causes are easily identified; for example, where species are restricted to scarce or isolated habitats, occur at the margins of their distribution areas, or are directly exposed to adverse human impact (see Karron 1987, Rosenzweig 1995, Kunin and Gaston 1997). More often, however, the study of morphological and functional characteristics of plant individuals—which allows assessments of their competitive and dispersal abilities, demographic characteristics, resistance to herbivory, relationships to the abiotic environment, and so forth—has shown that there is no simple answer to the question: What specific traits or combinations of traits are responsible for rarity? (Berg et al. 1994, Gustafsson 1994, Sætersdal 1994, Eriksson et al. 1995, Thompson et al. 1996, Witkowski and Lamont 1997, Webb and Peart 1999, Wolf et al. 1999; also see Chapter 6 for a discussion about the influences of filters in assembling communities). Again, compared to other, more obvious negative impact factors, such as competition or herbivory, the role of positive supportive interactions in determining plant species abundance is investigated much less often.

Early life stages, from seed germination to plant establishment, are important phases when the critical screening of colonizing diaspores takes place (Weiher and Keddy 1995, Kitajima and Tilman 1996). Following seed germination and the subsequent exhaustion of seed reserves, successful seedling establishment requires that plants acquire soil nutrients efficiently. Establishment also requires protection against pathogens, which in the majority of plant families is controlled, or at least influenced, by root symbioses with mycorrhizal fungi (Smith and Read 1997). More information is needed about the role of symbiotic relationships, including arbuscular mycorrhizal fungi, in early stages of plant life, because this may be an important determinant of the distribution and abundance of plant species.

Arbuscular Mycorrhiza and Plant Communities: More Evidence about the Role of the Invisible World

Arbuscular mycorrhizal fungi (AMF) communities are usually considered to be species-poor. Although Luoma et al. (1997) detected more than 200 distinct morphotypes of ectomycorrhiza on one single field site in Oregon, most studies on community richness of AMF did not mention more than 15 species (for example, Clapp et al. 1995, Stutz et al. 2000, Hildebrandt et al. 2001, Franke-Snyder et al. 2001). Studies that found more species (up to 40) did so by frequent field observations or by trapping in pot cultures (E. Allen et al. 1995, Egerton-Warburton and E. Allen 2000, Bever et al. 2001).

Fungi that form arbuscular mycorrhizal associations with the majority of land plants were originally assigned to the order Glomales within the Zygomycota (Morton and Benny 1990). However, publication of Gerdemann and Trappe's (1974) treatise redescribing the systematics of AMF led to consideration of differing species of AMF. An important result was that many of the groupings that Gerdemann and Trappe organized as genera (*Gigaspora, Acaulospora, Glomus*) are now recognized as belonging to different families, having diverged more than 300 million years ago. There are at least three clades of importance to us here: the Gigasporaceae (*Gigaspora, Scutellospora*) clade, which is morphologically distinct; the *Acaulospora* clade; and the *Glomus* clade (Morton and Redecker 2001). Walker and Trappe (1993) have identified 167 epithetes within the Glomeromycota so far. (The term *epithete* is used instead of species because the species concept in AMF is very controversial; see also the "Problems in Handling Arbuscular Mycorrhizal Fungi" section later in this chapter.)

Recent works have shown that the AMF can be unequivocally separated from all other major fungal groups in a monophyletic clade, based on molecular, morphological, and ecological characteristics. Consequently, they were removed from the polyphyletic Zygomycota and placed into a new monophyletic phylum, the Glomeromycota (Schüßler et al. 2001, Schwarzott et al. 2001), within which two new families, the Archaeosporaceae and Paraglomeraceae, both basal groups, have been described (Morton and Redecker 2001).

Both fossil and molecular evidence supports the idea that AMF are as old as the land flora and probably coevolved with the first land plants (Pirozynski and Malloch 1975, Simon et al. 1993, Remy et al. 1994, Redecker et al. 2000, Heckman et al. 2001). The AMF apparently were essential for the early land plants to scavenge phosphorus from the soil. Moreover, individual AMF and their communities may differ in their functional characters; for example, in efficiency of transporting phosphorus to plants (Jakobsen et al. 1992, 2001). Physiological and life cycle traits, too, are known to differ among AMF taxa (see Dodd et al. 2000).

The central role of symbiotic AMF in plant population dynamics—through their control of soil nutrient uptake, protection against root pathogens, and intra- and interplant species linkages—is now starting to be appreciated (Newsham et al. 1994, 1995a; Simard et al. 1997; Zobel et al. 1997; Watkinson 1998; van der Putten et al. 2001). Experiments conducted under greenhouse conditions have shown that the presence or absence of AMF may shift the competitive balance not only between mycorrhizal and a nonmycorrhizal species (E. Allen and M. Allen 1984) or between species with clearly different mycorrhizal dependency (Hetrick et al. 1989), but also between species of comparable mycorrhizal dependency, whether growing in even-aged cohorts (Allsopp and Stock 1992, Hartnett et al. 1993, Marler et al. 1999) or in systems containing adults and seedlings (Eissenstat and Newman 1990, Moora and Zobel 1996, Marler et al. 1999). The presence or absence of AMF may also have a major effect on the population structure of plant species in the field (Wilson et al. 2001). The presence of mycorrhiza in one generation may increase seedling competitive ability in the subsequent generations (Heppell et al. 1998). The effect of AMF also depends on the presence of a particular plant genotype or genotypes (Ronsheim and Anderson 2001).

In addition, one can observe the effect of the presence or absence of AMF on the species composition and diversity of the plant community, both under microcosm conditions (Grime et al. 1987) and in the field (Gange et al. 1993, Newsham et al. 1995c, Hartnett and Wilson 1999). Tree species, which were thought to be mainly ectomycorrhizal, were found to benefit from AMF colonization. Even low colonization rates seemed to have a high impact on the trees' nutrient demands (van der Heijden and Vosátka 1999, van der Heijden 2001). Egerton-Warburton and M. Allen (2001) analyzed the cost-benefit relationship in roots of *Quercus agrifolia* Nee. occupied by arbuscular mycorrhizal and ectomycorrhizal fungi and demonstrated a shift from an arbuscular mycorrhizal–dominated stage in young seedlings to an ectomycorrhizal-dominated stage in saplings.

Plants in most plant communities, however, are able to form symbiotic associations with one or more AMF species; nonmycorrhizal plant communities are very rarely found in nature (Smith and Read 1997). For this reason, the approach of comparing the presence or absence of AMF is gradually being replaced by the more informative study of the distribution and impact of natural AMF taxa (Clapp et al. 1995, van der Heijden et al. 1998b, Helgason et al. 1999, Bever et al. 2001). Although AMF can colonize roots of a taxonomically diverse range of plants, ecological specificity does occur in arbuscular mycorrhizal associations (McGonigle and Fitter 1990), and certain specific plant-AMF combinations can be more beneficial than others for both partners (Sanders 1993, van der Heijden et al. 1998b). Van der Heijden et al. (1998a) showed experimentally that the coexistence of vascular plant species in a microcosm was dependent on the species composition of the AMF community. These results lead to the conclusion that the occurrence and abundance of a vascular plant species in a particular community may depend on the presence of certain AMF species or combinations of species (Read 1998).

Are Plant and Fungal Communities Patchy?

Habitat fragmentation and its impact on the distribution of plant and animal species has now been acknowledged (Young et al. 1996, Hanski and Gilpin

1997, Tilman and Kareiva 1997). The significance of dispersal limitation for plant populations (Turnbull et al. 2000) and communities (Austin 1999, Huston 1999, Grace 2001) has been the object of many recent case studies and theoretical discussions. Both ongoing fragmentation of natural ecosystems in landscapes under human impact and dispersal limitations under fully natural conditions are important forces in structuring plant communities. One may ask: Is the distribution of AMF in natural ecosystems patchy, and does it have any significant impact on distribution of vascular plant species?

The first question to be answered, then, is: How are AMF dispersed under natural conditions? Several studies have shown that wind and water play important roles, but even animals of various sizes, from large mammals to small grasshoppers, might be vectors (M. Allen 1987, Warner et al. 1987).

Potentially concomitant variation in the species composition of AMF communities has been identified on the basis of the presence of spores in natural soils or in trap cultures. Considerable variation in AMF communities has been observed between plant communities (Johnson 1993, Merryweather and Fitter 1998b, Stutz et al. 2000) and within them (Rosendahl et al. 1989, Eom et al. 2000). These studies suggest that, at the least, the infection potential of soil in different plant communities may differ. Even seasonal dynamics may be important (Šmilauer 2001). Information about the variability of functionally active AMF communities in plant roots, however, is extremely scarce. Helgason et al. (1998, 1999) showed that AMF communities in plant roots differed between two contrasting types of ecosystems: an agricultural field and a deciduous woodland. There was also a considerable within-stand variation in AMF communities. A loss of AMF diversity due to agricultural activities (Helgason et al. 1998, Daniell et al. 2001) strongly suggests that variation in natural AMF communities may be partly due to habitat fragmentation, since intensively managed landscapes around and between natural ecosystems contain considerably smaller numbers of AMF taxa and do not function as efficient inoculation sources.

Thus, if specific compatible relationships between certain AMF and plant taxa are required for mutual symbiont survival, the loss of compatible AMF species or individuals may limit the distribution of a particular plant species. As a result, the availability of AMF taxa may influence the composition and function of the plant community. The impact of symbiotic microbes as a determinant of plant species abundance has been recognized only quite recently. For example, Thrall et al. (2000) demonstrated the differential dependency of rare and common tree species on nitrogen-fixing bacteria. More specifically, the potential importance of coavailability of plant species and AMF taxa has been discussed by Barroetavena et al. (1998). Öpik et al. (2003) and Moora et al. (2004) found that early establishment of a rare plant species, *Pulsatilla patens* (L.) Mill. (Ranunculaceae), and of a common species, *P. pratensis* (L.) Mill., depended upon different AMF taxa being unevenly distributed among fragments of natural landscapes. Thus, the establishment and performance of certain plant individuals in particular localities may really depend on the presence of the appropriate AMF taxa.

Occurrence and Function of AMF in Natural Ecosystems: A Case Study in the North American Sagebrush Steppe

Using assembly rules to facilitate restoration requires that we find such rules from observations of natural succession, coupled with experimental approaches including restoration research. Over the past three decades, the responses of sagebrush steppe communities to disturbance have been studied by Michael Allen and his research group (see corresponding articles cited in this section). The dominant species is an exotic introduced weed, *Salsola kali* L. This plant was introduced in the late 1800s from the Asian steppes and is considered a noxious weed. Many efforts have been made to control it in restoration situations. Grasses, including *Elymus smithii* (Rydb.) Gould, a C₃ species, and *Bouteloua gracilis* (H. B. K.) Lag. ex Steud., a C₄ species, are desirable outcomes (E. Allen 1982). These species are prime forage for rangelands for both cattle and native ungulates (for example, bison). The dominant shrub is *Artemisia tridentata* Nutt., or basin big sagebrush. This shrub is considered an increaser with grazing, and it is often removed in rangelands managed for cattle. However, it is important for wildlife including pronghorns, birds, and invertebrates.

In general, succession in sagebrush steppe communities proceeds from bare soil to weedy annuals to bunchgrasses to shrubs. The soils tend to be relatively nutrient-rich, but plant production is limited by drought and cold. An important characteristic is that, following disturbance, available nutrients increase and bound organic nutrients decline (M. Allen and MacMahon 1985). All late seral plants in these systems form arbuscular mycorrhiza relationships; ectomycorrhizae are limited by drought, being present only in riparian ribbon or higher-elevation conifer forests.

Possible Involvement Levels of Arbuscular Mycorrhizae in Plant Succession

In the initial sagebrush steppe research, it was postulated that *S. kali* could affect the rate of establishment of native plants. Further, because *S. kali* is a nonmycotrophic (never forming mycorrhizae) plant (Stahl 1900, Nicolson

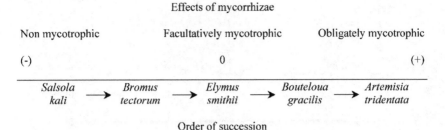

Order of succession

FIGURE 10.1 Plant successional model suggesting increasing dependence on mycorrhizal association in the sagebrush steppe. (After E. Allen 1984b.)

1960), it has been suggested that mycorrhizae might play an important role in succession and thus could be used to enhance vegetation recovery to the detriment of the nonmycotrophic *S. kali*. In addition, *S. kali* is nitrophilous and preferentially establishes and grows in nutrient-rich soils (E. Allen and Knight 1984). The observation that *S. kali* does not form mycorrhizal relationships was confirmed, as well as the finding that following disturbance, mycorrhizal fungi needed to build up inoculum density just as plants must colonize and expand their range (E. Allen and M. Allen 1980). Further, it was possible to demonstrate that AMF actually were parasitic on *S. kali*, inhibiting growth and, in some cases, actually killing seedlings (M. Allen et al. 1989). Further work also demonstrated that there was a gradient in responsiveness of plants to mycorrhizae; herbaceous grasses were facultatively mycorrhizal, and woody species were more responsive, approaching obligately mycorrhizal status (E. Allen 1984b). Based on these observations, a model of succession was postulated in which the initial seral stage was dominated by nonmycotrophic plants or by facultatively mycorrhizal species. With succession, as the mycorrhizal fungi became more prevalent, more responsive plants could establish and persist (Figure 10.1).

For assembly rules, it is important that the different groups of AMF also appear to form distinct functional groups. Hart and Reader (2002) associated AMF taxonomic groupings with colonization strategies. Newsham et al. (1995b) reported differing ecological functions among the different fungal groups. For example, the smaller *Glomus* spp. form many individual infections, whereas the Gigasporaceae tend to form larger external mycelium (Wilson and Tommerup 1992). With few exceptions, smaller *Glomus* spores disperse readily, including by wind in the arid steppe regions (M. Allen et al. 1993). *Acaulospora, Scutellospora, Gigaspora,* and some large-spored *Glomus* species disperse slowly, largely by animal vectors (M. Allen 1988, M. Allen et al. 1993). Thus, it was postulated that just as different plants and

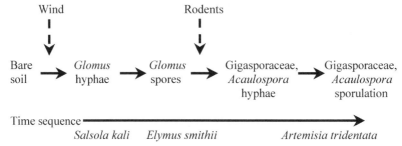

FIGURE 10.2 Fungal succession model with roles of animals and plants. Wind is the major vector for small *Glomus* spores and rodents are the major vector for the larger-spored species (M. Allen et al. 1993).

animals migrate using different vectors, and because invasion is related to successional status (for example, Diamond 1975), a similar process should occur for AMF. This led to the construction of a second model of succession based on the dispersal ability of the different AMF fungi (Figure 10.2). According to this model, AMF with small spores dispersed by wind would be prevalent in early-successional stages, whereas AMF with large spores dispersed by rodents would follow later.

Finally, we must consider some basic competitive outcomes. E. Allen and Knight (1984) demonstrated that S. *kali* competed with grasses for water and nutrients, reducing grass establishment. However, grass density and mass increased under S. *kali* when the soils contained AMF (E. Allen and Knight 1984). The presence of *Glomus* AMF aided the competitive ability and subsequent establishment of the grasses (E. Allen and M. Allen 1984, 1986). If S. *kali* was removed, however, grasses readily established but were replaced partially by A. *tridentata* over time (E. Allen 1988).

The mechanisms were not limited simply to competition. AMF were actually found to infect S. *kali*. These fungi would colonize and obtain enough carbon from the plant for sporulation, but they did not appear to provide nutrients in exchange (M. Allen and E. Allen 1990). Further, the fungi appeared to trigger an immune response that resulted in growth depressions and seedling mortality (M. Allen et al. 1989).

Conceptual Synthesis

Survey results showed that all major AMF groups can be found in association with sagebrush and grasses in the sage steppe region (E. Allen et al. 1995). Thus, the question becomes whether there are differential fungus-

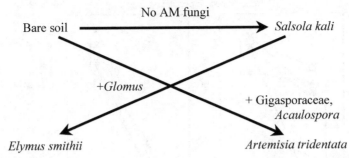

FIGURE 10.3 Postulated synthesized succession model in the sagebrush steppe showing different outcome of vegetation structure depending on availability and type of arbuscular mycorrhizal (AM) fungi.

plant responses leading to the patterns of succession observed and whether these responses can be used in restoration. It was postulated that *S. kali* would compete with grasses and that without AMF, *S. kali* would do better, but with AMF, the grasses would do best. It was expected that *Glomus* spp. would initially invade and facilitate the grasses, which in turn would support greater AMF biomass, including the larger *Acaulospora* and Gigasporaceae spores as they invaded. These fungi, in turn, would enhance growth of shrubs such as *A. tridentata* more than the growth of the smaller, more ephemeral *Glomus* spp. (Figure 10.3).

If this model is accurate, one could begin to manipulate the resulting community in several ways. First, by adding AMF, one could reduce the cover and persistence of *S. kali*. Second, a diversity of AMF could favor a complex mix of grasses and shrubs (as well as other species). Finally, by creating a patchy distribution of inoculum types, one could reconstruct a patchy but species-rich plant community capable of sustaining a higher diversity of animals. Alternatively, if the goal was to achieve range of grasses in some sections and a mixture of predominantly shrub cover for wildlife in another, initial inoculations could facilitate such a pattern.

Experimental Validation and Added Complexities

Several studies in this vegetation type and in others support such a model. E. Allen (1984a, 1984b) noted that the addition of AMF increased plant diversity by promoting the establishment of later seral forbs and grasses in the matrix of an early seral environment. Grime et al. (1987) reported that adding AMF increased plant diversity by facilitating arbuscular mycorrhizal plants in the competitive mix of arbuscular mycorrhizal and nonmy-

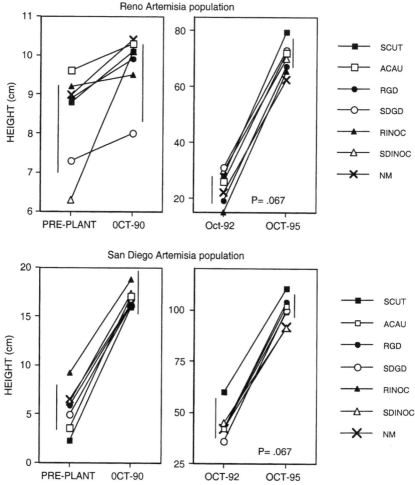

FIGURE 10.4 Growth responses of *Artemisia tridentata* to AM fungi using a reciprocal transplant experiment. The experimental design is described in Weinbaum et al. 1996. Two sites, Sky Oaks Biological Station, 120 km from San Diego, and Beddell Flats, 40 km from Reno, Nevada, were used, along with their populations of *A. tridentata* and AM fungi including *Scutellospora calospora* (found at the Reno site; SCUT), *Acaulospora elegans* (found at the San Diego site, ACAU), *Glomus deserticola* (found at Reno, RGD, as well as in San Diego, SDGD), and whole soil inoculum (a mixed species inoculum from Reno, RINOC, and San Diego, SDINOC; NM: nonmycorrhizal soil). Data are from M. Allen et al. 1992; Hickson 1993; E. Allen and M. Allen, unpublished. The two sites were analyzed separately. In both cases, using a repeated measures ANOVA, both plant populations ($p < .0001$ for both plant populations and sites) and inoculum sources ($p = 0.067$ for Reno site; $p = 0.012$ for San Diego site, data not shown) were different (see relevant power analyzes in Klironomos et al. 1999).

cotrophic plants. Van der Heijden et al. (1998a) found that a mix of AMF promoted diversity of plants, especially forbs. Klironomos et al. (1999) looked at the spatial structure of AMF in soil. There were distinctive patterns separating *Glomus, Acaulospora,* and Gigasporaceae, related to the presence of shrubs versus forbs in a shrubland patch recovering from fire.

On a practical note, direct replacement or short-term storage and replacement of topsoil ensures the reapplication of AMF to a site and facilitates recovery of mycorrhizae and mycotrophic plants (E. Allen and M. Allen 1980, Miller and Jastrow 1992). Most restoration plans now require the use of topsoil, and some are beginning to require mycorrhizae (see also the section "Occurrence and Function of AMF in Degraded Ecosystems" later in this chapter).

The addition of *Glomus* inoculum to soils with very low AMF (caused by long-term storage of the respread topsoil) enhanced the competitive ability of the native grass *E. smithii* in the presence of *S. kali.* This pattern became especially noticeable as drought stress became important (E. Allen and M. Allen 1986).

We still know little about how important species composition of AMF is to restoration. In a 5-year study of AMF and the establishment and growth of *A. tridentata,* important differences among species of fungi were found. Two populations of *A. tridentata* were monitored on the edges of the Great Basin in the western United States at two sites, one near San Diego, California, and one near Reno, Nevada. Soils had been tilled and AMF removed using benomyl (Weinbaum et al. 1996). The worst condition for the shrub's survival and growth was to be planted with no mycorrhizae. *Glomus deserticola* Trappe, Bloss & J.A. Menge initially stimulated plant growth, but the larger-spored AMF actually inhibited some growth initially. However, as *Scutellospora calospora* (T.H. Nicolson & Gerd.) C. Walker & F.E. Sanders and *Acaulospora elegans* Trappe & Gerd. expanded the hyphal network, these fungi subsequently stimulated the growth of *A. tridentata* more than *G. deserticola* (Figure 10.4).

This study also showed the importance of ecotypic differentiation. There were significant differences between the *G. deserticola* collected from the San Diego and Reno sites. For example, the *A. tridentata* at the San Diego site grew best with the San Diego *G. deserticola,* whereas it grew at intermediate rates (fifth) with the Reno *G. deserticola.* Almost the reverse was true for the Reno plants. Exotic fungi did have lower survival rates than local inoculum (Weinbaum et al. 1996), but the exotic inoculum did not disappear during the study period.

Although these community interactions show a level of organization that can be useful in restoration studies, the environment often plays havoc with any potential assembly rules we derive. This affects the outcomes, often in unpredictable ways.

A second example of this unpredictability was found in the Kemmerer, Wyoming, inoculation studies. It was predicted that adding inoculum would increase the competitive ability of *E. smithii* over that of *S. kali*. However, the AMF directly parasitized the *S. kali*, taking carbon (M. Allen and E. Allen 1990) and killing individual root tips (M. Allen et al. 1989). This reduced the growth of *S. kali* and, incidentally, also reduced the ability of the plant remains to trap snow during the winter. Soil moisture was thus reduced, and *E. smithii* was subject to early drought stress (E. Allen and M. Allen 1988). In this way, AMF actually were detrimental to the plants, whereas normally the symbiosis would be mutualistic.

Lessons for Using Arbuscular Mycorrhizae in Restoration Management

The role of mycorrhizae as symbionts in succession is relatively well documented (for example, Janos 1980, M. Allen 1987). The presence or absence of the mutualism has remarkably consistent effects, shifting the initial seral vegetation states to later ones. Further, because the hyphae bind soil particles and immobilize nutrients (in both plant and fungal tissue), these associations have dramatic effects on nutrient cycling. Some of these effects have already been incorporated into restoration practices such as carefully managing soils; providing source areas for immigration; and, in extreme cases, inoculating seedlings or soils. In essence, these practices are incorporating the assembly rules at the functional level.

However, the assembly of mycorrhizal fungal communities, like the assembly of all communities, contains a large element of chance. Dispersals of plants and fungi are independent events that become interdependent. Some plants prefer specific fungi, but the preferences are not absolute, and so there is elasticity in the system.

As noted earlier, the environment plays a major role in determining the outcomes of interactions among organisms. Since all rules are subject to the chaotic system that comprises our environment, caution is necessary. However, that does not prevent us from adopting some rules to enhance restoration. From the study of mycorrhizae, these rules include

• Manage soil for mycorrhizae and, if possible, retain existing inoculum.

- Use local mycorrhizae whenever possible.
- Understand the environment where restoration efforts will occur.
- Develop inoculation approaches as a last, but often essential, resort.
- Manage for AMF diversity, recognizing that the sequence of fungal additions may depend on plant composition.

Occurrence, Function, and Practical Use of Arbuscular Mycorrhizal Fungi in Degraded Ecosystems

Industrial activities increase the numbers of waste sites and degrade natural ecosystems. Such sites are open to succession, but their dynamics are generally slow because of a degraded soil fertility. Soil microflora is involved in fertility degradation in two ways. On the one hand, fertility is lost partially because of the direct detrimental effects of industrial activities on populations of soil microorganisms normally involved in cycling and mobilization of trophic resources. On the other hand, once soil microorganisms are established, edaphic stresses negatively influence them (Esher et al. 1992, Langer and Günther 2001). Populations of AMF are essential for soil development and successful plant establishment, as we showed in the previous section of this chapter. The presence of AMF as obligate plant symbionts in the majority of plant species may reduce the negative effects of stress caused by a lack of nutrients or organic matter resulting from adverse soil structure or extreme pH (Sylvia and Williams 1992). Other specific roles of the AMF symbiosis on degraded sites are pathogen defense (Filion et al. 1999), increased drought and heavy metal resistance (Galli et al. 1994, Hildebrandt et al. 1999, Kaldorf et al. 1999, Augé et al. 2001), and enhanced tolerance of high salinity because AMF enlarge the absorption zone of roots (Hardie 1985). In this context, the elimination of multifunctional AMF populations could hamper plant establishment and survival considerably (Visser 1985, Pfleger et al. 1994).

Succession of Vegetation and AMF Populations in Degraded Ecosystems

Recent research has shown that plant community structure is actually determined by the suitability of plant rhizosphere microorganism systems to specific site conditions (Smith and Smith 1996). Succession patterns and the involvement of AMF observed on highly disturbed sites such as spoil banks in central Europe resulting from industrial or mining activities are highly consistent with those of the case study and the proposed models presented in the "Occurrence and Function of AMF in Natural Ecosystems" section earlier in this chapter. A spontaneous plant succession is relatively slow, and primarily annual ruderal plants, mostly nonmycotrophic species, invade these sites at the beginning of succession. Even these first invaders (for example, Chenopodiaceae) can influence fungal populations indirectly and positively by providing a niche for sporulation in the cavity of dead seeds (Rydlová and Vosátka 2000). Various dominant species colonizing spoil banks during primary succession exhibit various levels of mycorrhizal dependence, from strictly nonmycotrophic *Chenopodium* and *Sisymbrium* to facultatively mycotrophic grasses such as *Elymus, Arrhenatherum,* and *Calamagrostis* (Rydlová and Vosátka 2001). Nonmycotrophic plant species can also be colonized, but arbuscules are usually not found and the function of mycorrhiza is thus questionable (Janos 1980, Rydlová and Vosátka 2001). Miller (1979) concluded that the occurrence of mycorrhizal colonization in the Chenopodiaceae is relevant to the life strategy of the plant. For example, shrubby members of the Chenopodiaceae with a stress-tolerant life strategy associate with AMF, whereas ruderal annuals of the family are nonmycorrhizal.

Secondary colonizers of degraded sites, particularly grasses, belong to facultative mycotrophs as classified by Janos (1980). Generally, temperate grasses are less mycotrophic, as found by Hetrick et al. (1989), but some grasses have been found to be highly mycorrhizal on degraded soils (Vosátka et al. 1995). Newsham et al. (1995a) concluded from a series of field experiments that the reason grasses benefit from AMF might be increased resistance to root pathogens rather than enhancement of nutrient uptake. It could be assumed that mycotrophy of plants in all succession stages is highly dependent not only on the plant species but also on many abiotic environmental factors.

The succession of AMF communities is closely interrelated with the succession of plant communities, changes in nutrient availability, and the distribution of mycorrhizal propagules on the site. The question remains whether native AMF or nonindigenous isolates are better adapted to edaphic conditions of soils from degraded sites. Some results indicate that native AMF can develop and function better in the soils of their origin (Enkhtuya et al. 2000). In soils from the first succession stages of vegetation on degraded sites, however, the composition of the native AMF population should differ from that of latter stages; in addition, there could be marked differences in the respective effectiveness of these AMF populations.

Potentials for Application of AMF Inocula in Restoration

As mentioned earlier, regeneration succession on highly degraded sites mostly results in a low diversity of plant communities and associated microbial populations. To promote and accelerate succession in restoration processes and to develop sustainable revegetation practices, it is necessary to study the implementation possibilities of phytomicrobial complexes (plants–symbiotic fungi–beneficial rhizosphere bacteria) that are tolerant to various stresses. To accomplish this objective, it is essential to restore or reintroduce functional populations of beneficial AMF in the soil. Mycorrhizal symbiosis develops when sufficient numbers of AMF propagules are in the soil; however, the number of spores and root colonization are often reduced by soil disturbance (Waaland and E. Allen 1987, Helgason et al. 1998). Therefore, if AMF have deteriorated or been eliminated, they should be reintroduced by inoculation of soils. Different populations or geographical isolates of AMF have been found to show a high variability in their tolerance to edaphic stress, represented by the amount of contamination present at a site (Weissenhorn et al. 1993, Bartolome-Esteban and Schenck 1994). In most cases, AMF isolates adapted to specific local soil conditions are better able to survive and have the potential to stimulate plant growth. It is probable that such AMF ecotypes result from a long-term adaptation to soils with extreme properties (Sylvia and Williams 1992). Such isolates should be selected and used to produce inoculum for field applications. A collection of stress-adapted AMF isolated from soils of different types of degraded ecosystems in central Europe has been established. Some of the isolates are registered in an international culture collection—the European Bank of Glomeromycota (Dodd et al. 1994; see "Sequence Analysis" later)—and can be used in future revegetation of polluted sites.

Inoculation biotechnology can be used to support restoration and reestablishment of AMF populations, where selected efficient strains of AMF and/or indigenous strains are multiplied and introduced to the soil either directly or by transplanting preinoculated plants. For example, inoculation technology was used for the restoration of a *Leymus arenarius* (L.) Hochst. (= *Elymus arenarius* L.) grass cover on a spoil bank at Samphire Hoe (UK) comprising material recovered from Eurotunnel excavations. Grasses preinoculated with native AMF from the site showed better development throughout two vegetation seasons than did noninoculated plants. It is interesting that precultivation of plants in highly fertilized substrate did not help to ensure sustainability of revegetation at this harsh site with high salinity, low fertility, and erosion problems (Dodd et al. 2002). Most of the pot and field experiments have shown the potential of mycorrhizal inoculations to facilitate plant growth on sites with unfavorable conditions resulting either from industrial activities or from naturally harsh environments. However, as noted, inoculations can be limited or even fail when they are conducted in highly contaminated soils or under extremely harsh conditions in degraded ecosystems. This is illustrated by the following examples of field applications in degraded ecosystems.

Experiment 1: Inoculation of Lotus corniculatus on Colliery Spoil Bank (Betteshanger, UK)

A field trial was conducted to investigate the effects of inoculation with AMF on the establishment and growth of Lotus corniculatus L. plants on colliery spoil bank in Betteshanger, Kent, UK. There was a noninoculated treatment and a treatment inoculated with a mixture of five AMF species, using the commercial mycorrhizal product TerraVital-D (PlantWorks Ltd., UK). Seeds of Lotus corniculatus were sown, and 20 mL of inoculum was placed under each seed patch. Survival of plants and mycorrhizal colonization were registered after 6 months of growth.

No plant survived the 6-month period on the noninoculated plots, whereas about 90 percent of the plants survived in the mycorrhizal treatment. Mycorrhizal colonization of plants was 52 percent (Vosátka 2000).

Experiment 2: Inoculation of Leymus arenarius on Volcanic Field (Hekla, Iceland)

Ecosystem degradation and desertification are by far the most serious environmental problems in Iceland. Erosion is accelerated by volcanic activity and harsh weather conditions. Efforts are constantly being made to reclaim eroded lands by distributing seeds and fertilizers. However, the results of these efforts are quite variable, since it usually takes years of fertilization, followed by reseeding, to obtain permanent vegetation cover. For a field trial conducted on the volcanic fields of the Hekla volcano, seeds of Leymus arenarius (L.) Hochst. were sown (200 per m^2) onto plots that were noninoculated or inoculated with 1 L of soil from either established adjacent L. arenarius stands or the commercial mycorrhizal product TerraVital-D (PlantWorks Ltd., UK) consisting of five AMF isolates, including some adapted to arctic conditions and eroded soils.

Inoculation with TerraVital-D substantially increased survival of L. arenarius plants after sowing. In control plots, only 8 plants per plot emerged, whereas on two plots treated with soil inoculum from adjacent vegetated sites or with TerraVital-D, 26 and 48 plants emerged, respectively (Figure 10.5a). The plants from the inoculated plots were significantly taller and more developed (Figures 10.5b, and 10.5c). These preliminary results indicate that reclamation work in naturally degraded and disturbed volcanic fields of cold deserts in Iceland can benefit from the addition of appropriate AMF. Further research is needed, however, to determine the long-term effects of inoculation and to select the most effective inoculum types and modes of application.

Experiment 3: Field Trial on Spoil Bank (Cottbus, Germany)

Two tree species, *Sorbus aucuparia* L. and *Robinia pseudoacacia* L., and the leguminous shrub *Cytisus scoparius* (L.) Link were planted on lignite spoil banks near Cottbus, Germany. The trial was conducted by the Brandenburgische Technical University in Cottbus. Seeds were sown to randomized blocks with four replications of 4-m rows. Four AMF species—*Glomus claroideum* N.C. Schenck & G.S. Sm., *G. geosporum* (T.H. Nicolson & Gerd.) C. Walker, *G. mosseae* (T.H. Nicolson & Gerd.) Gerd. & Trappe, and *G. intraradices* N.C. Schenck & G.S. Sm.—were used, with two isolates of each species, one from polluted soils and one from unpolluted soils.

Neither mortality nor height of *C. scoparius* was affected by mycorrhizal inoculation (data shown in Vosátka et al. 1999). Furthermore, *S. aucuparia* and *R. pseudoacacia* died during the summer and autumn of 1997. There were extremely adverse soil conditions in the field; that is, soil pH fluctuating around 3 and an unfavorable sandy soil structure. The results showed that the mycorrhizal inoculation alone could not compensate for these conditions. An additional reason for the plant dieback could be that in areas of AMF inoculation, the phosphorus supply was adequate but the nitrogen supply was not (Weber, personal communication). This may explain why the leguminous plants survived longer at the site. However, they also became nitrogen-limited, were not able to store enough resources to survive the winter, and finally died. This field study illustrates that for restoring extreme sites, further soil management in addition to the use of properly selected AMF may be essential.

Restoration Ecology Seen Through the Eyes of a Mycologist

AMF are present even in soils with very harsh physical and chemical properties, and at least certain AMF taxa already recolonize soils of degraded or anthropogenic ecosystems at early stages of plant succession. It is feasible to inoculate mycorrhizal fungi alone or together with bacteria, to increase the health, growth, and quality of plants used for revegetation. Before starting revegetation, soil conditions should be checked and, if necessary, sparse

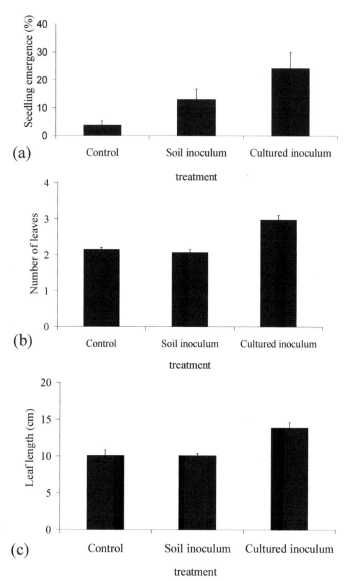

FIGURE 10.5 Survival rate (a), leaf length, (b) and number of leaves (c) of *Leymus arenarius* plants sown on 1-m^2 plots on a volcanic field in Iceland and inoculated with either soil inoculum from adjacent grass dune or commercial mycorrhizal product TerraVital-D (PlantWorks Ltd., UK).

nutrient pools should be compensated by fertilization to ensure basic essential conditions for plant establishment.

One of the strategies for the restoration of AMF populations could be based on an increase of native AMF by precropping with highly mycotrophic species. The resulting increased inoculum potential on such precultivated sites could be advantageous for a successful and consequent mycorrhization of plants used for revegetation. An alternative strategy is to use native fungi isolated from the site as inoculum. Further research should result in the development of a novel biotechnology for the inoculation of plant material for revegetation of low-grade industrial land. Application of appropriate phytomicrobial complexes, including stress-tolerant plants and beneficial plant-associated microbial consortia, bears the potential for a substantial increase in the effectiveness and sustainability of revegetation practices.

Distribution of AMF Communities in Nature: A Fungal Geobotany Is Needed

The examples presented illustrate the ecological importance of the structure and composition of native AMF communities associated with given plant species. The possibility of analyzing and using these structures constitutes a key to understanding spontaneously occurring ecosystem regeneration or to controlling managed restoration. In this context, poor basic knowledge of the distribution of AMF taxa under natural conditions is the main bottleneck to specifying their role in the functioning and dynamics of plant communities.

Limitations of Using Spores to Analyze AMF Communities

Traditionally, the taxonomy of AMF has been based on the morphology of their asexual soil-borne spores (Gerdemann and Trappe 1974, Morton and Benny 1990); these spores, however, cannot be traced to their host plant. Moreover, AMF communities *in planta* and spore communities in soil often differ in their species composition (for example, Clapp et al. 1995, Merryweather and Fitter 1998a, Turnau et al. 2001), due to sporulation and colonization differences between AMF taxa. Therefore, ecological studies of AMF based on analyses of spore diversity must be interpreted with caution. Most studies in the past were performed on spores, because their extraction from soil was relatively easy (Gerdemann and Nicolson 1963) and the only way to determine the different species was based on spore morphology.

In most cases, analyses of spore diversity that requires abundant quantities of vital spores were not performed directly on the spores collected in the field but were carried out after establishment of trap cultures under greenhouse conditions to gather pure isolates of the different spore types. This method, in which entrapping plants are inoculated with AMF structures from the field, is time- and space-consuming. Also, depending on the start-up method used (single spores, multiple spores, roots), the success rates can be quite variable. Additionally, the most effective mycorrhizal strains in the field might have a low tolerance to greenhouse conditions, and these strains might never be obtained in trap cultures. This phenomenon was shown in a phosphorus-polluted grassland in the vicinity of a former fertilizer plant in the central Saale Valley (Thuringia, Germany; for a detailed description, see Langer and Günther 2001). Although direct identification of AMF in roots from the field revealed dominance of a *Glomus* sp. related to *Glomus clarum* T.H. Nicolson & N.C. Schenck (AJ243275; 66–71 percent identity), identification by trapping maize plants under greenhouse conditions showed high infection rates by *Glomus intraradices*, which was quite rare in the field (Figure 10.6).

Approaches to Analyzing AMF Communities on the Mycorrhizal Roots Themselves

More recent studies have focused on direct detection of AMF in the roots of host plants. Cavagnaro et al. (2001) showed that the morphology of arbuscular mycorrhizae depends largely on the fungal partner rather than the plant partner. However, mycobiont identification based on features of AMF infection at best allows identification to the family level (Merryweather and Fitter 1998a). Specific staining techniques allow differentiation of some AMF species within certain groups (Vierheilig et al. 1998), but they are not adequate for general identification of AMF. For example, members of the Archaeosporaceae and Paraglomeraceae show weak or no staining with trypan blue, the most commonly used stain for mycorrhiza (Morton and Redecker 2001). Therefore, these techniques are of limited use under field conditions.

Within the whole spectrum of molecular methods, the high detection sensitivity of nested polymerase chain reaction (PCR) approaches is the most adaptable for amplifying target regions from the minute amounts of fungal DNA present in AM roots from the field (van Tuinen et al. 1998, Jacquot et al. 2000, Kjøller and Rosendahl 2000, 2001, Redecker 2000, Turnau et al. 2001). A problem with this method, however, is that DNA from all fungal

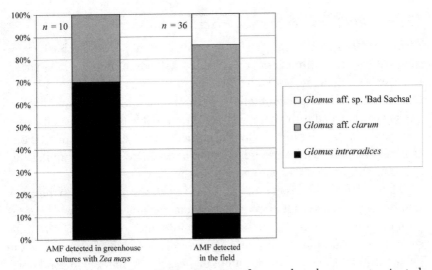

FIGURE 10.6 Survey of an AMF community from a phosphorus-contaminated grassland in the central Saale Valley 13 km north of Jena (Steudnitz, Thuringia, Germany; see Chapter 13 for details of the site) directly found in the field soil and in trap cultures with maize.

taxa of the roots, rhizoplane, and rhizosphere—the majority of which do not belong to AMF—are amplified when such universal primers for fungi as ITS1, ITS4, or ITS5 are used (White et al. 1990). To avoid this problem, most studies previously cited used primers with a narrow specificity for one to a few AMF taxa. This approach allows detection of specific mycobionts but fails to screen the entire diversity of AMF present in the field, which is necessary for ecological studies or for characterizing AMF populations adapted to specific stress conditions and to be used in restoration.

To resolve the conflict between amplification mainly of contaminants (non-AMF) and detection reduced to a narrow spectrum of AMF, a strategy was recently proposed based on a nested PCR in which one primer displays a specificity restricted to the glomalean fungi plus some groups in the basidiomycetes (Renker et al. 2003). Furthermore, the method tried to eliminate the most of the residual contaminants by performing a restriction between both reactions of the nested PCR with an enzyme that specifically does not cut the target region of AMF fungi. The first assessment of this method offered promising prospects for screening a wide range of AMF on field roots, even if it cannot reveal certain taxa of the most basal Glomeromycota.

The Need to Consider the Soil Mycelium of AMF

Apart from these methodological issues, which should be resolvable by new molecular tools, another basic weakness of AMF community analyses based solely on morphological or molecular investigations on roots is that the ecological role of some species might be overestimated. Molecular approaches neglect the fact that the extraradical mycelium is at least as important as the inter- and intracellular mycelia for the nutrient supply of the host plant (Horton and Bruns 2001). Therefore, it might also be necessary to monitor the species distribution directly in the soil (Claassen et al. 1996).

These basic problems in monitoring AMF diversity may explain the great number of unresolved questions in field studies dealing with (1) species composition and community structure, (2) the influence of fungal community structure on plant biodiversity, and (3) the impact of fungal abundance on plant nutrition. Effective investigations of these topics under field conditions would require analyzing the AMF directly on roots and in the soil; to date, such analysis has been done only indirectly using spores, which makes no more sense than characterizing plant communities by analysis of seed banks in soils alone.

Problems in Handling Arbuscular Mycorrhizal Fungi: The Weakness of the Species Concept

Another source of complication, even for interpretation of the most recent molecular ecological studies, arises from biological traits of AMF themselves:

1. Hyphae of AMF have a siphonal structure, and there is evidence that genetically heterogeneous nuclei may coexist in mycelium, and even the multicopy regions of the ribosomal DNA (rDNA) of single nuclei are known to be heterogeneous (Sanders et al. 1996, Kuhn et al. 2001).

2. AMF do not display sexual reproduction (Sanders 1999); but in the genus *Glomus*, for example, somatic exchanges of nuclei were observed via anastomosis between hyphae, which originated not only from the same spore but also from different spores of the same isolate (Giovannetti et al. 1999, 2001). This indicates that beyond nutritional flow in AMF mycelial networks, a flow of genetic information also exists (Giovannetti et al. 2001).

Recent studies have also revealed that hybridization between different genera might be possible and that *Entrophospora infrequens* (I.R. Hall) R.N. Ames & R.W. Schneid. might be one of the outcomes (Rodriguez et al.

2001). Consequently, genetic variation within one population and among different isolates of one species is often quite high (Dodd et al. 1996, Lloyd-MacGilp et al. 1996, Antoniolli et al. 2000, Clapp et al. 2001). A prerequisite to any study of population structure and diversity is the clarification of the species concept for AMF (Dodd et al. 1996). However, based on the lack of sexual reproduction in AMF, it has been suggested that applying the species concept could be difficult and that it would be more appropriate to base the description of AMF biodiversity on genetic diversity (Sanders et al. 1996).

Toward a Molecular Ecological Approach to Understanding How AMF Rule the Assembly of Plants in Communities

So far, we have on the one hand reviewed facts and experiments that support the preeminent role of AMF in ruling the formation of plant communities. On the other hand, we have emphasized that the geobotany of AMF, which is urgently needed in this context, is hampered by biological traits of AMF (see the preceding section, "Problems in Handling Arbuscular Mycorrhizal Fungi"). To resolve this conflict, we must develop molecular ecological tools that will allow us to monitor AMF directly in the field accurately enough to demonstrate their role in ruling the assembly of plants.

This last section will focus on a critical review of some molecular biological techniques that have been used to study the genetic diversity of AMF, based on analyses of glycoproteins (Wright et al. 1987, Hahn 1993, Thingstrup et al. 1995), isozymes (Rosendahl 1989), fatty acid patterns (Jabaji-Hare 1988, Bentivenga and Morton 1994), PCR coupled with restriction fragment length polymorphisms (RFLP) of target regions within the rDNA (Sanders et al. 1996), random amplified polymorphic DNA-PCR (RAPD, Abbas et al. 1996, Lanfranco et al. 1995), or microsatellite PCR (Longato and Bonfante 1997, Zézé et al. 1997) (for a detailed description of most methods, see Weising et al. 1995). Recent studies indicate that the most adequate techniques are based on sequence analyses of target regions within the nuclear rDNA, which combine the advantages of high copy numbers, highly conserved sequence tracks for the annealing of the primers, and variable regions between the priming sites (Simon et al. 1992, Abbas et al. 1996, Lloyd-MacGilp et al. 1996, Helgason et al. 1999, Antoniolli et al. 2000). Works focus either on the 18S gene encoding for the ribosomal small subunit (SSU) or on the internal transcribed spacer (ITS) region, which encompasses the two nonencoding spacers ITS1 and ITS2 separated by the 5.8S gene of the nuclear rDNA. Some studies analyze the 28S gene of the rDNA, which encodes for the ribosomal large subunit (LSU) (van Tuinen et al. 1998; Kjøller and Rosendahl 2000, 2001). So far, the number of available sequence data for this last region is quite low, so it seems preferable to consider the SSU or the ITS in ecological studies. The ITS displays the greatest polymorphism. In Ascomycota and Basidiomycota, it is adequate to differentiate species and in some groups to resolve intraspecific variation, whereas the SSU polymorphism is more adequate at higher taxonomic levels (Bruns et al. 1992, Horton and Bruns 2001). Therefore, the ITS is generally considered to be the most convenient region for investigating the structure of natural fungi communities in the field (Buscot et al. 2000). Within the Glomeromycota, analysis of ITS is also well established and is used mainly to study inter- or even intraspecific variation (for example, Lloyd-MacGilp et al. 1996, Antoniolli et al. 2000). In contrast, studies of the SSU deal mainly with higher taxonomic units, although they can be used at the species level (Helgason et al. 1998, 1999; Daniell et al. 2001; Schwarzott et al. 2001; Schüßler et al. 2001). Once the chosen target region is amplified by simple or nested PCR, various approaches, reviewed below, can be used to study its polymorphism.

Restriction Fragment Length Polymorphism of the Internal Transcribed Spacer (ITS-RFLP)

RFLP of the ITS region is a powerful tool for characterizing the fungal partners of ectomycorrhizal symbiosis (Horton and Bruns 2001). In the Glomeromycota, the intraspecific genetic variability of the rDNA and the multinucleate state complicate its use, because even one single spore may display several restriction patterns (Figure 10.7). Therefore, it is necessary to test a large number of enzymes on several PCR products of one "species" from pure cultures to understand the variability of its ITS before being able to monitor this species in the field. This limitation makes this method less interesting, apart from the fact that the more common types of AMF in the field might not be detectable in trap cultures.

Terminal Restriction Fragment Length Polymorphism (T-RFLP)

T-RFLP is based on a PCR in which one of the primers is labeled fluorescently. The PCR products obtained are digested with one single restriction enzyme or a set of restriction enzymes. The restriction bands obtained at the labeled end of the amplification product are separated on a polyacrylamide gel and detected by laser-induced fluorescence (LIF). Initial results using this technique in AMF have been published by Tonin et al. (2001).

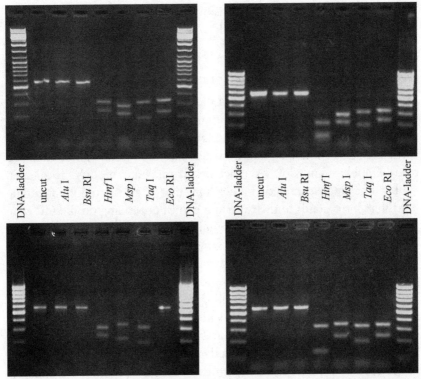

FIGURE 10.7 Four different PCR-RFLP patterns of maize roots inoculated with *Glomus intraradices* BEG 140 from one pot culture, illustrating the intraspecific variability of AMF. PCR products of the ITS were amplified with the primer pair ITS4/ITS5. Restriction fragments were separated on 2 percent agarose gels. Lanes 1 and 9 Gene Ruler DNA Ladder Mix.

Polymerase Chain Reaction–Single-Strand Conformation Polymorphism (PCR-SSCP)

PCR-SSCP is a reliable method for rapid and efficient detection of mutations and polymorphisms in genomic sequences. Target sequences are separated by polyacrylamide gel electrophoresis in a single-stranded state. Mutations are detected as mobility shifts resulting from changes in the conformation of single-stranded amplification products (Makino et al. 1992). This method has proved to be helpful in studies of genetic diversity for broad screening of different sequence types (Kjøller and Rosendahl 2000, 2001). However, as mentioned earlier, small changes in the genome are common in Glomeromycota, which might result in different patterns within the same "species." Therefore, this time-consuming method is not adequate for a broad screening of AMF in the field.

Denaturing Gradient Gel Electrophoresis (DGGE)

DGGE is a further development of the SSCP technique and is based on mobility differences of DNA fragments in a polyacrylamide gel, which result from both the base composition and the size of the product. DGGE polyacrylamide gels contain a gradient of a denaturant (usually formamide and urea). Double-stranded DNA molecules that migrate through the gradient gel become single-stranded at a position (that is, a denaturant concentration) that corresponds to their melting points, depending on the GC content of the molecules. When this point is reached, denaturation results in a sharp reduction of mobility. The fragment virtually "stops."

One disadvantage of these analyzing methods is that the different banding patterns they produce do not allow us to identify the detected species nor to assess the part of polymorphism resulting from intraspecific variations. Therefore, a prerequisite to investigations with these methods is a thorough determination of the species and their intraspecific variation by sequencing the used DNA target region. Another residual problem, which is omnipresent in soil microbiology, is the inability to exclude PCR products of nontarget species, which may greatly increase the number of different banding patterns.

Sequence Analysis

An important step for reaching the species level in molecular biological studies is the sequence analysis. In PCR products of AMF, each band may correspond to products with divergent sequences; resulting, for example, from the intraspecific polymorphism of DNA target regions. This makes it necessary to clone PCR products and to sequence a minimum number of clones systematically in preliminary work. SSCP was used in several recent studies to detect different patterns in PCR products. This allows one to reduce the number of samples to be sequenced (Kjøller and Rosendahl 2000, 2001; Clapp et al. 2001).

In recent years, a powerful tool has been established to analyze sequences: big sequence databases, such as GenBank and EMBL, with rapidly increasing amounts of sequence data sets for different fungal groups. Sequences of ITS and SSU rDNA of about one-third of the described Glomeromycota are already available. Using the databases in combination with programs such as BLAST (http://www.ncbi.nlm.nih.gov/; Altschul et al.

1997), the approximate taxonomic position of AMF can be determined on the basis of newly obtained sequences.

A large problem in using these databases is the uncertainty of the given species names, because the original determination of a taxon for which a sequence was deposed could be based on weak morphological observations. Another problem is misidentified sequence data belonging to contaminating fungi. An evaluation of the databases is urgently needed to remove such erroneous data, but this seems unlikely in view of the amount of new sequence data deposed daily. However, phylogenetic analysis including sequences of possible contaminants should be performed to identify an organism based on its sequence (Schüßler 1999).

In addition to these problems, the taxon sampling and sequencing must be completed to increase the chance of proper identification. To reach this goal, the available strains in the various culture collections (BEG—The International Bank for the Glomeromycota/European Bank of Glomeromycota, with 221 registered isolates; INVAM—International Culture Collection of Arbuscular and Vesicular-Arbuscular Mycorrhizal Fungi, with reference cultures of 77 species from all glomalean genera; GINCO— Glomeromycota In Vitro Collection, with isolates of 8 species available) should be screened completely. Alignments of sequence data should also be made available in Treebase (http://www.herbaria.harvard.edu/treebase/) or EMBL-Align (Multiple Sequence Alignment Database of European Molecular Biology Laboratory, http://www.ebi.ac.uk/embl/Submission/alignment.html) to allow fast access to such data sets for further studies.

Future Perspectives for the Use of Molecular Tools

A combination of molecular biology methods may be the most promising way to monitor the community structure and biodiversity of AMF in the field. Sequencing PCR products derived from nested PCR with mycorrhizal roots and also from products on spores will help bring new insights into "species" diversity. T-RFLP or DGGE will help to increase the number of analyzed samples, once a large number of roots and spores have been checked. Many of the gathered banding patterns should be addressable by the sequence data; then only samples of unknown origin detected at this level would have to be sequenced to determine their taxonomic status. This procedure will lead to an overview of biodiversity (sequencing) and community structure (T-RFLP/DGGE). For ecological studies in the future, development of microarrays or microchips might be very helpful to reduce the time needed to monitor AMF in field. By hybridization of given DNA fragments on the arrays or chips with amplified PCR products from the field, a fast and proper identification should be possible for large numbers of samples. Before reaching this stage in ecological studies, however, much work remains to be done. For example, all the unknown molecular strains should be sequenced to assess whether the commonly estimated diversity of only 150 to 200 AMF species worldwide is a reality, since it is based on the original classification system that used weak morphological characteristics. This work is an important prerequisite to systematic analysis of relationships between AMF diversity and plant communities.

Back to the Roots

Molecular AMF community studies of seminatural forests and arable fields in the United Kingdom that were based on sequencing of the 18S rRNA gene of Glomeromycota have revealed at least 18 AMF sequence groups detected from five natural and four crop plant species (Clapp et al. 1995; Helgason et al. 1998, 1999; Daniell et al. 2001) (Table 10.1). Only six of these taxa have shown close sequence similarity with reference species (*Glomus mosseae*, *G. geosporum*, *G. intraradices*, *Acaulospora scrobiculata* Trappe, *A. rugosa* J.B. Morton, *Scutellospora dipurpurescens* J.B. Morton & Koske). The taxonomic identity of the remaining fungi has not been established so far but awaits more extensive spore surveillance and trap-culturing studies in natural systems. Interestingly, Öpik et al. (2003) were able to detect six of these sequence types in trap seedlings and established native plants of two *Pulsatilla* spp. in boreal forest and grassland ecosystems in Estonia (see Table 10.1). Four other sequence types detected by these researchers, including the dominating sequence in natural plant roots at four sites, apparently represent new AMF sequence types (taxa).

Some ecological specificity of AMF is apparent from these data, since only 3 of 18 sequence types from natural plant roots have been detected in both arable and woodland systems; namely, *G. intraradices*, *A. scrobiculata*, and *S. dipurpurescens*, 5 from arable fields and 10 from woodland only. Five sequences by Öpik et al. (2003) represent woodland sequences and one, *G. intraradices*, is an arable land sequence detected from 2-month-old bait seedlings.

In addition to differences in AMF colonization patterns among plant species, temporal and spatial changes obviously occur in AMF community composition and species frequencies within an ecosystem type (Helgason et al. 1999, Daniell et al. 2001).

TABLE 10.1

AMF Taxa Detected by Two Research Groups in Different Ecosystems in the United Kingdom (UK) and Estonia (EST)[1]

AMF sequence group	Closest relative	Seminatural woodlands (UK)[2]	Arable fields (UK)[3]	Boreal forests and meadows (EST)[4]
Acau1	A. scrobiculata	+	+	+
Acau2[5]/MO-A1[6]	A. rugosa	+		
Acau3		+		
Acau4		+		
Acau5				
Glo1A	G. geosporum	+	+	
Glo1B	G. mosseae	+	+	
Glo2/MO-G5				+
Glo3/MO-G2				+
Glo4		+	+	
Glo5				
Glo6				
Glo7/MO-G6	G. intraradices	+		+
Glo8/MO-G3		+	+	+
Glo9/MO-G7				
Glo10		+	+	
Glo11				+
MO-G1				+
MO-G4				+
MO-G8				+
MO-G9				
Scut1	S. dipurpurescens	+	+	

[1] A. Acaulospora, G. Glomus, S. Scutellospora.
[2] Data from Helgason et al. (1998, 1999).
[3] Data from Helgason et al. (1998), Daniell et al. (2001).
[4] Data from Öpik et al. (2003).
[5] Sequence group designations after Helgason et al. (1998, 1999), Daniell et al. (2001).
[6] Sequence group designations after Öpik et al. (2003).

Another approach, using taxon-specific nested PCR of the 25S rRNA gene combined with root staining, has been used recently to detect and quantify AMF species *in planta* found as spores in the vicinity of roots of *Fragaria vesca* L. in a zinc waste site in Poland (Turnau et al. 2001). Five AMF taxa were identified—*Archaeospora gerdemannii* (S.L. Rose, B.A. Daniels & Trappe) J.B. Morton & D. Redecker, *Glomus mosseae*, *G. intraradices*, *G. claroideum*, and *Paraglomus occultum* (C. Walker) J.B. Morton & D. Redecker—the first of them being the most efficient colonizer. However, 12 percent of colonized roots contained fungi other than the morphologically recognized species. It is worth noting that the *Archaeospora* and *Paraglomus* sequences are probably not detectable with the primers used by these researchers.

Conclusions

Many questions still remain regarding the impact of arbuscular mycorrhizal fungi (AMF) on assembly in ecosystems and on the structure, dynamics, and restoration of plant communities. Nevertheless, promising results have been obtained recently by combining classical ecology, morphology, and molecular biology in field studies. Especially in the field of restoration ecology, successful applications have been conducted, promoting acceptance of the positive influence of AMF on plant establishment and growth on degraded soils. In addition, the influence of AMF on plant community structure, one of the most important questions, has been addressed by several field studies. Further investigation is needed to determine the importance of AMF more precisely.

Acknowledgments

The authors thank the following collaborators: Dr. J. Vráblíková (Jan Evangelista Purkyně University, Ústí nad Labem, Czech Republic), Ú. Óskarsson (Soil Conservation Service, Hella, Iceland), and Dr. E. Weber (Brandenburg Technical University, Cottbus, Germany). We acknowledge data provided by mycorrhizal inoculum producers PlantWorks Ltd., UK, and Symbio-M Ltd., Czech Republic. We are also grateful to Prof. K. Turnau (University of Kraków, Poland), and Dr. V. M. Temperton (Max Planck Institute for Biogeochemistry, Jena, Germany), who kindly reviewed the manuscript and made helpful suggestions.

Some studies were partly supported by the European COST Action 8.38, Maj and Tor Nessling Foundation (Finland); by the Estonian Science Foundation (Grant 4579); and by the *Deutsche Forschungsgemeinschaft* (DFG, German Research Council), Graduate Research Group GRK 266.

REFERENCES

Abbas, J. D.; Hetrick, B. A. D.; and Jurgenson, J. E. 1996. Isolate specific detection of mycorrhizal fungi using genome specific primer pairs. *Mycologia* 88:939–946.

Allen, E. B. 1982. Water and nutrient competition between *Salsola kali* and two native grass species (*Agropyron smithii* and *Bouteloua gracilis*). *Ecology* 63:732–741.

Allen, E. B. 1984a. The role of mycorrhizae in mined land diversity. In *Proceedings of the Third Biennial Symposium on Surface Coal Mine Reclamation on the Great Plains.* Billings, Mont., March 19–21, 273–295.

Allen, E. B. 1984b. VA mycorrhizae and colonizing annuals: implications for growth, competition, and succession. In VA *Mycorrhizae and Reclamation of Arid and Semi-arid Lands*, ed. S. E. Williams and M. F. Allen, 42–52. Scientific Report SA1261. Laramie: University of Wyoming Agricultural Experiment Station.

Allen, E. B. 1988. Some trajectories of succession in Wyoming sagebrush grassland: implications for restoration. In *The Reconstruction of Disturbed Arid Lands: An Ecological Approach*, ed. E. B. Allen, 89–112. Boulder, Colo.: Westview Press.

Allen, E. B.; and Allen, M. F. 1980. Natural reestablishment of vesicular-arbuscular mycorrhizae following stripmine reclamation in Wyoming. *Journal of Applied Ecology* 17:139–147.

Allen, E. B.; and Allen, M. F. 1984. Competition between plants of different successional stages: mycorrhizae as regulators. *Canadian Journal of Botany* 62:2625–2629.

Allen, E. B.; and Allen, M. F. 1986. Water relations of xeric grasses in the field: interactions of mycorrhizas and competition. *New Phytologist* 104:559–571.

Allen, E. B.; and Allen, M. F. 1988. Facilitation of succession by the nonmycotrophic colonizer *Salsola kali* (Chenopodiaceae) on a harsh site: effects of mycorrhizal fungi. *American Journal of Botany* 75:257–266.

Allen, E. B.; Allen, M. F.; Helm, D. J.; Trappe, J. M.; Molina, R.; and Rincon, E. 1995. Patterns and regulation of mycorrhizal plant and fungal diversity. *Plant and Soil* 170:47–62.

Allen, E. B.; and Knight, D. H. 1984. The effects of introduced annuals on secondary succession in sagebrush-grassland, Wyoming. *Southwestern Naturalist* 29:407–421.

Allen, M. F. 1987. Reestablishment of mycorrhizas on Mount St. Helens: migration vectors. *Transactions of the British Mycological Society* 88:413–417.

Allen, M. F. 1988. Reestablishment of VA mycorrhizas following severe disturbance: comparative patch dynamics of a shrub desert and a subalpine volcano. *Proceedings of the Royal Society of Edinburgh B* 94:63–71.

Allen, M. F.; and Allen, E. B. 1990. Carbon source of VA mycorrhizal fungi associated with Chenopodiaceae from a semiarid shrub-steppe. *Ecology* 71:2019–2021.

Allen, M. F.; Allen, E. B.; Dahm, C. N.; and Edwards, F. S. 1993. Preservation of biological diversity in mycorrhizal fungi: importance and human impacts. In *International Symposium on Human Impacts on Self-Recruiting Populations*, ed. G. Sundnes, 81–108. Trondheim: The Royal Norwegian Academy of Sciences.

Allen, M. F.; Allen, E. B.; and Friese, C. F. 1989. Responses of the non-mycotrophic plant *Salsola kali* to invasion by VA mycorrhizal fungi. *New Phytologist* 111:45–49.

Allen, M. F.; Clouse, S. D.; Weinbaum, B. S.; Jenkins, S.; Friese, C. F.; and Allen, E. B. 1992. Mycorrhizae and the integration of scales: from molecules to ecosystems. In *Mycorrhizal Functioning*, ed. M. F. Allen, 488–516. New York: Chapman and Hall.

Allen, M. F.; and MacMahon, J. A. 1985. Impact of disturbance on cold desert fungi: comparative microscale dispersion patterns. *Pedobiologia* 28:215–224.

Allsopp, N.; and Stock, W. D. 1992. Density dependent interactions between VA mycorrhizal fungi and even-aged seedlings of two perennial Fabaceae species. *Oecologia* 91:281–287.

Altschul, S. F.; Madden, T. L.; Schäffer, A. A.; Zhang, J.; Zhang, Z.; Miller, W.; and Lipman, D. J. 1997. Gapped BLAST and PSI-BLAST: a new generation of protein database search programs. *Nucleic Acids Research* 25:3389–3402.

Antoniolli, Z. I.; Schachtman, D. P.; Ophel-Keller, K.; and Smith, S. E. 2000. Variation in rDNA ITS sequences in *Glomus mosseae* and *Gigaspora margarita* spores from a permanent pasture. *Mycological Research* 104:708–715.

Augé, R. M.; Stodola, A. J. W.; Tims, J. E.; and Saxton, A. M. 2001. Moisture retention properties of a mycorrhizal soil. *Plant and Soil* 230:87–97.

Austin, M. P. 1999. The potential contribution of vegetation ecology to biodiversity research. *Ecography* 22:465–484.

Barroetavena, C.; Gisler, S. D.; Luoma, D. L.; and Meinke, R. J. 1998. Mycorrhizal status of the endangered species *Astragalus applegatei* Peck as determined from a soil bioassay. *Mycorrhiza* 8:117–119.

Bartolome-Esteban, H.; and Schenck, N. C. 1994. Spore germination and hyphal growth of arbuscular mycorrhizal fungi in relation to soil aluminum saturation. *Mycologia* 86:217–226.

Bentivenga, S. P.; and Morton, J. B. 1994. Stability and heritability of fatty acid methyl ester profiles of glomalean endomycorrhizal fungi. *Mycological Research* 98:1419–1426.

Berg, Å.; Ehnström, B.; Gustafsson, L.; Hallingbäck, T.; Jonsell, M.; and Weslien, J. 1994. Threatened plant, animal, and fungus species in Swedish forests: distribution and habitat associations. *Conservation Biology* 8:718–731.

Bever, J. D.; Schultz, P. A.; Pringle, A.; and Morton, J. B. 2001. Arbuscular mycorrhizal fungi: more diverse than meets the eye, and the ecological tale of why. *BioScience* 51:923–931.

Brown, V. K. 1994. Herbivory: a structuring force in plant communities. In *Individuals, Populations and Patterns in Ecology*, ed. S. R. Leather, A. D. Watt, N. J. Mills, and K. F. A. Walters, 299–308. Andover, UK: Intercept Ltd.

Brown, V. K.; and Gange, A. C. 1990. Insect herbivory below ground. *Advances in Ecological Research* 20:1–57.

Bruns, T. D.; Vilgalys, R.; Barns, S. M.; Gonzalez, D.; Hibbett, D. S.; Lane, D. J.; Simon, L.; Stickel, S.; Szaro, T. M.; Weisburg, W. G.; and Sogin, M. L. 1992. Evolutionary relationships within the fungi: analysis of nuclear small subunit rRNA sequences. *Molecular Phylogenetics and Evolution* 1:231–241.

Buscot, F.; Munch, J. C.; Charcosset, J. Y.; Gardes, M.; Nehls, U.; and Hampp, R. 2000. Recent advances in exploring physiology and biodiversity of ectomycorrhizas highlight the functioning of these symbioses in ecosystems. *FEMS Microbiology Reviews* 24:601–614.

Cavagnaro, T. R.; Gao, L.-L.; Smith, F. A.; and Smith, S. E. 2001. Morphology of arbuscular mycorrhizas is influenced by fungal identity. *New Phytologist* 151:469–475.

Chesson, P. L. 1986. Environmental variability and the coexistence of species. In *Community Ecology*, ed. J. Diamond and T. J. Case, 240–256. New York: Harper and Row.

Claassen, V. P.; Zasoski, R. J.; and Tyler, B. M. 1996. A method for direct soil extraction and PCR amplification of endomycorrhizal fungal DNA. *Mycorrhiza* 6:447–450.

Clapp, J. P.; Rodriguez, A.; and Dodd, J. C. 2001. Inter- and intra-isolate rRNA large subunit variation in *Glomus coronatum* spores. *New Phytologist* 149:539–554.

Clapp, J. P.; Young, J. P. W.; Merryweather, J. W.; and Fitter, A. H. 1995. Diversity of fungal symbionts in arbuscular mycorrhizas from a natural community. *New Phytologist* 130:259–265.

Connell, J. H. 1978. Diversity in tropical rain forests and coral reefs. *Science* 199:1302–1310.

Daniell, T. J.; Husband, R.; Fitter, A. H.; and Young, J. P. W. 2001. Molecular diversity of arbuscular mycorrhizal fungi colonising arable crops. *FEMS Microbiology Ecology* 36:203–209.

Diamond, J. M. 1975. Assembly of species communities. In *Ecology and Evolution of Communities*, ed. M. L. Cody and J. M. Diamond, 342–444. Cambridge, Mass.: Harvard University Press.

Dodd, J. C.; Boddington, C. L.; Rodriguez, A.; Gonzalez-Chavez, C.; and Mansur, I. 2000. Mycelium of arbuscular mycorrhizal fungi (AMF) from different genera: form, function and detection. *Plant and Soil* 226:131–151.

Dodd, J. C.; Dougall, T. A. G.; Clapp, J. P.; and Jeffries, P. 2002. The role of arbuscular mycorrhizal fungi in plant community establishment at Samphire Hoe, Kent, UK: the reclamation platform created during the building of the Channel tunnel between France and the UK. *Biodiversity and Conservation* 11:39–58.

Dodd, J. C.; Gianinazzi-Pearson, V.; Rosendahl, S.; and Walker, C. 1994. European Bank of *Glomales*: an essential tool for efficient international and interdisciplinary collaboration. In *Impact of Arbuscular Mycorrhizas on Sustainable Agriculture and Natural Ecosystems*, ed. G. Gianinazzi and H. Schüepp, 41–46. Basel, Switzerland: Birkhäuser Verlag.

Dodd, J. C.; Rosendahl, S.; Giovannetti, M.; Broome, A.; Lanfranco, L.; and Walker, C. 1996. Inter- and intraspecific variation within the morphologically-similar arbuscular mycorrhizal fungi *Glomus mosseae* and *Glomus coronatum*. *New Phytologist* 133:113–122.

Egerton-Warburton, L. M.; and Allen, E. B. 2000. Shifts in arbuscular mycorrhizal communities along an anthropogenic nitrogen deposition gradient. *Ecological Applications* 10:484–496.

Egerton-Warburton, L. M.; and Allen, M. F. 2001. Endo- and ectomycorrhizas in *Quercus agrifolia* Nee. (Fagaceae): patterns of root colonization and effects on seedling growth. *Mycorrhiza* 11:283–290.

Eissenstat, D. M.; and Newman, E. I. 1990. Seedling establishment near large plants: effects of vesicular-arbuscular mycorrhizas on the intensity of plant competition. *Functional Ecology* 4:95–99.

Enkhtuya, B.; Rydlová, J.; and Vosátka, M. 2000. Effectiveness of indigenous and non-indigenous isolates of arbuscular mycorrhizal fungi in soils from degraded ecosystems and man-made habitats. *Applied Soil Ecology* 14:201–211.

Eom, A.-H.; Hartnett, D. C.; and Wilson, G. W. T. 2000. Host plant species effects on arbuscular mycorrhizal fungal communities in tallgrass prairie. *Oecologia* 122:435–444.

Eriksson, Å.; Eriksson, O.; and Berglund, H. 1995. Species abundance patterns of plants in Swedish semi-natural pastures. *Ecography* 18:310–317.

Esher, R. J.; Marx, D. H.; Ursic, S. J.; Baker, R. L.; Brown, L. R.; and Coleman, D. C. 1992. Simulated acid rain effects on fine roots, ectomycorrhizae, microorganisms, and invertebrates in pine forests of the southern United States. *Water, Air, and Soil Pollution* 61:269–278.

Filion, M.; St-Arnaud, M.; and Fortin, J. A. 1999. Direct interaction between the arbuscular mycorrhizal fungus *Glomus intraradices* and different rhizosphere microorganisms. *New Phytologist* 141:525–533.

Franke-Snyder, M.; Douds, D. D.; Galvez, L.; Phillips, J. G.; Wagoner, P.; Drinkwater, L.; and Morton, J. B. 2001. Diversity of communities of arbuscular mycorrhizal (AM) fungi present in conventional versus low-input agricultural sites in eastern Pennsylvania, USA. *Applied Soil Ecology* 16:35–48.

Galli, U.; Schüepp, H.; and Brunold, C. 1994. Heavy metal binding by mycorrhizal fungi. *Physiologia Plantarum* 92:364–368.

Gange, A. C.; Brown, V. K.; and Sinclair, G. S. 1993. Vesicular-arbuscular mycorrhizal fungi: a determinant of plant community structure in early succession. *Functional Ecology* 7:616–622.

Gerdemann, J. W.; and Nicolson, T. H. 1963. Spores of mycorrhizal *Endogone* species extracted from soil by wet sieving and decanting. *Transactions of the British Mycological Society* 46:235–244.

Gerdemann, J. W.; and Trappe, J. M. 1974. The Endogonaceae in the Pacific Northwest. *Mycologia Memoir* 5:1–76.

Giovannetti, M.; Azzolini, D.; and Citernesi, A. S. 1999. Anastomosis formation and nuclear and protoplasmic exchange in arbuscular mycorrhizal fungi. *Applied and Environmental Microbiology* 65:5571–5575.

Giovannetti, M.; Fortuna, P.; Citernesi, A. S.; Morini, S.; and Nuti, M. P. 2001. The occurrence of anastomosis formation and nuclear exchange in intact arbuscular mycorrhizal networks. *New Phytologist* 151:717–724.

Grace, J. B. 2001. The roles of community biomass and species pools in the regulation of plant diversity. *Oikos* 92:193–207.

Grime, J. P.; Mackey, J. M.; Hillier, S. H.; and Read, D. J. 1987. Floristic diversity in a model system using experimental microcosms. *Nature* 328:420–422.

Gustafsson, L. 1994. A comparison of biological characteristics and distribution between Swedish threatened and non-threatened forest vascular plants. *Ecography* 17:39–49.

Hahn, A.; Bonfante, P.; Horn, K.; Pausch, F.; and Hock, B. 1993. Production of monoclonal antibodies against surface antigens of spores from arbuscular mycorrhizal fungi by an improved immunization and screening procedure. *Mycorrhiza* 4:69–78.

Hanski, I. A.; and Gilpin, M. E., eds. 1997. *Metapopulation Biology.* San Diego: Academic Press.

Hardie, K. 1985. The effect of removal of extraradical hyphae on water uptake by vesicular-arbuscular mycorrhizal plants. *New Phytologist* 101:677–684.

Hart, M. M.; and Reader, R. J. 2002. Taxonomic basis for variation in the colonization strategy of arbuscular mycorrhizal fungi. *New Phytologist* 153:335–344.

Hartnett, D. C.; Hetrick, B. A. D.; Wilson, G. W. T.; and Gibson, D. J. 1993. Mycorrhizal influence on intra- and interspecific neighbor interactions among co-occurring prairie grasses. *Journal of Ecology* 81:787–795.

Hartnett, D. C.; and Wilson, G. W. T. 1999. Mycorrhizae influence plant community structure and diversity in tallgrass prairie. *Ecology* 80:1187–1195.

Heckman, D. S.; Geiser, D. M.; Eidell, B. R.; Stauffer, R. L.; Kardos, N. L.; and Hedges, S. B. 2001. Molecular evidence for the early colonization of land by fungi and plants. *Science* 293:1129–1133.

Helgason, T.; Daniell, T. J.; Husband, R.; Fitter, A. H.; and Young, J. P. W. 1998. Ploughing up the wood-wide web? *Nature* 394:431.

Helgason, T.; Fitter, A. H.; and Young, J. P. W. 1999. Molecular diversity of arbuscular mycorrhizal fungi colonising *Hyacinthoides non-scripta* (bluebell) in a seminatural woodland. *Molecular Ecology* 8:659–666.

Heppell, K. B.; Shumway, D. L.; and Koide, R. T. 1998. The effect of mycorrhizal infection of *Abutilon theophrasti* on competitiveness of offspring. *Functional Ecology* 12:171–175.

Hetrick, B. A. D.; Wilson, G. W. T.; and Hartnett, D. C. 1989. Relationship between mycorrhizal dependence and competitive ability of two tallgrass prairie grasses. *Canadian Journal of Botany* 67:2608–2615.

Hickson, L. 1993. The effects of vesicular-arbuscular mycorrhizae on morphology, light harvesting, and photosynthesis of *Artemisia tridentata* ssp. *tridentata*. Thesis. San Diego State University.

Hildebrandt, U.; Janetta, K.; Ouziad, F.; Renne, B.; Nawrath, K.; and Bothe, H. 2001. Arbuscular mycorrhizal colonization of halophytes in Central European salt marshes. *Mycorrhiza* 10:175–183.

Hildebrandt, U.; Kaldorf, M.; and Bothe, H. 1999. The zinc violet and its colonization by arbuscular mycorrhizal fungi. *Journal of Plant Physiology* 154:709–717.

Horton, T. R.; and Bruns, T. D. 2001. The molecular revolution in ectomycorrhizal ecology: peeking into the black-box. *Molecular Ecology* 10:1855–1871.

Huntly, N. 1991. Herbivores and the dynamics of communities and ecosystems. *Annual Review of Ecology and Systematics* 22:477–503.

Huston, M. A. 1979. A general hypothesis of species diversity. *American Naturalist* 113:81–101.

Huston, M. A. 1999. Local processes and regional patterns: appropriate scales for understanding variation in the diversity of plants and animals. *Oikos* 86:393–401.

Jabaji-Hare, S. 1988. Lipid and fatty acid profiles of some vesicular-arbuscular mycorrhizal fungi: contribution to taxonomy. *Mycologia* 80:622–629.

Jacquot, E.; van Tuinen, D.; Gianinazzi, S.; and Gianinazzi-Pearson, V. 2000. Monitoring species of arbuscular mycorrhizal fungi *in planta* and in soil by nested PCR: application to the study of the impact of sewage sludge. *Plant and Soil* 226:179–188.

Jakobsen, I.; Abbott, L. K.; and Robson, A. D. 1992. External hyphae of vesicular-arbuscular mycorrhizal fungi associated with *Trifolium subterraneum* L. 2. Hyphal transport of ^{32}P over defined distances. *New Phytologist* 120:509–516.

Jakobsen, I.; Gazey, C.; and Abbott, L. K. 2001. Phosphate transport by communities of arbuscular mycorrhizal fungi in intact soil cores. *New Phytologist* 149:95–103.

Janos, D. P. 1980. Mycorrhizae influence tropical succession. *Biotropica* 12:56–64.

Janos, D. P. 1982. Tropical mycorrhizas, nutrient cycles, and plant growth. In *Tropical Rain Forest: Ecology and Management*, ed. S. L. Sutton, T. C. Whitmore, and A. C. Chadwick, 327–345. Oxford: Blackwell Science.

Johnson, N. C. 1993. Can fertilization of soil select less mutualistic mycorrhizae? *Ecological Applications* 3:749–757.

Kaldorf, M.; Kuhn, A. J.; Schröder, W. H.; Hildebrandt, U.; and Bothe, H. 1999. Selective element deposits in maize colonized by a heavy metal tolerance conferring arbuscular mycorrhizal fungus. *Journal of Plant Physiology* 154:718–728.

Karron, J. D. 1987. The pollination ecology of co-occurring geographically restricted and widespread species of *Astragalus* (Fabaceae). *Biological Conservation* 39:179–193.

Kitajima, K.; and Tilman, D. 1996. Seed banks and seedling establishment on an experimental productivity gradient. *Oikos* 76:381–391.

Kjøller, R.; and Rosendahl, S. 2000. Detection of arbuscular mycorrhizal fungi (Glomales) in roots by nested PCR and SSCP (Single Stranded Conformation Polymorphism). *Plant and Soil* 226:189–196.

Kjøller, R.; and Rosendahl, S. 2001. Molecular diversity of glomalean (arbuscular mycorrhizal) fungi determined as distinct *Glomus* specific DNA sequences from roots of field grown peas. *Mycological Research* 105:1027–1032.

Klironomos, J. N.; Rillig, M. C.; and Allen, M. F. 1999. Designing belowground field experiments with the help of semi-variance and power analyses. *Applied Soil Ecology* 12:227–238.

Kuhn, G.; Hijri, M.; and Sanders, I. R. 2001. Evidence for the evolution of multiple genomes in arbuscular mycorrhizal fungi. *Nature* 414:745–748.

Kunin, W. E.; and Gaston, K. J. 1993. The biology of rarity: patterns, causes and consequences. *Trends in Ecology and Evolution* 8:298–301.

Kunin, W. E.; and Gaston, K. J., eds. 1997. *The Biology of Rarity: Causes and Consequences of Rare-Common Differences.* London: Chapman and Hall.

Lanfranco, L.; Wyss, P.; Marzachì, C.; and Bonfante, P. 1995. Generation of RAPD-PCR primers for the identification of isolates of *Glomus mosseae*, an arbuscular mycorrhizal fungus. *Molecular Ecology* 4:61–68.

Langer, U.; and Günther, T. 2001. Effects of alkaline dust deposits from phosphate fertilizer production on microbial biomass and enzyme activities in grassland soils. *Environmental Pollution* 112:321–327.

Lloyd-MacGilp, S. A.; Chambers, S. M.; Dodd, J. C.; Fitter, A. H.; Walker, C.; and Young, J. P. W. 1996. Diversity of the internal transcribed spacers within and among isolates of *Glomus mosseae* and related mycorrhizal fungi. *New Phytologist* 133:103–111.

Longato, S.; and Bonfante, P. 1997. Molecular identification of mycorrhizal fungi by direct amplification of microsatellite regions. *Mycological Research* 101:425–432.

Luoma, D. L.; Eberhart, J. L.; and Amaranthus, M. P. 1997. Biodiversity of ectomycorrhizal types from southwest Oregon. In *Conservation and Management of Native Plants and Fungi*, ed. T. N. Kaye, A. Liston, R. M. Love, D. L. Luoma, R. J. Meinke, and M. V. Wilson, 249–253. Corvallis: Native Plant Society of Oregon.

MacArthur, R. H. 1972. *Geographical Ecology: Patterns in the Distribution of Species.* New York: Harper and Row.

MacArthur, R. H.; and Connell, J. H. 1966. *The Biology of Populations.* 2nd ed. New York: John Wiley and Sons.

Makino, R.; Yazyu, H.; Kishimoto, Y.; Sekiya, T.; and Hayashi, K. 1992. F-SSCP: fluorescent-based polymerase chain reaction–single-strand conformation polymorphism (PCR-SSCP) analysis. *PCR Methods and Applications* 2:10–13.

Marler, M. J.; Zabinski, C. A.; and Callaway, R. M. 1999. Mycorrhizae indirectly enhance competitive effects of an invasive forb on a native bunchgrass. *Ecology* 80:1180–1186.

McGonigle, T. P.; and Fitter, A. H. 1990. Ecological specificity of vesicular-arbuscular mycorrhizal associations. *Mycological Research* 94:120–122.

Merryweather, J.; and Fitter, A. 1998a. The arbuscular mycorrhizal fungi of *Hyacinthoides non-scripta.* I. Diversity of fungal taxa. *New Phytologist* 138:117–129.

Merryweather, J.; and Fitter, A. 1998b. The arbuscular mycorrhizal fungi of *Hyacinthoides non-scripta.* II. Seasonal and spatial patterns of fungal populations. *New Phytologist* 138:131–142.

Miller, R. M. 1979. Some occurrences of vesicular-arbuscular mycorrhiza in natural and disturbed ecosystems of the Red Desert. *Canadian Journal of Botany* 57:619–623.

Miller, R. M.; and Jastrow, J. D. 1992. The application of VA mycorrhizae to ecosystem restoration and reclamation. In *Mycorrhizal Functioning*, ed. M. F. Allen, 438–467. New York: Chapman and Hall.

Moora, M.; Öpik, M.; Sen, R.; and Zobel, M. 2004. Rare vs. common *Pulsatilla* spp. seedling performance with arbuscular mycorrhizal inoculum from contrasting native habitats. *Functional Ecology*, in press.

Moora, M.; and Zobel, M. 1996. Effect of arbuscular mycorrhiza on inter- and intraspecific competition of two grassland species. *Oecologia* 108:79–84.

Morton, J. B.; and Benny, G. L. 1990. Revised classification of arbuscular mycorrhizal fungi (Zygomycetes): a new order, Glomales, two new suborders, Glomineae and Gigasporineae, and two new families, Acaulosporaceae and Gigasporaceae, with an emendation of Glomaceae. *Mycotaxon* 37:471–491.

Morton, J. B.; and Redecker, D. 2001. Two new families of Glomales, Archaeosporaceae and Paraglomaceae, with two new genera *Archaeospora* and *Paraglomus*, based on concordant molecular and morphological characters. *Mycologia* 93:181–195.

Newsham, K. K.; Fitter, A. H.; and Watkinson, A. R. 1994. Root pathogenic and arbuscular mycorrhizal fungi determine fecundity of asymptomatic plants in the field. *Journal of Ecology* 82:805–814.

Newsham, K. K.; Fitter, A. H.; and Watkinson, A. R. 1995a. Arbuscular mycorrhiza protect an annual grass from root pathogenic fungi in the field. *Journal of Ecology* 83:991–1000.

Newsham, K. K.; Fitter, A. H.; and Watkinson, A. R. 1995b. Multi-functionality and biodiversity in arbuscular mycorrhizas. *Trends in Ecology and Evolution* 10:407–411.

Newsham, K. K.; Watkinson, A. R.; West, H. M.; and Fitter, A. H. 1995c. Symbiotic fungi determine plant community structure: changes in a lichen-rich community induced by fungicide application. *Functional Ecology* 9:442–447.

Nicolson, T. H. 1960. Mycorrhizae in the Gramineae. II. Development in different habitats, particularly sand dunes. *Transactions of the British Mycological Society* 43:132–145.

Öpik, M.; Moora, M.; Liira, J.; Kõljalg, U.; Sen, R.; and Zobel, M. 2003. Divergent arbuscular mycorrhizal fungal communities colonize roots of *Pulsatilla* spp. in boreal Scots pine forest and grassland soils. *New Phytologist* 160:581–593.

Pfleger, F. L.; Steward, E. L.; and Noyd, R. K. 1994. Role of AMF fungi in mine land revegetation. In *Mycorrhizae and Plant Health*, ed. F. L. Pfleger and R. G. Linderman, 47–82. St. Paul: APS Press.

Pirozynski, K. A.; and Malloch, D. W. 1975. The origin of land plants: a matter of mycotrophism. *BioSystems* 6:153–164.

Rabinowitz, D. 1981. Seven forms of rarity. In *The Biological Aspects of Rare Plant Conservation*, ed. H. Synge, 205–218. New York: John Wiley and Sons.

Read, D. J. 1998. Plants on the web. *Nature* 369:22–23.

Redecker, D. 2000. Specific PCR primers to identify arbuscular mycorrhizal fungi within colonized roots. *Mycorrhiza* 10:73–80.

Redecker, D.; Kodner, R.; and Graham, L. E. 2000. Glomalean fungi from the Ordovician. *Science* 289:1920–1921.

Remy, W.; Taylor, T. N.; Hass, H.; and Kerp, H. 1994. Four hundred-million-year-old vesicular arbuscular mycorrhizae. *Proceedings of the National Academy of Sciences USA* 91:11841–11843.

Renker, C.; Heinrichs, J.; Kaldorf, M.; and Buscot, F. 2003. Combining nested PCR and restriction digest of the internal transcribed spacer region to characterize arbuscular mycorrhizal fungi on roots from the field. *Mycorrhiza* 13:191–198.

Rodriguez, A.; Dougall, T.; Dodd, J. C.; and Clapp, J. P. 2001. The large subunit ribosomal RNA genes of *Entrophospora infrequens* comprise sequences related to two different glomalean families. *New Phytologist* 152:159–167.

Ronsheim, M. L.; and Anderson, S. E. 2001. Population-level specificity in the plant-mycorrhizae association alters intraspecific interactions among neighboring plants. *Oecologia* 128:77–84.

Rosendahl, S. 1989. Comparisons of spore-cluster forming *Glomus* species (Endogonaceae) based on morphological characteristics and isoenzyme banding patterns. *Opera Botanica* 100:215–223.

Rosendahl, S.; Rosendahl, C. N.; and Søchting, U. 1989. Distribution of VA mycorrhizal endophytes amongst plants from a Danish grassland community. *Agriculture, Ecosystems and Environment* 29:329–335.

Rosenzweig, M. L. 1995. *Species Diversity in Space and Time.* Cambridge: Cambridge University Press.

Rydlová, J.; and Vosátka, M. 2000. Sporulation of symbiotic arbuscular mycorrhizal fungi inside dead seeds of a non-host plant. *Symbiosis* 29:231–248.

Rydlová, J.; and Vosátka, M. 2001. Associations of dominant plant species with arbuscular mycorrhizal fungi during vegetation development on coal mine spoil banks. *Folia Geobotanica* 36:85–97.

Sætersdal, M. 1994. Rarity and species/area relationships of vascular plants in deciduous woods, western Norway: applications to nature reserve selection. *Ecography* 17:23–38.

Sanders, I. R. 1993. Temporal infectivity and specificity of vesicular-arbuscular mycorrhizas in co-existing grassland species. *Oecologia* 93:349–355.

Sanders, I. R. 1999. No sex please, we're fungi. *Nature* 399:737–739.

Sanders, I. R.; Clapp, J. P.; and Wiemken, A. 1996. The genetic diversity of arbuscular mycorrhizal fungi in natural ecosystems: a key to understanding the ecology and functioning of the mycorrhizal symbiosis. *New Phytologist* 133:123–134.

Schüßler, A. 1999. Glomales SSU rRNA gene diversity. *New Phytologist* 144:205–207.

Schüßler, A.; Schwarzott, D.; and Walker, C. 2001. A new fungal phylum, the *Glomeromycota*: phylogeny and evolution. *Mycological Research* 105:1413–1421.

Schwarzott, D.; Walker, C.; and Schüßler, A. 2001. *Glomus*, the largest genus of the arbuscular mycorrhizal fungi (Glomales), is nonmonophyletic. *Molecular Phylogenetics and Evolution* 21:190–197.

Simard, S. W.; Perry, D. A.; Jones, M. D.; Myrold, D. D.; Durall, D. M.; and Molina, R. 1997. Net transfer of carbon between ectomycorrhizal tree species in the field. *Nature* 388:579–582.

Simon, L.; Bousquet, J.; Lévesque, R. C.; and Lalonde, M. 1993. Origin and diversification of endomycorrhizal fungi and coincidence with vascular land plants. *Nature* 363:67–69.

Simon, L.; Lalonde, M.; and Bruns, T. D. 1992. Specific amplification of 18S fungal ribosomal genes from vesicular-arbuscular endomycorrhizal fungi colonizing roots. *Applied and Environmental Microbiology* 58:291–295.

Šmilauer, P. 2001. Communities of arbuscular mycorrhizal fungi in grassland: seasonal variability and effects of environment and host plants. *Folia Geobotanica* 36:243–263.

Smith, F. A.; and Smith, S. E. 1996. Mutualism and parasitism: diversity in function and structure in the "arbuscular" (VA) mycorrhizal symbiosis. In *Advances in Botanical Research*, vol. 22, ed. J. A. Callow, 1–43. London: Academic Press.

Smith, S. E.; and Read, D. J. 1997. *Mycorrhizal Symbiosis.* London: Academic Press.

Stahl, E. 1900. Der Sinn der Mycorrhizenbildung. *Jahrbücher für wissenschaftliche Botanik* 34:539–668.

Stutz, J. C.; Copeman, R.; Martin, C. A.; and Morton, J. B. 2000. Patterns of species composition and distribution of arbuscular mycorrhizal fungi in arid regions of southwestern North America and Namibia, Africa. *Canadian Journal of Botany* 78:237–245.

Sylvia, D. M.; and Williams, S. E. 1992. Vesicular-arbuscular mycorrhizae and environmental stress. In *Mycorrhizae in Sustainable Agriculture*, ed. G. J. Bethlenfalvay and R. G. Linderman, 101–124. Special Publication No. 54. Madison, Wisc: American Society of Agronomy.

Thingstrup, I.; Rozycka, M.; Jeffries, P.; Rosendahl, S.; and Dodd, J. C. 1995. Detection of the arbuscular mycorrhizal fungus *Scutellospora heterogama* within roots using polyclonal antisera. *Mycological Research* 99:1225–1232.

Thompson, K.; Hillier, S. H.; Grime, J. P.; Bossard, C. C.; and Band, S. R. 1996. A functional analysis of a limestone grassland community. *Journal of Vegetation Science* 7:371–380.

Thrall, P. H.; Burdon, J. J.; and Woods, M. J. 2000. Variation in the effectiveness of symbiotic associations between native rhizobia and temperate Australian legumes: interactions within and between genera. *Journal of Applied Ecology* 37:52–65.

Tilman, D. 1982. *Resource Competition and Community Structure*. Princeton: Princeton University Press.

Tilman, D. 1986. Evolution and differentiation in terrestrial plant communities: the importance of the soil resource: light gradient. In *Community Ecology*, ed. J. Diamond and T. J. Case, 344–380. New York: Harper and Row.

Tilman, D.; and Kareiva, P., eds. 1997. *Spatial Ecology*. Princeton: Princeton University Press.

Tonin, C.; Vandenkoornhuyse, P.; Joner, E. J.; Straczek, J.; and Leyval, C. 2001. Assessment of arbuscular mycorrhizal fungi diversity in the rhizosphere of *Viola calaminaria* and effect of these fungi on heavy metal uptake by clover. *Mycorrhiza* 10:161–168.

Turnau, K.; Ryszka, P.; Gianinazzi-Pearson, V.; and van Tuinen, D. 2001. Identification of arbuscular mycorrhizal fungi in soils and roots of plants colonizing zinc wastes in southern Poland. *Mycorrhiza* 10:169–174.

Turnbull, L. A.; Crawley, M. J.; and Rees, M. 2000. Are plant populations seed-limited? A review of seed sowing experiments. *Oikos* 88:225–238.

van der Heijden, E. W. 2001. Differential benefits of arbuscular mycorrhizal and ectomycorrhizal infection of *Salix repens*. *Mycorrhiza* 10:185–193.

van der Heijden, E. W.; and Vosátka, M. 1999. Mycorrhizal associations of *Salix repens* L. communities in succession of dune ecosystems. II. Mycorrhizal dynamics and interactions of ectomycorrhizal and arbuscular mycorrhizal fungi. *Canadian Journal of Botany* 77:1833–1841.

van der Heijden, M. G. A.; Boller, T.; Wiemken, A.; and Sanders, I. R. 1998a. Different arbuscular mycorrhizal fungal species are potential determinants of plant community structure. *Ecology* 79:2082–2091.

van der Heijden, M. G. A.; Klironomos, J. N.; Ursic, M.; Moutoglis, P.; Streitwolf-Engel, R.; Boller, T.; Wiemken, A.; and Sanders, I. R. 1998b. Mycorrhizal fungal diversity determines plant biodiversity, ecosystem variability and productivity. *Nature* 396:69–72.

van der Putten, W. H.; Vet, L. E. M.; Harvey, J. A.; and Wäckers, F. L. 2001. Linking above- and belowground multitrophic interactions of plants, herbivores, pathogens, and their antagonists. *Trends in Ecology and Evolution* 16:547–554.

van Tuinen, D.; Jacquot, E.; Zhao, B.; Gollotte, A.; Gianinazzi-Pearson, V. 1998. Characterization of root colonization profiles by a microcosm community of arbuscular mycorrhizal fungi using 25S rDNA-targeted nested PCR. *Molecular Ecology* 7:879–887.

Vierheilig, H.; Coughlan, A. P.; Wyss, U.; and Piché, Y. 1998. Ink and vinegar, a simple staining technique for arbuscular-mycorrhizal fungi. *Applied and Environmental Microbiology* 64:5004–5007.

Visser, S. 1985. Management of microbial processes in surface mined land reclamation in western Canada. In *Soil Reclamation Processes: Microbiological Analyses and Applications*, ed. R. L. Tade and D. A. Klein, 203–241. New York, Basel: Marcel Dekker.

Vosátka, M. 2000. A future role for the use of arbuscular mycorrhizal fungi in soil reme-diation: a chance for small-medium enterprises? *Minerva Biotechnologica* 13:69–72.

Vosátka, M.; Rydlová, J.; and Malcová, R. 1999. Microbial inoculations of plants for reveg-etation of disturbed soils in degraded ecosystems. In *Nature and Culture in Land-scape Ecology*, ed. P. Kovář, 303–317. Prague: Carolinum Press.

Vosátka, M.; Soukupová, L.; Rauch, O.; and Škoda, M. 1995. Expansion dynamics of *Calamagrostis villosa* and VA-mycorrhiza in relation to different soil acidification. In *Mountain National Parks and Biosphere Reserves: Monitoring and Management*, ed. J. Flousek and G. C. S. Roberts, 47–53. Proceedings of the International Conference, Špindleruv Mlýn, Czech Republic.

Waaland, M. E.; and Allen, E. B. 1987. Relationships between VA mycorrhizal fungi and plant cover following surface mining in Wyoming. *Journal of Range Management* 40:271–276.

Walker, C.; and Trappe, J. M. 1993. Names and epithets in the Glomales and Endogo-nales. *Mycological Research* 97:339–344.

Warner, N. J.; Allen, M. F.; and MacMahon, J. A. 1987. Dispersal agents of vesicular-arbuscular mycorrhizal fungi in a disturbed arid ecosystem. *Mycologia* 79:721–730.

Watkinson, A. R. 1998. The role of the soil community in plant population dynamics. *Trends in Ecology and Evolution* 13:171–172.

Webb, C. O.; and Peart, D. R. 1999. Seedling density dependence promotes coexistence of Bornean rain forest trees. *Ecology* 80:2006–2017.

Weiher, E.; and Keddy, P. 1995. The assembly of experimental wetland plant commu-nities. *Oikos* 73:323–335.

Weinbaum, B. S.; Allen, M. F.; and Allen, E. B. 1996. Survival of arbuscular mycorrhi-zal fungi following reciprocal transplanting across the Great Basin, USA. *Ecological Applications* 6:1365–1372.

Weising, K.; Nybom, H.; Wolff, K.; and Meyer, W. 1995. *DNA Fingerprinting in Plants and Fungi*. Ann Arbor, London, Tokyo: CRC Press.

Weissenhorn, I.; Leyval, C.; and Berthelin, J. 1993. Cd-tolerant arbuscular mycorrhizal (AM) fungi from heavy-metal polluted soils. *Plant and Soil* 157:247–256.

White, T. J.; Bruns, T.; Lee, S.; and Taylor, J. 1990. Amplification and direct sequencing of fungal ribosomal RNA genes for phylogenetics. In *PCR Protocols: A Guide to Methods and Applications*, ed. M. A. Innis, D. H. Gelfand, J. J. Sninsky, and T. J. White, 315–322. San Diego: Academic Press.

Whittaker, R. H.; and Levin, S. A. 1977. The role of mosaic phenomena in natural com-munities. *Theoretical Population Biology* 12:117–139.

Wilson, G. W. T.; Hartnett, D. C.; Smith, M. D.; and Kobbeman, K. 2001. Effects of mycorrhizae on growth and demography of tallgrass prairie forbs. *American Journal of Botany* 88:1452–1457.

Wilson, J. M.; and Tommerup, I. C. 1992. Interactions between fungal symbionts: VA mycorrhizae. In *Mycorrhizal Functioning*, ed. M. F. Allen, 199–248. New York: Chapman and Hall.

Witkowski, E. T. F.; and Lamont, B. B. 1997. Does the rare *Banksia goodii* have inferior vegetative, reproductive or ecological attributes compared with its widespread co-occurring relative *B. gardneri? Journal of Biogeography* 24:469–482.

Wolf, A.; Brodmann, P. A.; and Harrison, S. 1999. Distribution of the rare serpentine sunflower, *Helianthus exilis* (Asteraceae): the roles of habitat availability, dispersal limitation and species introductions. *Oikos* 84:69–76.

Wright, S. F.; Morton, J. B.; and Sworobuk, J. E. 1987. Identification of a vesicular-arbuscular mycorrhizal fungus by using monoclonal antibodies in an enzyme-linked immunosorbent assay. *Applied and Environmental Microbiology* 53:2222–2225.

Young, A.; Boyle, T.; and Brown, T. 1996. The population genetic consequences of habitat fragmentation for plants. *Trends in Ecology and Evolution* 11:413–418.

Zézé, A.; Sulistyowati, E.; Ophel-Keller, K.; Barker, S.; and Smith, S. 1997. Intersporal genetic variation of *Gigaspora margarita*, a vesicular arbuscular mycorrhizal fungus, revealed by M13 minisatellite-primed PCR. *Applied and Environmental Microbiology* 63:676–678.

Zobel, M.; Moora, M.; and Haukioja, E. 1997. Plant coexistence in the interactive environment: arbuscular mycorrhiza should not be out of mind. *Oikos* 78: 202–208.

Chapter 11

Modeling of Plant Community Assembly in Relation to Deterministic and Stochastic Processes

Gerrit W. Heil

Community ecology is concerned with the properties of sets of species at given locations in time and space. There are three major theories of community assembly: deterministic, stochastic, and alternative stable states (ASS). In the deterministic model, a community is seen as the inevitable consequence of physical and biotic factors (Clements 1916). In the stochastic model, the community composition and structure are essentially the results of a random process (Van der Maarel and Sykes 1993). The ASS theory is intermediate between the first two (see Chapter 3). Over the years, there has been an increasing focus on the use of individual-based models in ecology, as documented by various authors (for example, Huston et al. 1988, Hogeweg and Hesper 1990, DeAngelis and Gross 1992, Bart 1995). Huisman and Weissing (1999) discussed theories on competition within communities and the (limited) predictive capabilities of such mechanistic competition models. Huisman and Weissing (1999) showed the importance of chaotic fluctuations in species abundances. Their model was able to reproduce the overall behavior of the observed process quite well, suggesting that a stochastic approach can help provide insight into the functioning of community dynamics.

The importance of disturbance in community dynamics has been recognized only relatively recently (White and Pickett 1985, Chesson and Huntly 1997, White and Jentsch 2001). Chaotic disturbances interrupt succession processes in plant communities, often making it difficult to understand how they function (Hobbs 1999; also see Chapter 5). Consequently, the endpoint of a successional process is not a predictable, uniform outcome; on the contrary, several community states are possible, depending on biotic and abiotic conditions (Noble and Slatyer 1980, Hobbs 1994). These multiple community states may be stable for long periods of time, and distinct thresholds may exist that limit the transition from one state to another. Anand and Orlóci (1997) pointed out that community dynamics in the early state are linear but then break down into a "noisy" state. They showed that adding small amounts of quasi-random perturbation to a simple linear model can turn the normally well-behaved dynamics into a deterministically chaotic one.

A key to improving the predictive capabilities of such stochastic models might be to switch from dynamics describing the flux of individuals to the process of community assembly involving a series of filters that sieve species out of the regional pool (Díaz et al. 1988, Díaz et al. 1999; see also Part II, "Ecological Filters as a Form of Assembly Rule"). Additionally, there is a strong move away from working at the species level toward using functional groups of species or guilds (Grime 1979, Hobbs 1997, Wilson 1999). However, in moving from population dynamics up to the level of the community, there is a risk of generating dynamic systems too complex to provide insight into important community processes at the coarse scale of a landscape. Law (1999) suggested that it could be appropriate to think eventually in terms of a filtering process with the state variable $P(C_i,t)$ being the probability that the community is in state C_i at time t. The probability per unit time of moving from state C_i to state C_j, $w(C_j|C_i)$, can be seen as a stochastic Markov process depending on the probability per unit time that each species will arrive at the site and the set of resident species that results from each arrival.

In this chapter, I discuss the combination of a deterministic model and a stochastic model of community assembly. In the model, topography (slope angle) and distance to roads are used as stochastic filters. As an example of how deterministic and stochastic processes might interact in community assembly, the model has been applied to the plant community dynamics of the Izta-Popo National Park, Mexico.

Methods

Close to Mexico City in the southernmost stretch of the Sierra Nevada, the snowcapped volcanoes of Iztaccíhuatl and Popocatépetl rise to elevations of well over 5400 m and form the backbone of what is commonly known as Izta-Popo National Park. The park is located between 18°59′00″ N and 19°16′25″ N latitude and 98°34′54″ W and 98° 42′08″ W longitude. Extensive forestry in the region mainly involves cutting and collecting wood, cutting tree branches for torches or for utensils, collecting mushrooms, and hunting. Although these (often clandestine) activities seem to be small-scale, their adverse effects on the forest have been substantial (García et al. 1992).

The boundaries of the Izta-Popo National Park follow the 3600-m elevation topographic line (Figure 11.1). The area of the park is 25,679 hectares (ha). From the establishment of the park in 1935 until 1948, the 3000-m topographic line delineated the park boundaries, which encompassed 59,913 ha. Proposals have been made recently to lower the boundaries of the park to 2500 m, which would encompass 118,792 ha (SEDESOL et al. 1992). A management plan has been developed that consists of conservation and management, public use and recreation, and administration (Trigo et al. 1995). As part of this plan, an analysis has been carried out for designing buffer zones around protected areas within the park (Ridgley and Heil 1998).

Vegetation and Classification

The vegetation in the Izta-Popo National Park varies greatly with elevational and microclimatic changes across the slopes. The tree line at this latitude is around 4000 m (Velazques et al. 2001). The vegetation of the area can be characterized roughly into the following types:

- *Abies* forest: High evergreen forest with *Abies religiosa* as the dominant tree species, although other conifer tree species can be present.
- *Pinus* forest: More open than *Abies*; evergreen forest mostly with *Pinus hartwegii* as the dominant species.
- Mixed coniferous forest with *Abies* and *Pinus*.
- Open coniferous forest and secondary communities that develop where original forest vegetation is destroyed by cutting, grazing, or fire.
- Seminatural grasslands: Homogeneous communities of bushy graminae of 60–120 cm height at 2500–4300 m, and natural alpine grasslands, paramos, or alpine zacatonals (tropical bunch tussock grasslands from 4300 to 4500 m elevation).
- Bare soil or rock: Ground covered with hardly any vegetation.

Satellite images were used to create land-cover maps through a supervised classification (Trigo et al. 2003). Landsat Thematic Mapper (TM) images of the Izta-Popo area from January 1986 and February 1997 were processed according to standard procedures (Jensen 1986, Belward and Valenzuela 1991, Matson and Ustin 1991, Lilesand and Kiefer 1994). The ground resolutions of the Landsat TM images are 30 by 30 m. For practical reasons of data analysis and processing, and also to facilitate overlaying with existing maps such as digital road maps and digital elevation models (DEMs), the TM images were resampled to a 50-m by 50-m resolution. This resampling hardly affects the classification of the TM images, because the

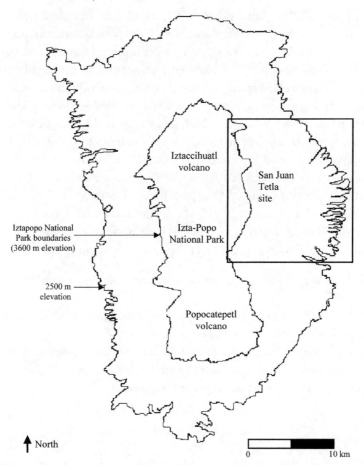

FIGURE 11.1 Location of the Izta-Popo National Park, its environment, and the modeling site San Juan Tetla. The park lies approximately 50 km southeast of Mexico City.

classification is only to the level of dominant plant communities (as stated earlier). For the supervised classification of the TM images, ground control points were assessed at 223 different positions in the whole area of the Izta-Popo National Park and buffer zones. For all ground control points, the geo-referenced position was obtained with a global positioning system (GPS), and annotations were made about the dominant vegetation.

Modeling

I used a matrix to summarize community transitions. Matrix modeling is a widely employed method in which the transitions among communities are used to build a simplified description of community dynamics (Van Hulst

1979, Caswell 1989, Orlóci et al. 1993). If we consider plant community dynamics as a cycle of specific phases, then the various plant communities are part of a permanent cycle of a species pool comprising the different community types (Rosenzweig 1995). Each community can be invaded by species from one or more of the other communities. Consequently, the union of these communities could constitute a permanent cycle of this species pool.

Model construction requires both knowledge of the analytical tools that eventually will be applied to the model and attention to the data and the manipulations needed to transform the data to the required form. The land cover of parts of the maps of 1986 and 1997 was used for cross-tabulation to quantify land-cover changes. One part of the map—the San Juan Tetla site (see Figure 11.1)—was withheld to validate the model. On the basis of a cross-tabulation, a matrix was created as the basis for the computation of the community dynamics (Figure 11.2). This matrix was constructed as follows.

1. The projection interval has been determined to be from 1986 to 1997.
2. Depending on elevation, each community can shift with a certain probability into one of the other communities.
3. Each probability is labeled by a coefficient; the coefficient a_{ij} on the transition from n_i to n_j gives the total cover of community i at time t + 1 per community in stage j at time t. So,

$$n_i (t + 1) = \sum_{j=1}^{S} a_{ij} \, n_j(t)$$

These coefficients are the transition probabilities taken from the cross-tabulation of the maps of Izta-Popo National Park and buffer zones (see Figure 11.2). The equation above can be rewritten as the matrix projection equation

$$n_i (t + 1) = An_i(t)$$

where n_i is the cover of plant community i and A is the projection matrix of the plant community.

These transitions were implemented in the PCRaster software (Van Deursen 1995). The architecture of the system permits the integration of environmental modeling functions, such as transitions, with classical Geographical Information System (GIS) functions (for example, Van Deursen and Heil 1993, Heil and Van Deursen 1996). For the numerical implementation of the simulation model, the calculated map with the actual community was classified according to the dominance of the community composition.

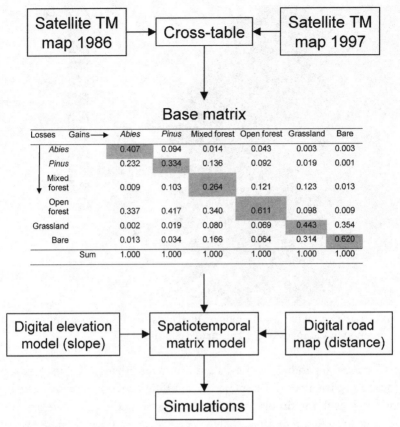

FIGURE 11.2 Diagram of the modeling procedure. The cross-table of the two maps is used as input for the base matrix, which is used in the Markov model. The digital elevation model and road map are used as distance operators for a disturbance filter. The base matrix shows transitions between land-cover classes from 1986 to 1997. Losses show the ratio of how much of a certain community was lost to other communities, gains show how much was gained from other communities, and the diagonal (gray) shows the proportion that remained the same.

Having used most of the Izta-Popo area for developing the transition matrix, model performance was evaluated using the San Juan Tetla area (see Figure 11.1). Starting with land cover from the 1986 map, simulation results were compared against the 1997 map of the same area. The comparison was executed using a geostatistical test according to the Kappa Index of Agreement (KIA; Burrough 1986, Eastman 1993). Next, the model was used for different simulations in which the impact from disturbance varied. Disturbance does not take place homogeneously over the area but is merely a fil-

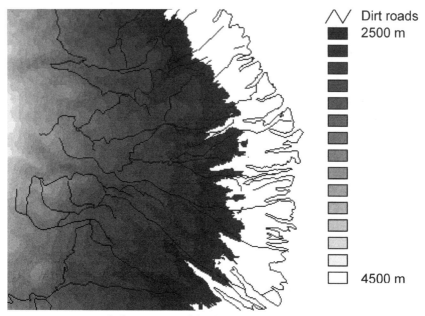

FIGURE 11.3 Digital elevation model (DEM) based on the topographical map and the map of the roads of the San Juan Tetla site.

tering process depending on the distance from disturbance vectors. Spatially localized disturbances of the plant communities by human activities were considered as the main disturbances. It was assumed that people are confined within a certain distance that they are willing to travel to obtain wood. It was also assumed that steep slopes would hinder people from crossing the landscape. Thus, both distance from roads and steepness of slope determine the intensity of disturbance; that is, the farther away from a road and the steeper a slope, the fewer the disturbances. The two disturbance factors were obtained from a digitized road map and a DEM, respectively (Figure 11.3). Although different scenarios can be applied, it was assumed that people's behavior does not vary with resource scarcity, resource monetary value, or human population size. However, data availability constrained our choice of criteria to digitally available sources. Maps of the two criteria were used in a spatial modeling environment to calculate the probability of land-cover changes. Transitions from the matrix were converted to annual transition probabilities, so calculations could be executed in time steps of 1 year. Simulation runs were executed that differed in the impact based on different maximum distance and maximum slope as correlates for disturbance. After each run, the resulting map was compared to the San Juan Tetla land-cover map of 1997. Using this approach, a kind of optimization could be performed, from which conclusions were drawn as to how far people are will-

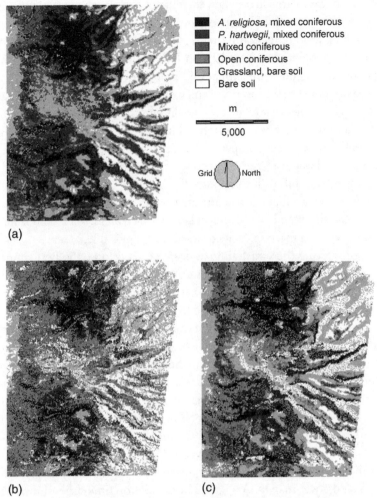

A. religiosa, mixed coniferous
P. hartwegii, mixed coniferous
Mixed coniferous
Open coniferous
Grassland, bare soil
Bare soil

m

5,000

Grid | North

(a)

(b)   (c)

FIGURE 11.4 Maps of the land cover in (a) 1986 and (b) 1997, and (c) a map resulting from the simulation. The different gray tones represent different land-cover types dominated by *Abies religiosa*, *Pinus hartwegii*, mixed coniferous forest of *Abies* and *Pinus*, open secondary forest, grassland, or bare soil.

ing to go and what maximum slope people are willing to tolerate to get their resources from the forest.

Results

Figure 11.4 shows the land-cover maps of the validation site San Juan Tetla in 1986 and 1997. The result of the cross-tabulation between the two maps of the Izta-Popo area, excluding the San Juan Tetla site, is shown in the Figure 11.2

matrix. In this matrix, losses show the ratio of how much of a certain community transitioned to a different community, gains show the ratio of how much was gained from other communities, and the diagonal shows the proportion that remained constant from 1986 to 1997. The numbers in the vertical and horizontal directions are similar and reflect the different communities; for example, *Abies* lost 0.23 to *Pinus* and gained 0.094 from *Pinus*. The matrix in Figure 11.2 shows that relatively large amounts of the *Abies* forest, *Pinus* forest, and mixed forest were transformed into open forest (about 33 percent, 42 percent, and 34 percent, respectively). Open forest mainly remained open forest; grassland was to a large extent converted into bare soil; and bare soil mostly remained bare soil, with some transitioning back into grassland. From these results, it appears that because the area under pressure continues to increase every year, the attendant opening of large forested areas, soil modification, and ensuing erosion make forest recovery almost impossible.

Considering the constraints of distance and slope, the best simulation results are shown in Figure 11.4. From the series of runs, it can be concluded that the maximum distance people are willing to go is 950 m, and the maximum slope that people are willing to climb is 35°.

Correlation between simulation results and the 1997 land-cover map was significant ($p < 0.05$) but rather low (KIA = 0.45). However, the overall changes of the plant communities in these simulations into other plant communities are in agreement and are in the same direction as the actual changes found in the land-cover maps from 1986 to 1997. In addition, when the *Abies*, *Pinus*, and mixed forest classes are combined into one class, the simulated output matches the 1997 land-cover map much better (KIA = 0.71). Also, both the 1997 land-cover map and the simulated map (of either combined or separate forest classes) were rather highly correlated with the 1986 land-cover map (KIA ≈ 0.6). The simulation time series shows that the changes are toward a dominance of open forest and grassland (Figure 11.5a). Even when the results of a long-term (50-year) simulation that included the simulated effects related to roads and slope were compared with the land-cover map of 1997, the spatial pattern was still strongly related to the pattern of the land-cover map of 1997 (KIA = 0.45).

When only the basic transitions in the simulation model are applied—that is, without the simulated effects of distance to roads and slope—the similarity with the land-cover map of 1997 became nonsignificant (KIA < 0.25), and the pattern of the vegetation changed toward uniform end states, strongly dominated by continuous patches of open forest, bare soil, and grassland (Figure 11.5b).

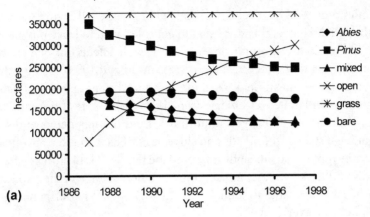

(a)

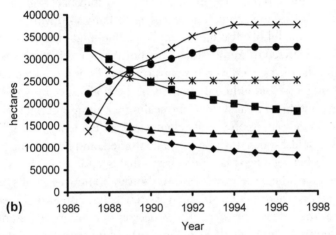

(b)

FIGURE 11.5 Time series of the simulations of the model including (a) and excluding (mean field, b) a spatially explicit disturbance filter caused by human impact. In (a), the effects of human impacts are restricted to distances within 950 m from a road and with slopes of less than 35‰. In (b), human impact is evenly distributed over the area.

Discussion and Conclusions

The basic difference between the modeling approach used in this study, being a nonmechanistic empirical model based on chance processes, and the mechanistic nonstochastic models on which ecological theory is mostly based, is that our purpose was not to uncover underlying mechanisms of community assembly but to analyze the adverse effects on the community assembly process of relatively small-scale filter processes that may seem to be insubstantial. Hence, this study had to consider what to leave out for modeling at the plant community level. Clearly, scale affects our perception of community pattern, process, and mechanism (Anand 1994). Using only the nonconstrained, mean-field transitions produced much poorer agreement with observed land-cover change from 1986 to 1997, resulting in a land-cover pattern that converged uniformly toward homogeneous end states. Including distance from roads and slope as correlates to intensity of disturbances by humans significantly improved the results. It is likely that other factors also influence human impacts on vegetation change in the park, and if data become available to model those factors, model performance would likely become even better.

Although the simulation model could not be validated firmly (see Jörgensen 1988), the model is rather persuasive and transparent in its simplicity, and results of the simulation model agree fairly well with the data. Obviously, in a model like this one, complicated population-dynamic processes are summarized in simple stochastic transitions and filters are based on subjective decisions. However, the construction of this model represents significant progress, and although analysis of the stochastic processes still leaves many questions unanswered, the scaling seems appropriate for our purpose: to model plant community changes at the landscape level. With this model, a landscape as a whole could be studied, and interconnections and feedbacks between two main processes—that is, community transitions and disturbance—became visible. More thorough knowledge of species dynamic processes at a stand scale, such as growth and dispersal of species, will uncover the underlying ecological processes of community assembly, but such details are not necessary for understanding transitions between community types.

Community assembly determines what happens where, given environmental and biotic constraints (filters). Restoration defines a goal and a time frame for management measures, and at this point any threshold in the system becomes important (also see Chapter 3). If we want to know the outcome of community restoration measures, we need to have a model that incorporates constraints on community membership and reasons for changes. In other words, we need to be able to make predictions to the greatest extent possible. The probabilistic Markov models of succession have been criticized for being too rigid (Usher 1981, Hobbs 1983). In particular, the requirement that transition probabilities be specified as a constant value over time conflict with the stochastic reality of year-to-year changes in the dynamics of natural systems. Anand and Orlóci (1997) suggested that the global pattern of state transition from determinism to apparent "noise" could be explained, in the long term, by a specific combination of deterministic and random processes; that is, the stationary Markov chain plus quasi-random perturbation. Anand and Heil (2000) used this approach in an analysis of a recovery process of heathland community. Petraitis and Latham (1999) emphasized the importance of scale and disturbance in the origin and persistence of alternative community assemblages. They suggested that alternative assemblages could arise in two ways: one through the natural process of succession and the other through a major disturbance event. Our results nicely distinguish these two cases.

This chapter showed that a combination of a deterministic and a stochastic Markov model can predict observed changes fairly well. That patterns and processes are scale-dependent is certainly not a new idea in the field of community ecology. The nature of this scale dependence is poorly understood, however. Here, I used empirical data to demonstrate that a process at a finer spatial scale (that is, constraints on human impacts) can help us understand the dynamics at a higher spatial scale (that is, vegetation change in the broader landscape).

To use the model for restoration purposes, we need to recouple biotic and abiotic influences on community development and define stepwise goals that encapsulate critical points in the development process (also see Chapter 4). Depending on the restoration goals, possible outcomes can be simulated by adjusting the transition probabilities according to the effects of specific restoration measures (for example, passive restoration, interseeding, and so forth; Samuels and Lockwood 2002). Future work will be directed toward the application of this model to analyze the potential impact of ecological restoration in the Izta-Popo area.

Acknowledgments

I am grateful to Betty Verduyn (department of plant ecology), who processed the data of the original TM images. I am also grateful to Roland Bobbink and Peter White, who reviewed an earlier version of the manuscript. This project is part of the Izta-Popo project of the *El Hombre y su Ambiente* (Humans and Their Environment) department of Universidad Autonoma Metropolitana-Xochimilco (UAM-X), Mexico City, Mexico.

References

Anand, M. 1994. Pattern, process and mechanism: fundamentals of scientific inquiry applied to vegetation science. *Coenoses* 9:81–92.

Anand, M.; and Heil, G. W. 2000. Analysis of a recovery process: Dwingelose Heide revisited. *Community Ecology* 1(1): 65–72.

Anand, M.; and Orlóci, L. 1997. Chaotic dynamics in a multispecies community. *Ecological and Environmental Statistics* 4:337–344.

Bart, J. 1995. Acceptance criteria for using individual-based models to make management decisions. *Ecological Applications* 5:411–420.

Belward, A. S.; and Valenzuela, C. R., eds. 1991. *Remote Sensing and Geographical Information Systems for Resource Management in Developing Countries.*Vol. 1, *Remote Sensing.* Dordrecht, The Netherlands: Kluwer.

Burrough, P. A. 1986. *Principles of Geographical Systems for Land Resources Assessment.* Oxford: Clarendon Press.

Caswell, H. 1989. *Matrix Population Models: Construction, Analysis, and Interpretation.* Sunderland, Mass.: Sinauer Associates.

Chesson, P.; and Huntly, N. 1997. The roles of harsh and fluctuating conditions in the dynamics of ecological communities. *American Naturalist* 150(5): 520–534.

Clements, F. E. 1916. *Plant Succession.* Publication No. 242. Washington, D.C.: Carnegie Institution.

DeAngelis, D. L.; and Gross, L. J. 1992. *Individual-Based Models and Approaches in Ecology Populations, Communities and Ecosystems.* London: Chapman and Hall.

Díaz, S.; Cabido, M.; and Casanoves, F. 1998. Plant functional traits and environmental filters at a regional scale. *Journal of Vegetation Science* 9(1): 113–122.

Díaz, S.; Cabido, M.; and Casanoves, F. 1999. Plant functional traits, ecosystem structure and land-use history along a climatic gradient in central-western Argentina. *Journal of Vegetation Science* 10:651–660.

Eastman, J. R. 1993. *Idrisi Manual.* Worcester, Mass.: Clark University, Graduate School of Geography.

García, M.; Lopez-Paniagua, J.; and Rosas, M. 1992. Socioeconomicos. In *Programa de Manejo para el Parque Nacional Ixtaccìhuatl-Popocatepetl,* SEDESOL-Banco Mundial-UAM-X, 99–119. Mexico City: Secretaria de Desarrollo Social.

Grime, J. P. 1979. *Plant Strategies and Vegetation.* Chichester, England: John Wiley.

Heil, G. W.; and Van Deursen, W. P. A. 1996. Searching for patterns and processes: modelling of vegetation dynamics with Geographical Information Systems and Remote Sensing. *Acta Botanica Neerlandica* 45:543–556.

Hobbs, R. J. 1983. Markov models in the study of post-fire succession in heathland communities. *Vegetatio* 56:17–30.

Hobbs, R. J. 1994. Dynamics of vegetation mosaics: can we predict responses to global change? *Ecoscience* 1:346–356.

Hobbs, R. J. 1997. Can we use plant functional types to describe and predict responses to environmental change? In *Plant Functional Types: Their Relevance to Ecosystem Properties and Global Change,* ed. T. M. Smith, H. H. Shugart, and F. I. Woodward, 66–91. Cambridge: Cambridge University Press.

Hobbs, R. J. 1999. Restoration of disturbed ecosystems. In: Ecosystems of the World 16. *Ecosystems of Disturbed Ground,* ed. L. Walker, 673–687. Amsterdam: Elsevier.

Hogeweg, P.; and Hesper, B. 1990. Individual-oriented modelling in ecology. *Mathematics and Computer Modelling* 13:83–90.

Huisman, J.; and Weissing, F. J. 1999. Biodiversity of plankton by species oscillations and chaos. *Nature* 402:407–410.

Huston, M.; DeAngelis, D. L.; and Post, W. 1988. New computer models unify ecological theory. *BioScience* 38:682–691.

Jensen, J. R. 1986. *Introductory Digital Image Processing*. Englewood Cliffs, N.J.: Prentice-Hall.

Jörgensen, S. E. 1988. *Fundamentals of Ecological Modelling. Developments in Environmental Modelling*, vol. 9. Amsterdam: Elsevier.

Law, R. 1999. Theoretical aspects of community assembly. In *Advanced Ecological Theory: Principles and Applications*, ed. J. McGlade, 143–171. Malden, Mass.: Blackwell Science.

Lilesand, T. M.; and Kiefer, R. W. 1994. *Remote Sensing and Image Interpretation*. New York: John Wiley and Sons.

Matson, P. A.; and Ustin, S. L. 1991. The future of remote sensing in ecological studies. *Ecology* 72:1917-1945.

Noble, I. R.; and Slatyer, R. O. 1980. The use of vital attributes to predict successional changes in plant communities subject to recurrent disturbances. *Vegetatio* 43:5–21.

Orlóci, L.; Anand, M.; and He, X. S. 1993. Markov chain: a realistic model for temporal coenosere? *Biométrie-Proximétrie* 33:7–26.

Petraitis, P. S.; and Latham, R. E. 1999. The importance of scale in testing the origins of alternative community states. *Ecology* 80:429–442.

Ridgley, M.; and Heil, G. W. 1998. Multicriterion planning of protected-area buffer zone: an application to Mexicoís Izta-Popo National Park. In *Multicriteria Evaluation in Land-Use Management*, ed. E. Beinat and P. Nijkamp, 293–311. Dordrecht, The Netherlands: Kluwer.

Rosenzweig, M. L. 1995. *Species Diversity in Space and Time*. Cambridge: Cambridge University Press.

Samuels, C. L.; and Lockwood, J. L. 2002. Weeding out surprises: incorporating uncertainty into restoration models. *Ecological Restoration* 20(4): 262–269.

SEDESOL-Banco Mundial-UAM-X. 1992. *Programa de Manejo para el Parque Nacional Iztaccihuatl-Popocatepetl*. Mexico City: Secretaria de Desarrollo Social.

Trigo, N.; Bobbink, R.; and Heil, G. W. 1995. A unified framework for ecological monitoring of the Izta-Popo National Park, Mexico. In *Ecosystem Monitoring and Protected Areas*, ed. T. B. Herman, S. Bondrup-Nielsen, J. H. M. Willison, and N. W. P. Munro, 575–580. Wolfville, Nova Scotia: Science and Management of Protected Areas Association.

Trigo, N.; Heil, G. W.; Bobbink, R.; and Verduyn, B. 2003. Classification and mapping of the plant communities using field observations and remote sensing. In *Ecology and Man in Mexicoís Central Volcanoes Area*, ed. G. W. Heil and R. Bobbink. Dordrecht, The Netherlands: Kluwer.

Usher, M. B. 1981. Modelling ecological succession, with particular reference to Markovian models. *Vegetatio* 46:11–18.

Van der Maarel, E.; and Sykes, F. S. 1993. Small-scale plant species turnover in a limestone grassland: the carousel model and some comments on the niche concept. *Journal of Vegetation Science* 4:179–1988.

Van Deursen, W. P. A. 1995. Geographical Information Systems and dynamic models. Ph.D. diss., Department of Physical Geography, University of Utrecht.

Van Deursen, W. P. A.; and Heil, G. W. 1993. Analysis of heathland dynamics using a spatial distributed GIS model. *Scripta Geobotanica* 21:17–27.

Van Hulst, R. 1979. On the dynamics of vegetation: Markov chains as models of succession. *Vegetatio* 40:3–14.

Velazques, A.; Romero, F. J.; Rangel-Cordero, H.; and Heil, G. W. 2001. Effects of landscape changes on mammalian assemblages at Izta-Popo Volcanoes, Mexico. *Biodiversity and Conservation* 10:1059–1075.

White, P. S.; and Jentsch, A. 2001. The search for generality in studies of disturbance and ecosystem dynamics. *Annals of Botany* 62:399–450.

White, P. S.; and Pickett, S. T. A. 1985. Natural disturbance and patch dynamics, an introduction. In *The Ecology of Natural Disturbance and Patch Dynamics*, ed. S. T. A. Pickett and P. S. White, 3–13. New York: Academic Press.

Wilson, J. B. 1999. Assembly rules in plant communities. In *Ecological Assembly: Advances, Perspectives Retreats*, ed. E. Weiher and P. Keddy, 130–164. Cambridge: Cambridge University Press.

Chapter 12

Application of Stable Nitrogen Isotopes to Investigate Food-Web Development in Regenerating Ecosystems

JAN ROTHE AND GERD GLEIXNER

Stable isotope analysis has become an essential tool in ecology and environmental science to investigate the cycling of carbon and nitrogen at various spatial and temporal scales in order to understand the dynamics of organic matter (Lajtha and Michener 1994). The path organic matter takes through an ecosystem is defined by the ecosystem structure. Food webs can be regarded as the biotic backbone of the system, since they connect all organisms by trophic relationships. The food-web structure reflects energy flow, which is a major driving force of ecosystem development. Consequently, structural changes in food webs (for example, due to species loss or invasion) influence the pattern of organic matter and energy transfer and hence affect both the isotopic composition of single food-web constituents and the general isotopic patterns within the food web.

Recently, Vander Zanden et al. (1999) and Stapp et al. (1999) demonstrated that natural and anthropogenic disturbances cause deviations from system-specific isotope patterns. Therefore, improvement of ecosystem health during recovery from disturbances should also be evident in the isotopic picture. However, so far, this application has not been specifically investigated for regenerating ecosystems.

Ecosystem regeneration after disturbance constitutes the reestablishment of abiotic and biotic system components (structural recovery) and of their interactions (functional recovery). At each stage in the transition from a disturbed to an intact, healthy state (Hobbs and Norton 1996), the rules underlying community assembly ensure that the present ensemble of system components is able to maintain vital ecosystem functions, such as primary production, decomposition, and remineralization of organic matter. If these functions are not maintained, the system does not recover. The order in

245

which species assemble is governed by species traits and by the multifactorial complex of environmental conditions acting as a dynamic filter (see Chapter 6). Stepwise assembly requires dynamic adaptation of the food web in terms of structure and functioning and leads to increasing complexity and stability of the trophic network. For each trophic link, both its actual interaction strength and its relative stability contribute to the isotope signal built up in the consumer. Therefore, in stabilized trophic relationships, the isotopic difference between consecutive trophic levels should be rather constant. However, the species involved might reveal an isotope shift related to isotopic changes of their food sources during ecosystem regeneration.

In this study, we used stable nitrogen isotopes to describe the progress in ecosystem development after disturbance, which is a key issue of restoration ecology (Hobbs and Norton 1996). Based on a few macroinvertebrate species, we monitored temporal changes of the food-web structure at four differently degraded grassland sites. We also considered functional aspects related to food-web complexity and stability. Because the systems investigated are far too complex to uncover the causalities between changing isotope patterns and regeneration processes from a few analyses, our study is a first attempt to find descriptive parameters that promise to be useful in the context of community assembly and ecological restoration. Community ecologists and restoration ecologists need appropriate tools to find patterns relevant to understanding the factors and processes involved in community development and to detect rules that govern the assembly of the system (see Chapter 1). In this chapter, therefore, we address the following questions: Can stable isotopes

• Distinguish developmental states reached by differently disturbed ecosystems?
• Reflect changes in the food-web structure and dynamics of regenerating ecosystems?
• Provide information on food-web and ecosystem complexity, functionality, and stability?

Stable Isotope Method

The element nitrogen (N), which is a major component of organic matter, has two stable isotopes: ^{15}N and ^{14}N. Due to isotope discrimination in the attainment of equilibrium in physical reactions (thermodynamic isotope effects) and in biochemical reactions (kinetic isotope effects), the isotopic composition of organic matter depends on (1) the isotopic composition of the reactants, (2) the biochemical pathways involved, (3) the reaction kinetics, and (4) the physical and chemical conditions during organic matter synthesis, transformation, and degradation (Wada et al. 1995, Gleixner et al. 2001). The combination of all four factors leads to a dynamic stable isotope fingerprint in every biogenic material. Isotope ratios are expressed in conventional delta (δ) notation in parts per thousand (‰) relative to international standard materials:

$$\delta^{15}N\ (‰) = (R_{sample}/R_{standard} - 1) \times 1000\ ‰ \qquad (12.1)$$

where R is the ratio of the heavier to the lighter isotope (^{15}N:^{14}N) of a sample or a standard. International standards—for example, atmospheric N_2 (AIR) for N—are provided by the International Atomic Energy Agency (Coplen et al. 1992). Commonly, isotope ratios are determined using an isotope ratio mass spectrometer (IRMS) in combination with online and offline coupled sample preparation systems (Brand 1996).

Stable Isotopes in Food Webs

The relative abundance of stable isotopes in living organisms depends on the isotopic composition of their food sources and their internal fractionation. Generally, internal fractionation leads to an enrichment of the heavier isotope in consumers relative to their diet. For ^{15}N, the enrichment is caused by preferential excretion of ^{15}N-depleted metabolites such as ammonia. Regardless of the habitat, Minagawa and Wada (1984) reported an average ^{15}N enrichment of $3.4 \pm 1.1‰$ per trophic level. This difference in $\delta^{15}N$ values is also referred to as trophic level shift (TLS). As an empirical parameter, TLS is used to assign organisms to distinct trophic levels (Ponsard and Arditi 2000, Rothe and Gleixner 2000, Scheu and Falca 2000).

Generally, the ^{15}N shift of $3.4‰$ reflects conditions in which the consumers were able to equilibrate isotopically with their diet according to the thermodynamic and kinetic constraints governing isotope fractionation. In fact, the actual isotopic difference between consumer and diet often deviates from $3.4‰$ in either direction, depending on such factors as food quality, food quantity, age, and sex (Focken 2001) and also on the time of adaptation to the food source. Nevertheless, previous work has confirmed that stable N isotopes provide a time-integrated measure of food-web relationships because they indicate the trophic position of a consumer at a given time (Fantle et al. 1999, Stapp et al. 1999, Vander Zanden et al. 1999). The comparison of trophic positions of certain species among ecosystems,

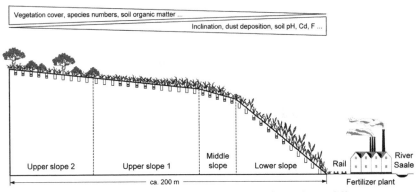

FIGURE 12.1 Steudnitz research area. Profile view along the pitfall trap transect, divided into four sections. Within each section, animal material from five pitfall traps (indicated as half-filled) was used to represent distinct segments of the transect (see text). General environmental gradients present at the beginning of ecosystem regeneration are indicated in the top panel.

however, requires that we take into account the system-specific isotope background determined by the isotopic composition of the organic matter input at the base of the food web (Ponsard and Arditi 2000). Moreover, food-web studies in terrestrial ecosystems need to consider the soil compartment, even if the main focus is on the aboveground food web (AFW). Although changes in soil properties during regeneration will have equivalent impact on the entire ecological community (Scheu et al. 1999), changes in biotic ecosystem components feed back to the soil (Scheu 1997). Thus, soil isotopic signatures may reflect the history of structural and functional ecosystem development (Bardgett et al. 1998, Bengtsson et al. 1998, Mikola et al. 2001).

Study Site

We studied the regeneration of a grassland ecosystem (Figure 12.1; for general site characteristics, see Heinrich 1984 and Chapter 13) that had been disturbed for 30 years by dust and gaseous emissions originating from an adjacent phosphate fertilizer plant (Metzner et al. 1997). Major parts of the lower slope were almost completely degraded (Heinrich et al. 2001). Restrictions in the cycling of organic matter caused by dissimilarities between primary production and decomposition hampered the reestablishment of ecosystem structure and functioning in the first years of regeneration.

Sampling and Sample Preparation

In 1990, the Institute of Ecology (University of Jena) installed a permanent transect of 40 pitfall traps over 200 m to monitor faunal development along the former pollution gradient. For the present study, we used data from 20 of these traps to define four distinct, differently impacted segments: (1) the most-disturbed, steep lower slope (LS) next to the polluter; (2) the flatter middle slope (MS) just below the edge of the valley shoulder; (3) the lower part of the slightly inclined upper slope (US1); and (4) the least-impacted, upper part of the upper slope already invaded by woody plants (US2). Within each segment, we pooled data from the five adjacent trap locations near the center of the segment (see Figure 12.1)

From 1991 to 1996 and in 2000, pitfall traps were emptied every 14 days to sample epigeic (surface) arthropods. In addition, at LS, US1, and US2, the fauna in the upper parts of the vegetation was sampled by sweep-netting. Animals were preserved in 70 percent ethyl alcohol. After taxonomic identification, animals caught on three successive sampling dates were pooled by species and site to obtain sufficient amounts of animal dry mass. Each pooled sample contained 20 to 50 individuals.

For isotope analysis, we selected 14 species from the most dominant species to represent different taxonomic categories, trophic levels, and functional groups. Taxa selected included woodlice, which are important macrodecomposers sensitive to potentially harmful elements such as zinc, cadmium, copper, and lead (Beyer and Anderson 1985, Hopkin 1990, Jones and Hopkin 1998); their recolonization is thought to have a pronounced effect on quantity and quality of soil organic matter and thus on ecosystem development in general. In addition to woodlice, several phytophagous and zoophagous species of both, beetles and bugs, were selected to include the trophic levels of herbivores and carnivores. The final selection (Table 12.1) comprised samples of spring, summer (in 2000 only), and autumn catches, respectively, of two woodlouse (1991–1996 and 2000), seven beetle (1991, 1996, and 2000), and five bug species (1992, 1994–1996, and 2000, at US1 and US2 only). Nomenclature of the species investigated follows Gruner (1966) for woodlice, Köhler and Klausnitzer (1998) for beetles, and Wagner (1967) for bugs.

In 1994, 1998, and 2000, around each pitfall trap, we took 10 soil cores from the A-horizon (0–10 cm) for a pooled sample. Soil was air dried, sieved to < 2 mm, cleared of residual plant material, and stored in paper bags. The five pooled samples from each of the four sites along the slope were analyzed separately and the data were pooled again. Also, in 2000, separate samples of the most abundant plant species and pooled litter samples were taken

TABLE 12.1

Animal Species Analyzed from the Steudnitz Research Area, Their
Taxonomic Relationship, Main Ecological Function, Mode of Nutrition,
and Locations of Sampling; Sampled Between 1991 and 1996 and in 2000

Animal group / Species	Order, class / Family	Main function[1]	Nutrition[1]	Slope site[2]
Woodlice	Isopoda, Crustacea			
Armadillidium vulgare	Armadillidiidae	Decomposer	Saprophagous[3]	1–4
Trachelipus rathkei	Trachelipidae	Decomposer	Saprophagous[3]	1–4
Beetles	Coleoptera, Insecta			
Agriotes murina	Elateridae	Leaf chewer	Phyto-/saprophagous	1–4
Otiorhynchus raucus	Curculionidae	Leaf chewer	Phytophagous	1–4
Amara aulica	Carabidae	Predator	Zoo-/phytophagous	1–4
Anisodactylus binotatus	Carabidae	Predator	Zoophagous	1–4
Calathus melanocephalus	Carabidae	Predator	Zoophagous	1–4
Harpalus affinis	Carabidae	Predator	Zoo-/phytophagous	1
Poecilus cupreus	Carabidae	Predator	Zoophagous	1–4
Bugs	Heteroptera, Insects			
Amblytylus nasutus	Miridae	Plant-sap sucker	Phytophagous	1
Myrmus miriformis	Coreoidea	Plant-sap sucker	Phytophagous	1, 3, 4
Notostira elongata	Miridae	Plant-sap sucker	Phytophagous	1, 3, 4
Trigonotylus coelestialium	Miridae	Plant-sap sucker	Phytophagous	1
Nabis cf. brevis	Nabidae	Predator	Zoophagous	1, 3, 4

[1]Of the adult (in the case of beetles).
[2]1 = lower slope (LS), 2 = middle slope (MS), 3 = lower part of the upper slope (US1), 4 = upper part of the upper slope (US2).
[3]Including necrophagy and coprophagy; that is, all types of dead organic material.

at the four sites. Plants were divided into shoots and roots, dried at 35°C, and stored in paper bags.

Isotope Analysis

Prior to isotope analysis, ethyl alcohol was removed from the animal material by decanting the liquid and volatilizing the remnants in a fume hood. Plant, animal, and soil samples were freeze dried, ground, and stored in airtight glass bottles at −10°C. Samples equivalent to 150 µg N were weighed into tin capsules and combusted in an EA 1110 Elemental Analyzer (ThermoQuest, 20090 Rodano, Italy). The resulting N_2 gas was analyzed for ^{15}N content in a DeltaPlusXL isotope ratio mass spectrometer (Finnigan MAT, 28127 Bremen, Germany). The analytical precision was ± 0.2‰. Working standards were acetanilide and caffeine, calibrated against international standard IAEA-N1. Accuracy and repeatability of measurements were ensured according to Werner and Brand (2001). Isotope ratios were expressed as $\delta^{15}N$ values (see Equation 12.1).

Calculations and Statistics

In 2000, the differences of $\delta^{15}N$ values between the analyzed animal species and their likely food sources were calculated separately for all possible combinations at each site, considering trophic links between carnivores and herbivores/detritivores, herbivores and plant shoots, detritivores and plant shoots, and detritivores and litter. Intratrophic-level predation was excluded. The resulting data were divided into $\delta^{15}N$ classes of 1‰ width.

We used the mean value of the frequency distribution of classified $\delta^{15}N$ differences between trophic links to evaluate the developmental state of the food web after 10 years of regeneration. This analysis is based on the idea that the stability and functionality of food webs are accomplished by the whole array of trophic links. Since the theoretical isotopic equilibration of a consumer with its diet is reflected by a shift in the $\delta^{15}N$ value of 3.4 ± 1.1‰ (Minagawa and Wada 1984), we hypothesize that if the observed food webs are developmentally advanced, they will exhibit a high proportion of stable trophic links with mean ^{15}N shifts of between 3‰ and 4‰. To test this hypothesis, the distribution data were subjected to bootstrap analysis (1000 replications; S-PLUS 6.0, Insightful Corp. 2001) to test the reliability of the observed mean. The bias between observed and empirical means was below 0.002‰. In addition, we tested if the distribution mean values observed in 2000 resulted from developmental processes; that is, we calculated the distribution mean values in 1991, 1996, and 2000 for the trophic interactions of the predators. Linear regression was applied to analyze temporal trends (SPSS for Windows 11.0, SPSS Inc. 2001).

In regard to the temporal ^{15}N dynamics from 1991 to 2000, the isotopic difference between soil and species representing the AFW was used to characterize changes in the coupling between aboveground and belowground parts of the ecosystem with ongoing regeneration. This difference should decrease the more AFW and soil food web (SFW) become structurally and functionally connected. Depending on the degree of functional integration, the isotopic signatures of the soil and of the aboveground components should show similar dynamics. Temporal trends and differences in the dynamics (that is, the degree of parallelism) of the measured $\delta^{15}N$ values (annual means of two or three seasonal records) were tested by generalized linear models (GLM) based on univariate analysis of variance (ANOVA). The interaction between species and time, that is, the duration of regeneration, was included (SPSS for Windows 11.0, SPSS Inc. 2001). Data sets of four beetle and three bug species were omitted because of incompleteness.

Results and Discussion

We will now discuss results from our analysis as they pertain to the developmental state of the food web, food-web structure, dynamics of the entire food web, stability of the food web, and development of soil relative to the aboveground food web.

Developmental State of the Food Web

In 2000, after 10 years of regeneration, the frequency distribution of $\delta^{15}N$ differences between consumer species and their likely diet varied clearly among the investigated sites (Figure 12.2). Distribution mean values (given as bootstrap means with lower and upper bounds of the 95 percent confidence interval) were highest at the least-disturbed sites US1 (3.65‰; 3.11–4.18; N = 41) and US2 (3.57‰; 3.22–3.93; N = 57), intermediate at the most-disturbed site LS (3.01‰; 2.53–3.46; N = 68), and lowest at intermediately disturbed site MS (2.13‰; 1.58–2.69; N = 30). The overall mean was 3.18‰ (N = 185). The means were statistically identical at US1 and US2, but the means found at LS and MS did not fall into the 95 percent confidence interval found at the other sites.

From 1991 to 1996, the mean ^{15}N shift between predators and their possible prey strongly increased at all four sites (Figure 12.3); thereafter, it remained constant. The linear trend was significant in the case of LS ($p = 0.01$) and US2 ($p = 0.02$). The increase in the isotopic shift between adjacent trophic levels agreed with our hypothesis that an advance in regeneration should be reflected in the stable isotope pattern of the assembling community. That ^{15}N shifts remained constant between 1996 and 2000 indicates that community food webs had already stabilized by 1996.

The mean of the frequency distribution of classified $\delta^{15}N$ differences between consumer species and their likely diet (see Figure 12.2) served as a descriptive variable for evaluating the developmental state of food webs. This parameter differentiates between sites along a pollution gradient representing different advanced stages in ecosystem regeneration. The site-specific ecosystem complexity that can be estimated from vegetation and macroinvertebrate data (Heinrich et al. 2001, Perner and Voigt, University of Jena unpublished) corresponds with the calculated distribution mean values. At the least-disturbed sites US1 and US2, the means of the ^{15}N shifts (3.65‰

FIGURE 12.2 Frequency distribution (class width: 1‰) of the isotopic difference ($\delta^{15}N$) between food-web components and their assumed food source at the four sites of the Steudnitz study area in 2000. Mean values of the distribution are indicated by arrows, the 95 percent confidence interval of the means by vertical dashed lines. Filling patterns illustrate the contribution of the three different types of consumers.

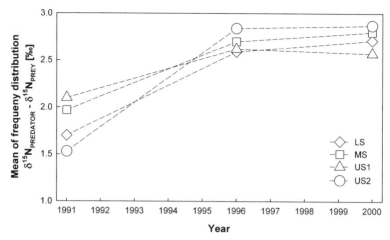

FIGURE 12.3 Temporal variation of the mean of $\delta^{15}N$ differences between predators and potential prey, spanning 10 years of ecosystem regeneration after disturbance. Calculations based on samples from seven beetle, five bug, and two woodlouse species at four differently impacted sites (LS: lower slope, MS: middle slope, US1: lower part of upper slope, US2: upper part of upper slope).

and 3.57‰, respectively) were closest to the expected shift of 3.4‰ (see Minagawa and Wada 1984). In a study of macroinvertebrates in the litter of undisturbed woodlands, based on a total of almost 300 samples of detritivorous and predatory species taken in three seasons at three sites, Ponsard and Arditi (2000) found an average ^{15}N shift between detritivores and litter (L-horizon) of 3.5 ± 1.2‰. Between predators and detritivores, the ^{15}N shift (3.6 ± 1.0‰) significantly differed from the 3.4‰ target value only in one of the nine cases. These findings corroborate our hypothesis that the distribution mean value should be close to 3.4‰ at the sites that are most advanced with respect to food-web regeneration.

At LS, the number of grass and herb species increased from 2 in 1991 to 50 in 2000 (Heinrich et al. 2001). This strong increase in plant diversity corresponds with a mean ^{15}N shift of 3.0‰, close to the expected shift of 3.4‰; that is, the community dynamics at LS led to a faster stabilization of trophic relationships compared to the dynamics at MS. At MS, biodiversity was lowest, as was the mean ^{15}N shift (2.1‰, see Figure 12.2). The vegetation was strongly dominated by the grass *Agropyron repens*. Due to its low palatability for macrodecomposers, the litter of *A. repens* was only partly degraded, and it accumulated to a thick layer on the soil surface (Heinrich 1984). The barrier of undecayed litter presumably caused a decoupling between soil and aboveground processes, which in turn reduced the diversity of trophic interactions and the speed of regeneration. As a result, the developmental state was less advanced at MS than at LS and was even less advanced than at US1 and US2.

Considering the correspondence between vegetation diversity and food-web linkage, in conjunction with the evidence that the mean trophic shifts for ^{15}N found in 2000 resulted from developmental processes (see Figure 12.3), the system at the upper slope (sites US1 and US2) can be regarded as recovered from disturbance. We hypothesize that community assembly has reached an "optimum state"insofar as the addition of new species would not necessarily improve functionality and stability of the food web but would only increase complexity by introducing further functional redundancy.

Food-Web Structure

Food-web dynamics over 10 years were reflected by the development of animal δ^{15}N values (Figure 12.4). At all four sites, the general trophic order of the selected species was identical throughout the time period. The woodlouse *Armadillidium vulgare* and the phytophagous beetle *Otiorhynchus raucus* were consumers at the base of the food web. The woodlouse *Trachelipus rathkei* was always about 2‰ enriched in ^{15}N compared with *A. vulgare* and represented a distinctly higher level within the trophic hierarchy. The next level was composed of predatory organisms (that is, the beetle *C. melanocephalus* and the bug *Nabis brevis*) but also included one beetle species (*Agrypnus murina*) assumed to be predominantly phytophagous. The plantsap-sucking bug *Notostira elongata* was found at US2 at the level of herbivores, but at US1 at the level of carnivores.

Our ^{15}N analysis detected differences in food-web structure of the four differently polluted sites (see Figure 12.4). Although stable-isotope analysis cannot provide a complete picture of the trophic links between individual species in food webs, this technique does allow a general and reliable evaluation of the trophic structure of a community (Ponsard and Arditi 2000, Scheu 2002). Moreover, ^{15}N signatures provide insight into the nutritional behavior of the species analyzed. For example, the δ^{15}N values of woodlice indicated a clear niche differentiation in food sources at all four sites, which was stable in time. Over 10 years, the ^{15}N signals of *A. vulgare* and *T. rathkei* differed consistently by 2‰, suggesting that *A. vulgare* fed mainly on fresh litter and plant biomass whereas *T. rathkei* used materials more enriched in ^{15}N, such as degraded litter, feces, and carcasses; that is, detritus including associated microbes. This difference is in accordance with field studies

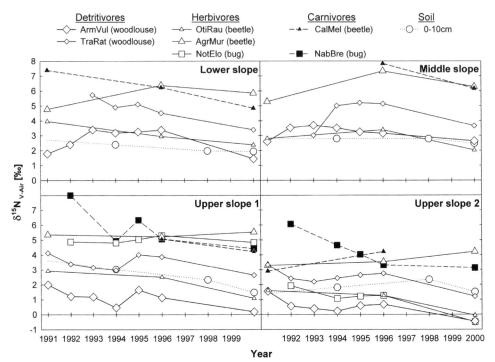

FIGURE 12.4 δ^{15}N values of the main consumer species and soil. δ^{15}N values (extrapolated from three data points) at the four sites of the Steudnitz research area between 1991 and 2000. If seasonal data of animal species were available, the arithmetic mean of two to three values was used (standard deviations always < 0.8‰). Data points for δ^{15}N values of soil are represented by circles, those of animals by polygons (open symbols for detritivorous and phytophagous species, solid symbols for zoophagous species). Dotted lines connect data points of soil measurements, solid lines data points of detritivorous and phytophagous species, dashed lines data points of zoophagous species. ArmVul: *Armadillidium vulgare*, TraRat: *Trachelipus rathkei*, AgrMur: *Agrypnus murina*, OtiRau: *Otiorhynchus raucus*, CalMel: *Calathus melanocephalus*, NotElo: *Notostira elongata*, NabBre: *Nabis brevis*.

(Schmidt et al. 1997, Scheu and Falca 2000) showing that differences in the feeding mode of taxonomically related species are traceable using [15]N:[14]N ratios. Hence, Scheu and Falca (2000) suggested distinguishing primary and secondary decomposers to describe the continuum of decomposer species according to the preferential food sources.

The high δ^{15}N values at the level of carnivores often found in the beetle *A. murina* contrasted with its a priori classification as phytophagous. We assume that apart from plant material, this large, polyphagous elaterid beetle used ^{15}N-enriched food sources extensively. Besides a possible switch to a saprophagous feeding mode (including dead animal biomass), zoophagy should be considered as an alternative nutritional strategy in A. *murina*, an observation that has rarely been reported. This hypothesis is corroborated by a C:N ratio (4.4 ± 0.2) only slightly above that of obligate-carnivorous beetles such as *Calathus melanocephalus*, *Anisodactylus binotatus*, and *Poecilus cupreus* (4.3 ± 0.2). Hence, we propose that in the Steudnitz system, adults of A. *murina*—contrary to the common classification (see Table 12.1)—were predominantly acting as predators and therefore showed higher $\delta^{15}N$ values than detritivores and herbivores (see Figure 12.4). In addition, depending on the time of sampling relative to the date of emergence, the isotopic signature of an adult insect can show an influence from the larval diet. Consequently, one should consider possible effects of shifting functional type, such that if holometabolic species are selected for food-web reconstruction, it would be better to select species such as O. *raucus* (herbivorous generalist) and C. *melanocephalus* (carnivorous generalist) that retain a common functional type of nutrition in larva and adult.

The differing ^{15}N signals of the plantsap-sucking bug N. *elongata* at US1 and US2 (see Figure 12.4) relate to variations in the isotopic composition of the accessible diet. In our case, N *elongata* preferred immature seeds of A. *repens* (W. Voigt, University of Jena, personal communication), which we found to be up to 4‰ enriched in ^{15}N compared with other abundant grasses in Steudnitz. Although A. *repens* was present at US1, it was lacking at US2. Thus, at US1, the presence of A. *repens* allowed the apparently anomalous ^{15}N enrichment observed in N. *elongata*, resulting in a carnivore-like ^{15}N signal, whereas at US2 this bug exploited food sources with lower ^{15}N content similar to those of the phytophagous beetle O. *raucus*. This example shows that feeding preferences can be useful in tracing food sources and helping to avoid misleading conclusions regarding trophic-level membership. Therefore, lacking knowledge of specific food preferences within functional groups, generalists should be selected to obtain an unbiased picture of the trophic structure of the studied community, because generalists integrate over a number of food sources.

Dynamics of the Entire Food Web

Based on the ^{15}N data of the animals and the soil, analysis of variance confirmed that LS ($p = 0.013$), US1 ($p = 0.001$), and US2 ($p = 0.007$) experienced shifts in the entire system toward lower $\delta^{15}N$ values with time. At MS, no relationship between isotope pattern and time of regeneration was detectable (GLM, $p = 0.269$). However, the hump-shaped curves (for example, in T. *rathkei* and A. *murina*) show that after a general shift to higher $\delta^{15}N$ values, the whole system returned to a ^{15}N level similar to the initial situation in 1991.

The δ^{15}N values of the species investigated indicated that all trophic levels were affected by regeneration processes; that is, the entire food web showed an isotopic response to changing environmental conditions (see Figure 12.4). We argue that the system ^{15}N shifts at Steudnitz were controlled by two factors: the performance of N_2-fixing plants and the efficiency of organic matter decomposition.

At US1 and US2, the δ^{15}N values of the basal food-web components decreased while the abundance of two species of the genus *Vicia* (Fabaceae) was increasing or permanently high (Heinrich et al. 2001). *Vicia* is able to fix atmospheric N_2 with δ^{15}N values close to 0‰ (Nadelhoffer and Fry 1994). The continuous input of fresh N with δ^{15}N values below the mean ecosystem value caused the observed decrease of animal and soil δ^{15}N values at US1 and US2. The high and almost constant soil δ^{15}N values at LS and MS are due to the absence of legumes (see Figure 12.4). Because the availability of N influences plant species composition, biomass production, and the performance of soil microbes, it affects community assembly with respect to diversity and strength of trophic interactions. Our study demonstrates that food-web regeneration may benefit from the presence of N_2-fixing plants. In fact, we found the ecosystems at US1 and US2 to be most advanced in regeneration because of the effective supply of N, which accelerated developmental processes. Chapter 16 discusses the role of N_2-fixing plants in ecological restoration and comes to a similar conclusion: single plant species may become key components in community assembly (also see Chapter 4).

Another factor governing ^{15}N dynamics was the presence of macrodecomposers. Due to their sensitivity to heavy metals and high salt concentrations (Beyer and Anderson 1985, Hopkin et al. 1986, Paoletti and Hassall 1999), woodlice were almost absent from Steudnitz before 1990. Heinrich (1984) observed that primary production exceeded the decomposition rate, leading to litter accumulation on the soil surface. The woodlouse species investigated (*A. vulgare* and *T. rathkei*) did not attain significant abundances until 1994, but since then they have facilitated the decomposition of accumulated and fresh litter. Likewise, the improvement of soil conditions supported the recovery of the soil microflora (Langer and Günther 2001). Both factors ensured the input of new ^{15}N-depleted compounds into the soil, which was also recognizable at LS and MS (see Figure 12.4). Because the lack of decomposers may considerably hamper developmental progress due to restrictions in the cycling of matter, site preparation in restoration projects needs to consider this factor.

Stability of Trophic Relationships

Concerning the similarity in the ^{15}N dynamics of species of various trophic levels, at MS ($p = 0.555$) and US1 ($p = 0.137$) the model showed that the isotope signals of the analyzed species did not behave differently over time. Instead, the lack of significant differences in the interactions between species ^{15}N signals and year of sampling suggests a high level of stability in the processes and factors governing the isotope dynamics at both sites. At LS, the species-time interaction was almost significant ($p = 0.074$); that is, the similarity in the ^{15}N patterns of the analyzed species was lower at LS than at MS and US1. At US2, the species-time interaction was highly significant ($p = 0.002$). However, when omitting the beetle species *C. melanocephalus* and *A. murina* from the analysis (both showing positive ^{15}N trends; see Figure 12.4), it becomes evident that the $\delta^{15}N$ values of the four species at the base of the food web shifted almost in parallel with time (GLM, $p = 0.163$).

From 1991 to 2000 at each of the four Steudnitz sites, the trophic hierarchy was rather constant (see Figure 12.4); that is, the principal food-web structure was already established in the initial phase of regeneration. We suggest that the attainment of functional stability in trophic relationships can be identified from analyzing the ^{15}N dynamics of species representing different trophic levels (see Figure 12.4). For example, at US1 and US2, the $\delta^{15}N$ values of most AFW components ran parallel during the last years. This indicates that in less than 10 years in both systems at the upper slope, regeneration processes have compensated the negative effects of disturbance with respect to the relative stability of the food web and its functioning.

Development of the Soil Relative to the Aboveground Food Web

Along the pollution gradient from LS to US2, the soil ^{15}N content shifted consistently toward that of higher trophic levels of the AFW (see Figure 12.4). At LS, the $\delta^{15}N$ values of the soil were clearly below (~1‰), and at MS slightly below, the $\delta^{15}N$ values of *A. vulgare* and *O. raucus*. At US1 and US2, the soil $\delta^{15}N$ values were >1‰ above the values of *A. vulgare* and slightly to well above the values of *O. raucus*.

Over 10 years of regeneration, at all four sites the ^{15}N dynamics of the soil were similar to those of the AFW species (GLM, $p > 0.05$). However, between 1998 and 2000 at LS and MS, animal $\delta^{15}N$ values decreased further while the soil remained rather constant, perhaps reflecting the ongoing integration of AFW and SFW at these most-impacted sites. In comparison, the ^{15}N patterns at US1 and US2 suggest that the coupling of aboveground and soil processes at the least-impacted sites had already stabilized between 1991 and 1996.

The relative shift of soil δ^{15}N values toward those of AFW species increased along the pollution gradient in Steudnitz. At US2, one of the least-impacted sites, soil δ^{15}N values of the Ah-horizon were intermediate between those of primary consumers (primary decomposers and herbivores) and higher trophic levels (secondary decomposers and predators). Similarly, in intact forest ecosystems, the δ^{15}N value of the Ah-horizon was found to be higher (1.5‰ on average) compared with detritivores but lower (1.9‰ on average) compared with predators, both present in the litter layer (Ponsard and Arditi 2000). From this, we hypothesize that the δ^{15}N difference of AFW components and soil reflects the level of functional integration between AFW and SFW, which should be most advanced under undisturbed conditions. The level of integration influences the functionality, efficiency, and stability of the entire food web. Although stable interactions between SFW and AFW are required for such essential ecosystem functions as nutrient cycling (including decomposition and remineralization), disturbance may lead to severe interruptions in the cycling of elements, at least in part decoupling ecosystem components from one another (Asner et al. 1997). The ^{15}N data of Steudnitz imply an increasing level of integration of the trophic network with a decreasing degree of disturbance. If our hypothesis is valid, one should be able to estimate the developmental state of an ecosystem on its trajectory from disturbed toward recovered in relation to reference ecosystems.

The temporal development of the soil δ^{15}N values relative to those of detritivores and herbivores describes the interdependence of SFW and AFW based on structural and functional links, which constitute bottom-up control of the aboveground community by soil animals as well as top-down forces by generalist predators benefiting from belowground energy supply (Scheu 2001). Stable relations between aboveground and belowground processes should cause the ^{15}N signals of AFW species and the soil to run parallel with time, as demonstrated for the least-degraded Steudnitz sites US1 and US2.

So far, our interpretation of stable isotope data is a first attempt to scale up from species-specific measurements to ecosystem properties. More evidence is needed to develop this approach into a universal tool in ecological restoration.

Conclusions: Stable Isotopes in Ecosystem Research

The potential of stable isotopes to assess developmental progress can be used to describe natural succession as well as to evaluate the success of restoration measures. Since the rules creating isotope patterns are universal, systems that are not yet known can be approached. Previous knowledge about the attributes and history of the system under study is an asset but is not imperative. Various types of ecosystems with differing species compositions can be compared. Even a small sample size allows one to get an impression of the structure and functional performance of a community. Considering the amount of work needed to gather the data necessary to study community structure and its variation over time, the isotope methods are faster and easier to use than classical ones (Ponsard and Arditi 2000). To compare the system under restoration with a reference system, however, the "isotopic baseline"defined by the organic matter input at the base of the trophic network must be included. In addition, isotope analysis of the soil is required to get information about the functional coupling between aboveground and belowground compartments (Scheu 2001).

In our study, the good correspondence of spatial and temporal isotope patterns with abiotic and community development provided arguments that isotope signatures can be linked to such ecosystem properties as stability, functionality, and complexity and that these signatures can be translated into "developmental states."Both steps demand high levels of abstraction and carry the risk of an oversimplistic interpretation of the complex reality. To account for this complexity, comprehensive studies of various ecosystems (for example, Ponsard and Arditi 2000, Scheu and Falca 2000) should provide statistical evidence for a causal relationship between ecosystem state and descriptive variables derived from isotope data; for example, the mean value of $\delta^{15}N$ differences between consumer and diet as suggested above.

Empirical findings such as the "3.4‰-rule"encourage one to use the information offered by isotope patterns. These patterns are produced by the same forces that drive the piece-by-piece assembly of species into a harmonic entity, the community—despite (or because of?) the presence of thresholds and environmental filters (see Part II, "Ecological Filters as a Form of Assembly Rule"). The driving forces are thermodynamics and reaction kinetics of biogeochemical processes (Schmidt et al. 1995, Wada et al. 1995, Gleixner and Schmidt 1997) acting in the framework set by environmental conditions. These conditions control the flow of energy and matter and introduce structure into the systems, creating the "dynamic stable-isotope fingerprint"in biogenic material that can be used in ecological studies. Since the patterns and processes studied span several spatial and temporal scales, the search for assembly rules must cover various levels of complexity. Different resolution and hence different insight can be extracted from raw data by suitable data reduction methods and statistical procedures. Using such tools to screen our isotope data for hidden patterns and synoptic variables allowed us to focus on such community attributes as stability and complexity, which could not be assessed easily otherwise. In this context, our view on the development of regenerating ecosystems from the perspective of stable isotopes may serve as an encouraging example.

With regard to assembly rules in the context of restoration ecology, stable nitrogen isotopes offer great potential to describe food-web development in regenerating ecosystems and to gain insights into the mechanisms and principles governing structural and functional reorganization of communities after disturbance.

Acknowledgments

This research was supported by the *Deutsche Forschungsgemeinschaft* (DFG, German Research Council) as part of the Graduate Research Group on Analysis of the Functioning and Regeneration of Disturbed Ecosystems (GRK 266). We are grateful to Willy Brand and the ISOLAB team at Max Planck Institute for Biogeochemistry, Jena, for isotope measurements and to Wolfgang Heinrich for sharing his knowledge about the Steudnitz ecosystem. Winfried Voigt and Jörg Perner supplied most of the animal material as well as information about the faunal development in Steudnitz and the biology of the various animal taxa. Jürgen Einax provided archived soil samples. Finally, we give special thanks to Jens Schumacher for support in statistical analysis.

REFERENCES

Asner, G. P.; Seastedt, T. R.; and Townsend, A. R. 1997. The decoupling of terrestrial carbon and nitrogen cycles: human influences on land cover and nitrogen supply are altering natural biogeochemical links in the biosphere. *BioScience* 47:226–233.

Bardgett, R. D.; Wardle, D. A.; and Yeates, G. W. 1998. Linking above-ground and below-ground interactions: how plant responses to foliar herbivory influence soil organisms. *Soil Biology and Biochemistry* 30:1867–1878.

Bengtsson, J.; Lundkvist, H.; Saetre, P.; Sohlenius, B.; and Solbreck, B. 1998. Effects of organic matter removal on the soil food web: forestry practices meet ecological theory. *Applied Soil Ecology* 9:137–143.

Beyer, W. N.; and Anderson, A. 1985. Toxicity to woodlice of zinc and lead oxides added to soil litter. *Ambio* 14:173–174.

Brand, W. A. 1996. High precision isotope ratio monitoring techniques in mass spectrometry. *Journal of Mass Spectrometry* 31:225–235.

Coplen, T. B.; Krouse, H. R.; and Böhlke, J. K. 1992. Reporting of nitrogen-isotope abundances. *Pure and Applied Chemistry* 64:907–908.

Fantle, M. S.; Dittel, A. I.; Schwalm, S. M.; Epifanio, C. E.; and Fogel, M. L. 1999. A food web analysis of the juvenile blue crab, *Callinectes sapidus*, using stable isotopes in whole animals and individual amino acids. *Oecologia* 120:416–426.

Focken, U. 2001. Stable isotopes in animal ecology: the effect of ration size on the trophic shift of C and N isotopes between feed and carcass. *Isotopes in Environmental and Health Studies* 37:199–211.

Gleixner, G.; Kracht, O.; Schmidt, H.-L.; and Schulze, E.-D. 2001. Isotopic evidence for the origin and formation of refractory organic substances. In *Refractory Organic Substances in the Environment*, ed. F. H. Frimmel and G. Abbt-Braun, 146–162. Weinheim, Germany: Wiley-VCH.

Gleixner, G.; and Schmidt, H. L. 1997. Carbon isotope effects on the fructose-1,6-bisphosphate aldolase reaction, origin for non-statistical ^{13}C distributions in carbohydrates. *Journal of Biological Chemistry* 272:5382–5387.

Gruner, H.-E. 1966. Krebstiere oder Crustacea. V. Isopoda. 2. Lieferung. In *Die Tierwelt Deutschlands*, ed. F. Dahl, M. Dahl, and F. Peus, 150–380. Jena, Germany: Gustav Fischer Verlag.

Heinrich, W. 1984. Über den Einfluß von Luftverunreinigungen auf Ökosysteme III: Beobachtungen im Immissionsgebiet eines Düngemittelwerkes. *Wissenschaftliche Zeitschrift der Friedrich-Schiller-Universität Jena, Naturwissenschaftliche Reihe* 33:251–290.

Heinrich, W.; Perner, J.; and Marstaller, R. 2001. Regeneration und Sekundärsukzession: 10 Jahre Dauerflächenuntersuchungen im Immissionsgebiet eines ehemaligen Düngemittelwerkes. *Zeitschrift für Ökologie und Naturschutz* 9:237–253.

Hobbs, R. J.; and Norton, D. A. 1996. Towards a conceptual framework for restoration ecology. *Restoration Ecology* 4:93–110.

Hopkin, S. P. 1990. Species-specific differences in the net assimilation of zinc, cadmium, lead, copper and iron by the terrestrial isopods *Oniscus asellus* and *Porcellio scaber*. *Journal of Applied Ecology* 27:460–474.

Hopkin, S. P.; Hardisty, G. N.; and Martin, M. H. 1986. The woodlouse *Porcellio scaber* as a biological indicator of zinc, cadmium, lead and copper pollution. *Environmental Pollution, Series B: Chemical and Physical* 11:271–290.

Jones, D. T.; and Hopkin, S. P. 1998. Reduced survival and body size in the terrestrial isopod *Porcellio scaber* from a metal-polluted environment. *Environmental Pollution* 99:215–223.

Köhler, F.; and Klausnitzer, B. 1998. Verzeichnis der Käfer Deutschlands. *Entomologische Nachrichten und Berichte, Beihefte* 4:1–185.

Lajtha, K.; and Michener, R. H., eds. 1994. *Stable Isotopes in Ecology and Environmental Research*. Oxford: Blackwell Scientific.

Langer, U.; and Günther, T. 2001. Effects of alkaline dust deposits from phosphate fertilizer production on microbial biomass and enzyme activities in grassland soils. *Environmental Pollution* 112:321–327.

Metzner, K.; Friedrich, Y.; and Schäller, G. 1997. Bodenparameter eines Immisionsgebietes vor und nach der Schließung eines Düngemittelwerkes (1979–1997). *Beiträge zur Ökologie* 3:51–75.

Mikola, J.; Yeates, G. W.; Wardle, D. A.; Barker, G. M.; and Bonner, K. I. 2001. Response of soil food-web structure to defoliation of different plant species combinations in an experimental grassland community. *Soil Biology and Biochemistry* 33:205–214.

Minagawa, M.; and Wada, E. 1984. Stepwise enrichment of ^{15}N along food chains: further evidence and the relation between $\delta 15N^{15}N$ and animal age. *Geochimica et Cosmochimica Acta* 48:1135–1140.

Nadelhoffer, K. J.; and Fry, B. D. 1994. Nitrogen isotope studies in forest ecosystems. In *Stable Isotopes in Ecology and Environmental Research*, ed. K. Lajtha and R. H. Michener, 22–44. Oxford: Blackwell Scientific.

Paoletti, M. G.; and Hassall, M. 1999. Woodlice (Isopoda: Oniscidea): their potential for assessing sustainability and use as bioindicators. *Agriculture Ecosystems and Environment* 74:157–165.

Ponsard, S.; and Arditi, R. 2000. What can stable isotopes ($\delta^{15}N$ and $\delta^{13}C$) tell about the food web of soil macro-invertebrates? *Ecology* 81:852–864.

Rothe, J.; and Gleixner, G. 2000. Do stable isotopes reflect the food web development in regenerating ecosystems? *Isotopes in Environmental and Health Studies* 36:285–301.

Scheu, S. 1997. Effects of litter (beech and stinging nettle) and earthworms (*Octolasion lacteum*) on carbon and nutrient cycling in beech forests on a basalt-limestone gradient: a laboratory experiment. *Biology and Fertility of Soils* 24:384–393.

Scheu, S. 2001. Plants and generalist predators as links between the below-ground and above-ground system. *Basic and Applied Ecology* 2:3–13.

Scheu, S. 2002. The soil food web: structure and perspectives. *European Journal of Soil Biology* 38:11–20.

Scheu, S.; and Falca, M. 2000. The soil food web of two beech forests (*Fagus sylvatica*) of contrasting humus type: stable isotope analysis of a macro- and a mesofauna-dominated community. *Oecologia* 123:285–296.

Scheu, S.; Theenhaus, A.; and Jones, T. H. 1999. Links between the detritivore and the herbivore system: effects of earthworms and Collembola on plant growth and aphid development. *Oecologia* 119:541–551.

Schmidt, H. L.; Kexel, H.; Butzenlechner, M.; Schwarz, S.; Gleixner, G.; Thimet, S.; and Werner, R. A. 1995. Non-statistical isotope distribution in natural compounds: mirror of their biosynthesis and key for their origin. In *Stable Isotopes in the Biosphere*, ed. E. Wada, T. Yoneyama, M. Minagawa, T. Ando, and B. D. Fry, 17–35. Kyoto: Kyoto University Press.

Schmidt, O.; Scrimgeour, C. M.; and Handley, L. L. 1997. Natural abundance of ^{15}N and ^{13}C in earthworms from a wheat and a wheat-clover field. *Soil Biology and Biochemistry* 29:1301–1308.

Stapp, P.; Polis, G. A.; and Piñero, F. S. 1999. Stable isotopes reveal strong marine and El Niño effects on island food webs. *Nature* 401:467–469.

Vander Zanden, M. J.; Casselman, J. M.; and Rasmussen, J. B. 1999. Stable isotope evidence for the food web consequences of species invasions in lakes. *Nature* 401:464–467.

Wada, E.; Ando, T.; and Kumazawa, K. 1995. Biodiversity of stable isotope ratios. In *Stable Isotopes in the Biosphere*, ed. E. Wada, T. Yoneyama, M. Minagawa, T. Ando, and B. D. Fry, 17-35. Kyoto: Kyoto University Press.

Wagner, E. 1967. *Wanzen oder Heteropteren. II. Cimicomorpha*. Jena, Germany: Gustav Fischer Verlag.

Werner, R. A.; and Brand, W. A. 2001. Referencing strategies and techniques in stable isotope ratio analysis. *Rapid Communications in Mass Spectrometry* 15:501-519.

PART FOUR

Assembly Rules in Severely Degraded Environments

The following chapters consider how regeneration of severely degraded ecosystems requires us to focus on abiotic as well as biotic factors. Chapter 13 investigates the interaction between species dispersal ability and stress tolerance in community assembly over time in a degraded grassland. Chapter 14 considers similar issues in the same grassland but uses an experimental approach that centers on dispersal, stress tolerance, and facilitation. Chapter 15 analyzes vegetation patterns in and around a former mining dump to search for the effects of dispersal and site quality on the development of alternative stable states of vegetation. Chapter 16 reviews the overarching importance of nutrient relationships for restoring ecosystem function in terrestrial systems.

Chapter 13

The Roles of Seed Dispersal Ability and Seedling Salt Tolerance in Community Assembly of a Severely Degraded Site

MARKUS WAGNER

An early but much neglected experiment on plant community assembly was carried out by Ellenberg (1953, 1954), who grew a number of grasses under different water table regimes in monocultures and mixed cultures. His experiments—which illustrated the differences between the fundamental and realized niche—showed that in monoculture all species grew best at intermediate water tables, whereas in mixed cultures the same conditions resulted in an almost complete elimination of some species from the experimental community by competition. It is probably these more or less optimal environmental conditions that are thought of by those who take the view that species interactions govern assembly rules (for example, Lawton 1987, Wilson and Gitay 1995, Wilson 1999).

Restoration ecologists encounter such optimal conditions and the effects of interaction-mediated assembly rules on sites with increased soil fertility when reestablishment of high-diversity, seminatural vegetation fails due to competitive exclusion of slow-growing target species by high-yielding species (for a review, see Marrs and Gough 1989). In contrast, when restoration ecology aims to reclaim areas where environmental conditions are far from optimal for plant growth, the nature of plant-plant interactions often shifts more toward facilitation (Bertness and Callaway 1994, Pugnaire and Luque 2001; also see Chapters 14 and 16).

In harsh environments, however, a number of factors other than biotic interactions can also influence plant establishment and therefore plant community assembly. These factors are (1) the ability of diaspores to reach the site, which in turn depends on the distance to the source of the diaspores and their mode of dispersal; (2) the level of resources required for plant establishment and growth; and (3) the occurrence of toxic substances. Resource inadequacy or excess and the presence of toxic substances create conditions that can be defined as stress, and under severe stress, species are commonly excluded from a community by their inability to survive and reproduce (Grime 1979). Stress and all the other factors just listed can be subsumed under the concept of environmental filters. A simple filter model may consist of a sequence of historical ("can it reach the site?"), abiotic ("can it cope with the physicochemical conditions?"), and biotic ("can it compete and defend itself?") filters through which a candidate species from the regional pool of species must pass to enter the local community (Lambers et al. 1998; but also see Roughgarden and Diamond 1986). Further elaborations of the filter concept can be found in Part II, "Ecological Filters as a Form of Assembly Rule."

Bradshaw (Chapter 16) deals with the role of resource requirements and discusses the influence of nutrients on the assembly of plant communities in more detail. Here, I consider the role of dispersal and toxicity at a site that is still recovering from deposition and soil degradation resulting from emissions of an adjacent fertilizer factory. It is thought that one of the main factors initially limiting plant reestablishment at this site was the high level of soil salinity, which later diminished considerably. The following questions were investigated:

1. Did plant species with the capacity for long-distance wind dispersal colonize the site earlier than other species?
2. Did species with more salt-tolerant seedlings tend to colonize the site earlier? If yes, was there a stronger correlation for species with efficient long-distance wind dispersal than for those without it?

Study Site

The Steudnitz field site (Figure 13.1) is a southeast-facing slope of 35–40° with underlying limestone geology. It is situated about 10 km north of Jena, Germany, on the western side of the Saale River Valley (51°01′ N, 11°41′ E). Just 100 m south of the site, a factory that made fertilizers—predominantly phosphorus (P) but also phosphorus-potassium (P-K)—emitted large quantities of alkaline dust between 1957 and 1990. The composition of these dust emissions was determined by the raw materials. For P fertilizer production, apatite $Ca_5(PO_4)_3(OH, F, Cl)$, soda (Na_2CO_3), and quartz sand were used. Some additional emissions of hydrochloric acid arose from the P-K fertilizer production (Peter Hagenguth, former environmental officer of Dornburger Cement Company, personal communication).

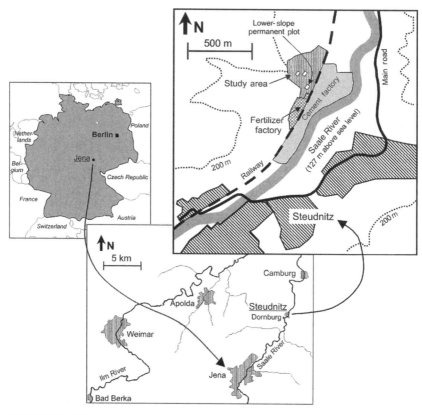

FIGURE 13.1 (a) Map and (b) oblique aerial photograph of the Steudnitz, Germany, study area (after Heinrich et al. 2001). Note that the slope rises from the Saale River to the 200-m elevation contour.

The emissions severely altered the soil conditions at the site, which lay in the prevailing wind direction from the factory. The original vegetation was replaced by species-poor pollution-tolerant assemblages, which persisted until the factory was shut down in 1990. One of these assemblages, a near-monoculture of the halophyte *Puccinellia distans* (Poaceae), together with a few stunted individuals of *Atriplex sagittata* (Chenopodiaceae), was located in the immediate vicinity of the factory on the lower slope (see Figure 13.2a; nomenclature follows Wisskirchen and Haeupler 1998). This species-poor community gave way to the subsequent assembly of a diverse ruderal plant community. A 20-m by 30-m permanent plot on this part of the slope, which in 1990 was inhabited only by the two above-mentioned species, was already inhabited by 55 species of higher plants and 9 moss species by 1999 (Heinrich et al. 2001; also see Figure 13.2b, which shows the slope in 2000). The development of species richness on the permanent plot is illustrated in Figure 13.3. No further increase in the number of herbaceous species was observed from 1999 onward, whereas the number of woody species continues to increase slowly.

Table 13.1 gives an overview of the most important soil parameters measured and how they have changed over time since the factory was closed. Particularly noteworthy is the decrease in total sodium (Na) by about 50 percent, strongly indicating high initial soil salinity. The initial soil pH of about 9 was slightly elevated compared to the current pH (about 8); the current pH is comparable to the pH of other unpolluted soils in the region (Heinrich 1984). Initially, there was also a strong enrichment of the pollutants cadmium (Cd) and fluorine (F) in the topsoil. Although Cd levels are still high, F levels have declined markedly. The high Cd levels result from the soil conditions at the field site, where high levels of carbonate and phosphate occurring at high pH have led to a permanent fixation and immobilization of Cd (McLaughlin et al. 1996). At the same time, a high pH favors fluoride solubility (for example, Larsen and Widdowson 1971).

Other changes caused by the emissions are persistently high levels of total and plant-available P and the elevated levels of potassium and magnesium (Mg). The amounts of organic matter and nitrogen in the soil, on the other hand, are still low. The soil structure in the vicinity of the factory is also modified, due to particle emissions the size of sand grains. The resulting coarse texture, in combination with the topography of the slope, favors a quick drying-out after rainfall, particularly in the summer.

It is obvious that the soil salinity changed considerably after the factory was closed, as indicated by the decrease in F and Na levels (see Table 13.1).

FIGURE 13.2 Vegetation in (a) 1991 and (b) 2000 at the Steudnitz, Germany, field slope, showing the phosphorus fertilizer factory and impacted slope. In 1991, vegetation was completely dominated by *Puccinellia distans*. By 2000, a diverse plant community of up to 60 species had developed.

The initially high salinity, however, was probably also caused by high levels of sulfate, phosphate, chloride, and, to a lesser extent, nitrate. These constituents were not measured in the soil but were found in abundance in soluble form in pollutant dust samples collected from sheltered locations at the factory site in 1997 (K. Metzner, unpublished data).

Decreasing levels of soil salinity and declining pH seem most likely to have widened the "mesh width"of the abiotic filter (see Chapter 6) after factory closure, gradually allowing for the colonization of the site by less tolerant species.

Methods

Permanent-plot vegetation data from Heinrich et al. (2001 and unpublished) provided baseline data for addressing the questions outlined in this chapter's introduction. Only data from the permanent plot on the lower slope were used

TABLE 13.1

Means (and Standard Deviations) of Soil Parameters at Lower-Slope Permanent Plot at Steudnitz, Germany, for 1990–1999.[1,2]

Soil factor	1990 ($n = 11$)	1996 ($n = 11$)	1999 ($n = 24$)
Plant nutrients			
Total N (%)	Not measured	0.07 (0.07)	0.11 (0.03)
Total P (g · kg^{-1})	83.9 (26.6)[3]	76.3 (26.3)[4]	Not measured
Available P (g · kg^1)	Not measured	8.6 (3.1)	4.4 (1.5)
Total K (g · kg^{-1})	6.7 (2.8)	4.0 (1.2)	4.9 (2.0)
Available K (mg · kg^{-1})	Not measured	281 (67)	344 (137)
Mg (g · kg^{-1})	4.5 (1.0)	4.0 (2.1)	8.5 (2.4)
Pollutants			
Cd (mg · kg^{-1})	10.8 (7.9)[3]	7.8 (2.5)	9.6 (4.3)
Na (g · kg^{-1})	16.1 (6.0)	12.8 (5.5)	7.9 (2.0)
F (g · kg^{-1})	8.2 (3.3)	3.2 (0.9)[4]	Not measured
Other parameters			
pH	9.14 (0.38)	7.9 (0.10)	8.09 (0.14)
Sand (%)	Not measured	58	54
Silt (%)	Not measured	34	36
Clay (%)	Not measured	8	10
Organic C (%)	Not measured	3.7 (1.0)	6.0 (0.6)

[1]1990 and 1996 data from Metzner et al. (1997); 1999 data from Wagner (unpublished).
[2]Unless otherwise stated, total soil contents are given.
[3]Values from 1991.
[4]Values from 1997 and $n = 3$.
[5]$n = 10$.

because the other two permanent plots, being farther away from the factory (see Figure 13.1), were only slightly degraded by emissions (Heinrich et al. 2001). Statistical tests were carried out using SPSS for Windows, version 10.0.

Role of Dispersal Type

To investigate the impact of dispersal ability, speed of colonization was compared between those species having a plumed diaspore, facilitating long-distance wind dispersal, and those with other dispersal adaptations or no adaptations. All 80 species that colonized the plot between 1992 and 2002 were classified according to this criterion, using the comprehensive seed morphology database by Müller-Schneider (1986). They were ranked according to the year they were first recorded on the permanent plot, and a Mann-Whitney test (exact test, one-tailed) was used to test whether plant species with the capacity for long-distance wind dispersal arrived earlier at the site than those lacking this capacity.

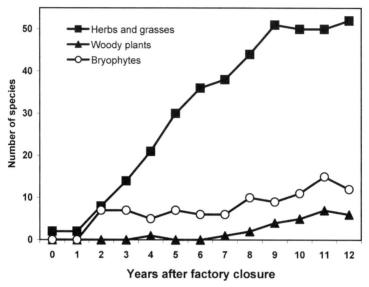

FIGURE 13.3 Development of species richness on the Steudnitz lower-slope permanent plot subsequent to fertilizer factory closure in 1990 (data from Heinrich et al. 2001 and unpublished).

Role of Salt Tolerance at the Early Seedling Stage

The effects of salt stress on seed germination and seedling radicle extension were tested under controlled conditions in December 2000 and December 2001, using seeds of species recorded on the permanent plot that were collected locally in the respective field season. Salt stress was generated by using sodium chloride (NaCl) solutions.

The two species initially present in the permanent plot, *P. distans* and *A. sagittata*, as well as 22 of the colonizing species (12 species with plumed seeds and 10 without; see Table 13.2 for details) were tested. Species from the pool of colonizers were chosen to represent the whole range from early colonization in 1992 to late colonization in 1999.

For *A. sagittata*, a species with dimorphic seeds, only black seeds were tested because in both years the plants produced few brown seeds. To break dormancy, seeds of two of the species, *A. sagittata* and *Chenopodium rubrum*, were subjected to cold-moist stratification (5 weeks in the dark at 5°C) before the experiment. The stratification requirements of the two genera to which these species belong are documented in the ECOFLORA database (Fitter and Peat 1994).

The experiment consisted of five different treatments: one distilled water control treatment and four different NaCl solution treatments with osmotic potentials of –0.4, –0.8, –1.2 and –1.6 MPa. The calculations required for preparation of the NaCl solutions were performed using a computer program developed by Michel and Radcliffe (1995).

For each species and treatment, there were four replicates. Each replicate consisted of 20 to 25 seeds of a species (see Table 13.2) sown into a 9-cm petri dish containing two layers of Whatman No. 1 filter paper and 12 mL of the test solution to moisten the filter paper. The petri dishes were kept in a controlled environment with 12 hours of light at 22°C and 12 hours of dark at 14°C. Daily temperature fluctuations of this magnitude are known to promote seed germination in a wide range of species (Grime et al. 1981).

To keep the solute potentials constant, petri dishes were sealed with Parafilm™ at the beginning and between counts, and at each count one-quarter of the solution was replaced. Counts were carried out for the first time 3 days after the onset of the experiment and then after 7, 14, 21, 28, 35, 42, and 49 days. On each occasion, the numbers of germinated seeds, seedlings with a radicle length ≥ 5 mm, and seedlings with a radicle length ≥ 10 mm were counted. Seedlings with a radicle length ≥ 10 mm were removed from the experiment. Data analysis in this chapter is based on the final count of seedlings reaching 5-mm radicle extension.

I formulated a salt-tolerance index (I_s) to characterize salt tolerance for each of the tested species at the early seedling stage:

$$I_s = \frac{(\sum_{i=1}^{4} n_i)}{(4 \times n_c)}$$

where n_i and n_c are the numbers of seedlings with ≥ 5-mm radicle extension in the four salt treatments and in the control treatment, respectively. Therefore, a species completely unaffected by salinity over the range of salt concentrations tested will have an I_s value of close to 1.0, whereas a very sensitive species germinating only in the distilled water treatment will have a value of 0. The resulting index values for each species are shown in Table 13.2.

The strength of association between the I_s of a species and the time needed for species establishment after factory closure was measured using Spearman rank correlation. This method is recommended when observations are in the form of indices (Fowler et al. 1998). The significance of the correlation was tested using a one-tailed test (at $p < 0.05$).

TABLE 13.2

Properties of Species Tested for Salt Tolerance in Petri Dish Experiments[1,2]

Species	First year[3]	Plumed (yes/no)[4]	Remarks	Seed weight (mg)[5]	Salt-tolerance index (I_s)	Evidence for salt tolerance (literature source)
Atriplex sagittata	1990	No	Only black seeds tested, stratified	1.40	0.43	Slightly salt-tolerant (Brandes 1999)
Puccinellia distans	1990	No	Glumes removed	0.23	0.91	Halophyte (Ellenberg 1992)
Artemisia vulgaris	1992	No		0.13	0.33	
Chenopodium rubrum	1992	No	Stratified	0.11	0.61	Slightly salt-tolerant (Ellenberg 1992)
Lactuca serriola	1992	Yes		0.57	0.34	
Erigeron acris	1993	Yes		0.09	0.19	
Senecio vulgaris	1993	Yes		0.22	0.21	
Taraxacum officinale	1993	Yes	20 seeds per replicate	0.68	0.37	Slightly salt-tolerant (Ellenberg 1992)
Cirsium vulgare	1994	Yes		2.67	0.31	
Conyza canadensis	1994	Yes		0.04	0.24	
Lepidium ruderale	1994	No		0.07	0.36	Slightly salt-tolerant (Rich 1991, Brandes 1999)
Senecio vernalis	1994	Yes		0.21	0.21	
Arenaria serpyllifolia	1995	No		0.03	0.10	
Epilobium tetragonum	1995	Yes	20 seeds per replicate	0.07	0.02	
Inula conyzae	1995	Yes		0.23	0.11	
Picris hieracioides	1995	Yes	20 seeds per replicate	0.64	0.29	
Achillea millefolium	1996	No		0.18	0.30	
Bromus japonicus	1996	No	20 seeds per replicate	2.55	0.57	Frequent in saline U.S. prairies (Cooper and Jean 2001)
Verbascum lychnitis	1996	No	20 seeds per replicate	0.08	0.06	
Solidago gigantea	1997	Yes		0.07	0.13	
Erophila verna	1998	No		0.02	0.00	
Anthemis tinctoria	1999	No	20 seeds per replicate	0.44	0.22	
Hieracium piloselloides	1999	Yes		0.23	0.11	
Thlaspi perfoliatum	1999	No		0.08	0.08	

[1]Results are summarized as salt-tolerance index values (see text).
[2]Unless otherwise indicated under Remarks, salt tolerance is based on 25 seeds per replicate.
[3]Data from Heinrich et al. (2001).
[4]Species are classified with respect to seed morphology (Müller-Schneider 1986).
[5]Air-dry seed weight measured from batches of 500 seeds.

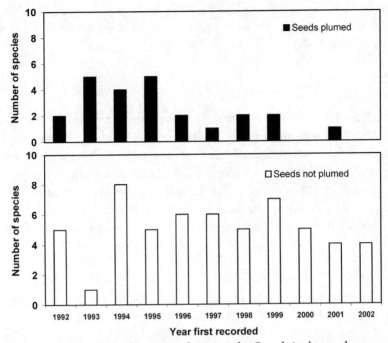

FIGURE 13.4 Numbers of species colonizing the Steudnitz lower-slope permanent plot between 1992 and 2002 (data from Heinrich et al. 2001 and unpublished). Species with plumed diaspores (black bars, $n = 24$) and those without (white bars, $n = 56$) are displayed separately (classification after Müller-Schneider 1986).

Results

Figure 13.4 illustrates the distribution of numbers of colonizers with and without plumed diaspores for the period from 1992 to 2002. The Mann-Whitney test (one-tailed) confirms that colonization by species with plumed seeds occurred significantly earlier than colonization by those without plumed seeds ($n_1 = 56$, $n_2 = 24$; $U = 423$; $p = 0.004$).

Because the mechanism of seed dispersal seems to have influenced the order of colonization at the slope, the Spearman rank correlation used to establish a link between salt tolerance and time of colonization was carried out separately for each group of species (those with a plumed seed morphology and those without). Spearman rank correlation yielded very similar and significant correlation coefficients for both groups (nonplumed species: $n = 10$, $r_s = -0.63$, $p = 0.026$; plumed species: $n = 12$, $r_s = 0.66$, $p = 0.010$). A test for significant difference between the two correlation coefficients was not carried out. Because the difference between the two coefficients and the sample sizes were rather small, the power of such a test would have been very small (Zar 1999).

Discussion

Results from the experiments and analyses presented here demonstrate how the species pool is limited both by seedling salt tolerance and by the dispersal ability of the various species. First, I will discuss the method used to characterize seedling salt tolerance. Next, I will discuss how diaspore morphology and seedling salt tolerance influence species arrival at the Steudnitz site. Finally, I will draw some conclusions about how these results may be generalizable to other severely degraded or high-stress environments.

Salt Tolerance Characterization

When interpreting the results of the petri dish experiment, it must be kept in mind that soil salinity at the field site was caused by a mixture of salts, including ions other than sodium and chloride. NaCl was selected for this study to allow comparison with results of other published work, the great majority of which also use this salt.

It is not uncommon to use germination (defined as a visible protrusion of the radicle from the seed coat; see Kitajima and Fenner 2000) for the characterization of salt tolerance (see Ungar 1995). Indeed, several authors have found a correlation between germination response and salinity gradient in the field (for example, Rozema 1975, Bakker et al. 1985, Mariko et al. 1992), such that halophytes, on average, seem to be able to germinate at somewhat higher salinities than glycophytes (Ungar 1995).

The relationship between salt tolerance and germination is blurred, however. Seeds of a number of highly salt-tolerant plants exhibit osmotically induced dormancy (Ungar 1995). There is evidence that in both glycophytes and halophytes, germination is often controlled by osmotic potential in the surrounding medium, rather than by specific ion effects (Schratz 1934). This has been confirmed by comparative experiments on the same species as in the present study, using either polyethylene glycol 6000 or NaCl for the simulation of corresponding osmotic potentials (Wagner unpublished data). In contrast, ion toxicity mainly affects the subsequent process of seedling growth, thus lending support to Rozema's (1975) hypothesis that the ability of seedlings to tolerate salt stress is a much better measure of salt tolerance than the ability to germinate. The present study, therefore, used counts of seedlings with radicles longer than 5 mm, rather than germination rates. This method is less arbitrary than it may seem. In the distilled water control treatment, even in those species with the smallest seeds, almost 100 percent of the germinated seeds managed a radicle extension > 5 mm. In contrast, even low salinity levels caused necrosis of the radicle tip in a number of glycophytes almost immediately after the onset of germination, preventing any further root growth. Therefore, seed size and seed reserves do not matter as much as one might expect.

A comparison of the calculated salt-tolerance index with the relevant literature shows that the species with $I_s > 0.35$ are in fact known to display some degree of salt tolerance in Central Europe (see Table 13.2). The only exception is *Bromus japonicus*, a rather large-seeded species that is typical of open, ruderal habitats but that is rare in Germany (Jäger and Werner 2002). According to Cooper and Jean (2001), however, it is an abundant species in saline prairie habitats in the northwestern Great Plains of North America. There, it is also found at oil industry saltwater blowout sites (Halvorson and Lang 1989).

Overall, comparison with the literature suggests that the I_s is useful for characterizing salt tolerance in glycophytes. Because one cannot rule out the possibility that seed size exerted some influence on the results, air-dry seed weights (determined by weighing of batches of 500 seeds from the local collections) are provided in Table 13.2.

Plant Community Assembly at the Steudnitz Field Site

Clear evidence was found that plumed species entered the plant community at the Steudnitz field site earlier than nonplumed species. This is no surprise, as the importance of efficient long-range dispersal in the early colonization of open habitats has been pointed out previously (for example, Bradshaw 1983, Fenner 1987a, Ash et al. 1994).

Order of colonization was also correlated with seedling salt tolerance. Figure 13.5 suggests that up to about 1994 (that is, 4 years after factory closure), salinity was an important component of the abiotic filter acting as a selective force upon species establishment, since only in 1995 did any of the most salt-sensitive ($I_s < 0.15$) species colonize. Correlations between salt tolerance and year of colonization were about equal for both plumed and nonplumed species. Assuming a higher probability of immediate colonization of suitable sites (that is, soil salinity falls below critical threshold) by plumed species, one would have expected a more predictable colonization order—and therefore a better correlation—in this group compared to the group of nonplumed species.

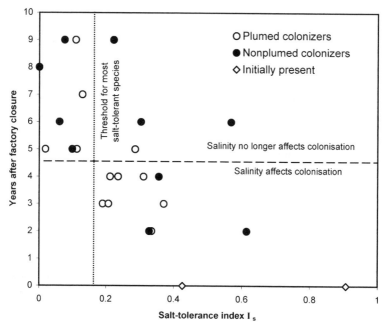

FIGURE 13.5 Scatter plot of time between factory closure and colonization versus seedling salt-tolerance index I_s (see text for details). The species present initially were *Puccinellia distans* ($I_s = 0.91$) and *Atriplex sagittata* ($I_s = 0.43$).

Regardless of the apparently simple patterns found in this study, the role of the abiotic filter is undoubtedly much more complex. There is a diverse array of soil factors, each of which can vary over different time scales. For instance, factors such as salt levels and pH have strongly declined, whereas other factors such as soil structure (water-holding capacity) and nutrient imbalance (excess phosphorus or nitrogen deficiency) change over much longer periods of time. Thus, the choice of potential restoration measures in a system recovering from degradation, such as the one investigated in this study, would depend on the timing of the intervention (see Chapter 5).

P. distans maintained its dominance until 1995 and then abruptly disappeared (Heinrich et al. 2001). Increasing competitive pressure may have triggered this disappearance, as *P. distans* is not normally restricted to saline soils. Grown in experimental monocultures, it grows more vigorously in the absence of salinity (Beyschlag et al. 1996) than it does even at moderate salinity levels. It is generally described as a weak competitor (Beyschlag et al. 1996) and can therefore be regarded as one of the many species described by Ellenberg (1954) that are relegated to suboptimal conditions because, under better conditions, superior competitors exclude them.

The complete collapse of the *P. distans* population in 1996 indicates that salinity probably played no further role at the investigated site, especially in light of the fact that woody species recolonized the slope in the next year and that those native to Germany exhibit a very low salt tolerance (Brandes 1999). The low water-holding capacity and the persistently low nitrogen content of the soil suggest, however, that at present the remaining components of the abiotic filter still exert some direct influence on plant community assembly at the site.

The Role of Stress in Plant Community Assembly

Wilson and Gitay (1995:369) define assembly rules as "ecological restrictions on the observed patterns of species presence or abundance that are based on the presence or abundance of one or more other species or groups of species (not simply the response of individual species to the environment)." In the present study, nonrandom colonization patterns could be demonstrated, implying the existence of assembly rules in this system (Lawton 1987), but the overall pattern could be explained without invoking biotic interactions. Therefore, the alternative concept for assembly rules outlined in Keddy (1992; also see Weiher and Keddy 1999) seems to be more useful in the context of this study. Keddy advocates that assembly rules should include any constraint on entry of species into the community. Based on this idea, he suggests the development of assembly rules based on species traits, which "could include morphological, physiological or ecological features"(Keddy 1992:159). This approach was also employed in the present study, using one morphological trait (adaptation to long-distance seed dispersal) and one physiological trait (salt tolerance at the early seedling stage).

Measuring the stress tolerance of seedlings not only is an easy method of characterization but also is likely to be an appropriate one. As stated by Angevine and Chabot (1979:189), "The germination of seeds and subsequent early growth of seedlings are not only essential phases of the life cycle of all higher plants, but represent periods of maximum vulnerability to physical changes in the environment and minimal potential for homeostatic response or physiological retrenchment." Mortality risk is usually much higher in the early postgermination stage than in later stages of the life cycle (Fenner 1987b). Therefore, in many cases, the bottleneck for species colonization in stressful habitats could lie in the ability of seedlings to withstand the adverse abiotic conditions in such habitats. The large seed mass (Hodkinson et al. 1998) and prolonged seedling shade tolerance (Grime and Jeffrey 1965) of species regularly establishing under light-deficient conditions are examples of adaptations to circumvent such bottlenecks. A large seed mass was also suggested as an adaptation for establishment in nutrient-poor soils (Milberg and Lamont 1997). However, there are also systems in which community composition is determined largely by the stress tolerance of adult plants. One example is salt marsh ecosystems, where windows of opportunity for seedling establishment are provided by an annual "low-salinity gap"and plant community zonation is determined largely by the salt and inundation tolerance of adult plants (Zedler and Beare 1986, Partridge and Wilson 1987).

Keddy's (1992) definition of assembly rules in general seems to be particularly useful in situations where stress plays a role in the exclusion of members of the local species pool from a community. Such situations occur regularly in the restoration of degraded habitats.

The importance of stress tolerance for community assembly is best illustrated in systems where stress levels change over time. Short-term extremes of environmental conditions can cause the species composition of communities to undergo selection, reflecting the different capacities of species to withstand these extremes. Examples are the dependence of floodplain meadow species composition on the duration of the spring flood (see Eliashevich 1982) and the dependence of dry grassland community structure on the extent of droughts in summer (Hopkins 1978, Ellenberg 1996). In the present study, it was the order of colonization that was influenced by the gradual amelioration of soil conditions.

Grime's (1979) conceptual model pointed out the importance of stress levels for species density and provided an explanation for the relationship between total biomass and species density in herbaceous vegetation. In this so-called humpbacked model of species density, community membership in productive habitats is restricted by competition for light, and membership in unproductive habitats is restricted mainly by extreme conditions of stress or disturbance.

Grime's model and more recent models that specify the relative importance of facilitation along stress gradients (Belcher et al. 1995, Callaway and Walker 1997) can be amalgamated into a simple conceptual community assembly model summarizing how the relative importance of the abiotic stress filter, interspecific facilitation, and competition could vary along a stress gradient (Figure 13.6). Exact curve shapes would depend on the identity or combination of stress factors particular to a system. The abiotic environment not only has a direct effect on community assembly via the abiotic stress filter but also has indirect effects through the modulation of biotic interactions. For example, the relative competitive ability of the species

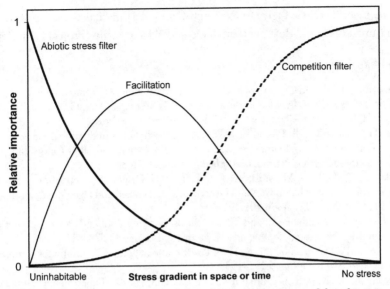

FIGURE 13.6. Conceptual model of the relative importance of the abiotic stress filter and of plant-plant interactions (facilitation and competition) for plant community assembly along a hypothetical stress gradient.

involved depends on the abiotic environment. Additionally, facilitation is more important in severe environments, whereas competition plays a bigger role under benign conditions (see Chapter 14 for more on facilitation). Although the model does not incorporate such other factors as disturbance, life history processes, and interactions with other biota (for these factors, see Walker and Chapin 1987; also see Chapters 10 and 17), it does illustrate that biotic interactions are only part of the whole story and that the usefulness of any definition of assembly rules relying mainly on such interactions is limited to habitats that are unaffected by high stress levels.

Acknowledgments

I thank Tony Bradshaw, Dietmar Brandes, and Vicky Temperton for carefully reviewing the manuscript. Through thorough proofreading and valuable discussions, Rosemary Setchfield helped to compile the rules of assembly for this paper. Discussions with Jens Schumacher, Tim Nuttle, and Gottfried Jetschke are also kindly appreciated. Wolfgang Heinrich provided unpublished permanent-plot data for the years 2000 and 2001. Jörg Perner and Patricia Purdy helped with Figure 13.1. This study was sponsored by the *Deutsche Forschungs-*

gemeinschaft (DFG, German Research Council), Graduate Research Group on Functioning and Regeneration of Degraded Ecosystems (Project GRK 266).

REFERENCES

Angevine, M. W.; and Chabot, B. F. 1979. Seed germination syndromes in higher plants. In *Topics in Plant Population Biology*, ed. O. T. Solbrig, S. Jain, G. B. Johnson, and P. H. Raven, 188–206. New York: Columbia University Press.

Ash, H. J.; Gemmell, R. P.; and Bradshaw, A. D. 1994. The introduction of native plant species on industrial waste heaps: a test of immigration and other factors affecting primary succession. *Journal of Applied Ecology* 31:74–84.

Bakker, J. P.; Dijkstra, M.; and Russchen, P. T. 1985. Dispersal, germination and early establishment of halophytes and glycophytes on a grazed and abandoned salt-marsh gradient. *New Phytologist* 101:291–308.

Belcher, J. W.; Keddy, P. A.; and Twolan-Strutt, L. 1995. Root and shoot competition intensity along a soil depth gradient. *Journal of Ecology* 83:673–682.

Bertness, M. D.; and Callaway, R. 1994. Positive interactions in communities. *Trends in Ecology and Evolution* 9:191–193.

Beyschlag, W.; Ryel, R. J.; Ullmann, I.; and Eckstein, J. 1996. Experimental studies on the competitive balance between two Central European roadside grasses with different growth forms. II. Controlled experiments on the influence of soil depth, salinity and allelopathy. *Botanica Acta* 109:449–455.

Bradshaw, A. D. 1983. The reconstruction of ecosystems. *Journal of Applied Ecology* 20:1–17.

Brandes, D. 1999. Flora und Vegetation salzbeeinflusster Habitate im Binnenland: eine Einführung. In *Vegetation salzbeeinflusster Habitate im Binnenland*, ed. D. Brandes, 7–12. Brunswick, Germany: Universitätsbibliothek TU Braunschweig.

Callaway, R. M.; and Walker, L. R. 1997. Competition and facilitation: a synthetic approach to interactions in plant communities. *Ecology* 78:1958–1965.

Cooper, S. V.; and Jean, C. 2001. Wildlife succession in plant communities natural to the Alkali Creek vicinity, Charles M. Russell National Wildlife Refuge, Montana. Report to the U.S. Fish and Wildlife Service. Helena: Montana Natural Heritage Program.

Eliashevich, N. V. 1982. Year-to-year variability of floodplain meadows. *The Soviet Journal of Ecology* 12:147–155.

Ellenberg, H. 1953. Physiologisches und ökologisches Verhalten derselben Pflanzenarten. *Berichte der Deutschen Botanischen Gesellschaft* 65:350–361.

Ellenberg, H. 1954. Über einige Fortschritte der kausalen Vegetationskunde. *Vegetatio* 5/6:199–211.

Ellenberg, H. 1996. *Vegetation Mitteleuropas mit den Alpen in ökologischer, dynamischer und historischer Sicht.* Stuttgart: Ulmer.

Ellenberg, H. 1992. Zeigerwerte der Gefäßpflanzen (ohne *Rubus*). In *Zeigerwerte von Pflanzen in Mitteleuropa*, ed. H. Ellenberg, H. E. Weber, R. Düll, V. Wirth, W. Werner, and D. Paulißen, 9–166. Göttingen: Scripta Geobotanica XVIII.

Fenner, M. 1987a. Seed characteristics in relation to succession. In *Colonization, Succession and Stability*, ed. A. J. Gray, M. J. Crawley, and P. J. Edwards, 103–114. Oxford: Blackwell Scientific.

Fenner, M. 1987b. Seedlings. *New Phytologist* 106 (suppl.): 35–47.

Fitter, A. H.; and Peat, H. J. 1994. The ecological flora database. *Journal of Ecology* 82:415–425.

Fowler, J.; Cohen, L.; and Jarvis, J. 1998. *Practical Statistics for Field Biology*. Chichester, England: John Wiley and Sons.

Grime, J. P. 1979. *Plant Strategies and Vegetation Processes*. Chichester, England: John Wiley and Sons.

Grime, J. P.; and Jeffrey, D. W. 1965. Seedling establishment in vertical gradients of sunlight. *Journal of Ecology* 53:621–642.

Grime, J. P.; Mason, G.; Curtis, A. V.; Rodman, J.; Band, S. R.; Mowforth, M. A. G.; Neal, A. M.; and Shaw, S. 1981. A comparative study of germination characteristics in a local flora. *Journal of Ecology* 69:1017–1059.

Halvorson, G. A.; and Lang, K. J. 1989. Revegetation of a salt water blowout site. *Journal of Range Management* 42:61–65.

Heinrich, W. 1984. Über den Einfluß von Luftverunreinigungen auf Ökosysteme. III. Beobachtungen im Immissionsgebiet eines Düngemittelwerkes. *Wissenschaftliche Zeitschrift der Friedrich-Schiller-Universität Jena, Naturwissenschaftliche Reihe* 33:251–289.

Heinrich, W.; Perner, J.; and Marstaller, R. 2001. Regeneration und Sekundärsukzession: 10 Jahre Dauerflächenuntersuchungen im Immissionsgebiet eines ehemaligen Düngemittelwerkes. *Zeitschrift für Ökologie und Naturschutz* 9:237–253.

Hodkinson, D. J.; Askew, A. P.; Thompson, K.; Hodgson, J. G.; Bakker, J. P.; and Bekker, R. M. 1998. Ecological correlates of seed size in the British flora. *Functional Ecology* 12:762–766.

Hopkins, B. 1978. The effects of the 1976 drought on chalk grassland in Sussex, England. *Biological Conservation* 14:1–12.

Jäger, E. J.; and Werner, K. 2002. *Exkursionsflora für Deutschland/begründet von Werner Rothmaler. Gefäßpflanzen: Kritischer Band*. Heidelberg, Germany: Spektrum.

Keddy, P. A. 1992. Assembly and response rules: two goals for predictive community ecology. *Journal of Vegetation Science* 3:157–164.

Kitajima, K.; and Fenner, M. 2000. Ecology of seedling regeneration. In *Seeds: The Ecology of Regeneration in Plant Communities*. ed. M. Fenner, 331–359. Wallingford, UK: CAB International.

Lambers, H.; Chapin, F. S., III; and Pons, T. L. 1998. *Plant Physiological Ecology*. New York: Springer.

Larsen, S.; and Widdowson, A. E. 1971. Soil fluorine. *Journal of Soil Science* 22:210–221.

Lawton, J. H. 1987. Are there assembly rules for successional communities? In *Colonization, Succession and Stability*, ed. A. J. Gray, M. J. Crawley, and P. J. Edwards, 225–244. Oxford: Blackwell Scientific.

Mariko, S.; Kachi, N.; Ishikawa, S.-I., and Furukawa, A. 1992. Germination ecology of coastal plants in relation to salt environment. *Ecological Research* 7:225–233.

Marrs, R. H.; and Gough, M. W. 1989. Soil fertility: a potential problem for habitat restoration. In *Biological Habitat Reconstruction*, ed. G. P. Buckley, 29–44. Chichester, England: John Wiley and Sons.

McLaughlin, M. J.; Tiller, K. G.; Naidu, R.; and Stevens, D. P. 1996. Review: the behaviour and environmental impact of contaminants in fertilizers. *Australian Journal of Soil Research* 34:1–54.

Metzner, K.; Friedrich, Y.; and Schäller, G. 1997. Bodenparameter eines Immissionsgebietes vor und nach der Schließung eines Düngemittelwerkes (1979–1997). *Beiträge zur Ökologie* 3:51–75.

Michel, B. E.; and Radcliffe, D. 1995. A computer program relating solute potential to solution composition for five solutes. *Agronomy Journal* 87:126–130.

Milberg, P.; and Lamont, B. B. 1997. Seed/cotelydon size and nutrient content play a major role in early performance of species on nutrient-poor soils. *New Phytologist* 137:665–672.

Müller-Schneider, P. 1986. Verbreitungsbiologie der Blütenpflanzen Graubündens. *Veröffentlichungen des Geobotanischen Institutes der ETH, Stiftung Rübel* 85:1–263.

Partridge, T. R.; and Wilson, J. B. 1987. Salt tolerance of salt marsh plants of Otago, New Zealand. *New Zealand Journal of Botany* 25:559–566.

Pugnaire, F. I.; and Luque, M. T. 2001. Changes in plant interactions along a gradient of environmental stress. *Oikos* 93:42–49.

Rich, T. C. G. 1991. *Crucifers of Great Britain and Ireland.* London: Botanical Society of the British Isles.

Roughgarden, J.; and Diamond, J. 1986. Overview: the role of species interactions in community ecology. In *Community Ecology*, ed. J. Diamond and T. J. Case, 333–343. New York: Harper and Row.

Rozema, J. 1975. The influence of salinity, inundation and temperature on the germination of some halophytes and nonhalophytes. *Oecologia Plantarum* 2:367–380.

Schratz, E. 1934. Beiträge zur Biologie der Halophyten. I. Zur Keimungsphysiologie. *Jahrbücher für Wissenschaftliche Botanik* 80: 112–142.

Ungar, I. A. 1995. Seed germination and seed-bank ecology in halophytes. In *Seed Development and Germination*, ed. J. Kigel and G. Galili, 599–628. New York: Marcel Dekker.

Walker, L. R.; and Chapin, F. S., III. 1987. Interactions among processes controlling successional change. *Oikos* 50:131–135.

Weiher, E.; and Keddy, P. A. 1999. Assembly rules as general constraints on community composition. In *Ecological Assembly Rules: Perspectives, Advances, Retreats*, ed. E. Weiher and P. Keddy, 251–271. Cambridge: Cambridge University Press.

Wilson, J. B. 1999. Assembly rules in plant communities. In *Ecological Assembly Rules: Perspectives, Advances, Retreats*, ed. E. Weiher and P. Keddy, 130–164. Cambridge: Cambridge University Press.

Wilson, J. B.; and Gitay, H. 1995. Limitations to species coexistence: evidence for competition from field observations, using a patch model. *Journal of Vegetation Science* 6:369–376.

Wisskirchen, R.; and Haeupler, H. 1998. *Standardliste der Farn- und Blütenpflanzen Deutschlands.* Stuttgart, Germany: Ulmer.

Zar, J. H. 1999. *Biostatistical Analysis.* Upper Saddle River, N.J.: Prentice Hall.

Zedler, J. B.; and Beare, P. A. 1986. Temporal variability of salt marsh vegetation: the role of low-salinity gaps and environmental stress. In *Estuarine Variability*, ed. D. A. Wolfe, 295–306. San Diego: Academic Press.

Chapter 14

Order of Arrival and Availability of Safe Sites: An Example of Their Importance for Plant Community Assembly in Stressed Ecosystems

VICKY M. TEMPERTON AND KERSTIN ZIRR

Ecological communities subjected to extreme disturbances, such as mining or industrial pollution, can be stripped of vegetation and can take on characteristics of primary succession (Bradshaw 1997). In such systems, the invasion of plant species during regeneration after the disturbance usually depends on the availability of propagule sources in the surrounding area, since a seed bank often no longer exists (Strykstra et al. 1998). The order of arrival of organisms into a community can play an important role in determining which of several possible assembly trajectories the community takes (priority effects; see Drake 1991). A number of studies have shown such effects, using organisms such as spiders (Ehmann and MacMahon 1996), acacia ants (Palmer et al. 2002), aquatic microorganisms (Drake 1991), and nurse plants (Cavieres et al. 1998). Connell and Slatyer (1977) were among the first to formulate a clear theory of succession that included the idea that the identity and environmental effects of the first organisms (that is, priority effects) in primary succession are important for a community's further development. They proposed three models, facilitation, tolerance, and inhibition, to describe, respectively: a positive effect of the presence of early-successional species on the subsequent performance of late-successional species; a tolerance (neither negative nor positive) of the first organisms' presence by late-successional newcomers; and a negative effect on establishment of both early- and late-successional newcomers by the early occupants. Connell and Slatyer postulated that the relative importance of facilitation versus tolerance or inhibition changes as a community goes through succession (see Figure 13.5 in Chapter 13). However, it is very difficult to show unequivocally the workings of facilitation, tolerance, or inhibition in ecological communities. There is a growing consensus that all three processes can work

285

simultaneously, and the relative importance of any one process can change with the spatial scale and patchiness of the system.

Disturbance—in particular its quality, intensity, and frequency—also affects membership in a community (White and Jentsch 2001; also see Chapter 17). The role of disturbance in driving the availability of safe sites for plant germination and establishment has been recognized only recently (Denslow 1985, Bakker 1987, Hobbs and Huenneke 1992, Burke and Grime 1996, Buckland et al. 2001). There is now an increasing awareness that disturbance can be a very useful tool in ecological restoration, particularly when the community is locked in an undesirable state, under a certain threshold, and needs a kick to move beyond this threshold into a new, desired stable state (see Hobbs and Norton 1996, Chambers 1997).

A good deal of attention has been paid to negative interactions between plants, especially to the role of competition (for example, Connell 1980, Kelt et al. 1995, Tilman 1997, Holl 1998). Only recently—and often in conjunction with increasing evidence from mycorrhizal research of the importance of positive interactions between organisms (for example, Van der Heijden et al. 1998; also see Chapter 10)—has there been a renewed interest in direct positive interactions between organisms (Hacker and Gaines 1997). Additionally, the existence of nurse plants—plants of one species that provide a favorable microclimate for another plant species to emerge from the seed bank and establish itself—has been recognized. This phenomenon has been noted mainly in extreme environments such as high alpine (Schlag and Erschbamer 2000) or arid habitats such as deserts; examples include cactus (for instance, *Echinocereus*, *Mammillaria*, and *Peniocereus* species) growing in the shade and accumulated soil of *Larrea tridentata* (Franco and Nobel 1989). The use of nurse crops to aid regeneration in ecological restoration projects is a well-established technique (Pywell et al. 1995), but little critical attention has been paid to their potential role in the recovery of vegetation in polluted ecosystems (see Bradshaw 1983, Pywell et al. 1995, Bradshaw 1997).

Plants growing in chemically polluted ecosystems must adapt to various imbalances in nutrient availability caused by the chemical pollution. In many cases, the capacity of a plant to establish itself and grow in a polluted site depends on its ability to deal with stress caused by the abiotic conditions on the site; that is, its capacity to find safe sites (Urbanska 1997). Grime's (1979) theoretical model of primary ecological plant strategies (CSR theory) categorizes plants as those that prevail under conditions of high competitive exclusion (competitors), high stress (stress tolerators), high disturbance (ruderals), or combinations of the three conditions. Lichtenthaler (1996) describes stress as "a state of the plant under the condition of a force applied."

It is known that stress leads to physiological changes during growth and affects developmental processes.

The response of a plant to abiotic conditions brings a new factor into the discussion of priority effects in community development. A plant species may arrive at a site first but be unable to deal with the site's abiotic environment. The availability of safe sites for germination and establishment is by no means guaranteed. Because of the limited number of safe sites, the concept of ecological filters (see Part II, "Ecological Filters as a Form of Assembly Rule") becomes useful. Filters allow only certain species from the local pool to thrive in a certain habitat with its set of environmental and living conditions. A process of deletion takes place, analogous to a filtering out of those organisms not adapted to the habitat conditions (Keddy 1992, Zobel 1992, Weiher and Keddy 1995). The next step, related to the search for assembly rules, is the issue of ecological prediction: If one knows the local species pool surrounding a habitat, to what extent can one predict the future composition of the community?

Filtering effects change over time, however. Certain plant species, such as annual ruderals, escape over time by surviving dormant in a persistent seed bank in the soil (Thompson and Grime 1979, Gutterman 2000), waiting for the right conditions, the right safe sites, to germinate and establish themselves again. The human-induced dust disturbance described at the Steudnitz site (see Chapters 10, 11, and 13) has changed the abiotic filters through which invading species must pass to establish successfully (see Chapters 6 and 16).

The Study System

A grassland slope with underlying limestone geology, situated close to a phosphate ($CaNaPO_4$) fertilizer factory in Steudnitz, north of Jena, Germany (51°09' N, 11°06' E), was polluted with phosphate dust emissions from 1957 until 1990 (see Chapter 13 for more details about the site and pollution conditions). Soil conditions on the lower slope (situated in the prevailing wind direction), particularly in areas closest to the factory, were massively altered by the emissions, which caused a gradient of pollution over distance from the factory. Since the shutdown of the factory, the ecosystem has been recovering at an astonishing rate (see Chapters 9, 12, and 13; also see Heinrich et al. 2001). As one moves up the hill, there is a succession of vegetation types from the ruderal, Asteraceae-dominated vegetation on the lower slope; through a belt of quack grass–dominated vegetation (*Elymus repens* (L.) Gould s. Str.) on the central slope; to a grassland with ruderal characteristics and encroaching bushes and trees on the upper slope. During the time of dust emissions from the factory, *E. repens* was one of two species able to grow near the factory. Since the factory shutdown, *E. repens* has started to disappear slowly, while false oat grass (*Arrhenatherum elatius* L.) is expanding (Heinrich et al. 2001). One might expect that, over time, the spatial succession present as one moves up the hill will be reflected in a temporal development of the lower slope into the central slope and the central slope into the upper slope, ending up with a woodland across the whole area. Evidence from the first 10 years of recovery does not suggest that this is actually happening, however. The lower slope seems to be skipping the grassland stage, with the invasion and successful establishment of pioneer tree species such as *Fraxinus excelsior* L., *Betula pendula* Roth., or Salix *spp.*, and since 1999 there has been no new increase in the number of herbaceous species.

The Experiments

To assess the roles of dispersal, facilitation, and stress tolerance in the assembly of the degraded grassland, we set up the following experiments. First, we investigated the effect of a nurse plant, *Picris hieracoides* L., on the emergence and establishment of six plant species present on the central and upper sections of the grassland on a degraded slope. The central question was whether the absence of the six selected species on the lower slope was due to remnant soil pollution and other abiotic conditions or to dispersal limitations that kept these species out of the lower slope area. In addition, we performed laboratory experiments on the two dominant grass species in the grassland, *E. repens* and *A. elatius*, to investigate the importance of order of arrival and water stress for plant performance.

We hypothesized that

- Disturbance and nurse plants are important providers of safe sites for emergence and establishment in the regenerating grassland.
- The performance of *E. repens* and *A. elatius* is affected by the additional factor of water stress.
- Order of arrival is important for the development of the plant community: the performance of a grass species that arrives after another grass species is affected negatively or positively by the presence of the first plant species (priority effect).

Plant Material

Plant nomenclature follows Wisskirchen and Haeupler (1998). We collected seeds of *Torilis japonica* (Houtt.) D.C., *Pastinaca sativa* L., *Inula conyzae*

(Griess.) Meikle, and *Achillea millefolium* L. directly from plants at the study site. This was not possible for *A. elatius*, *E. repens*, and *Dactlyis glomerata* L., so we obtained seeds from a north German seed-cultivation business (Deutscher Saatgut Verein, Lippstadt). We used transplanted adult rosettes of *P. hieracoides* plants growing at the Steudnitz site as nurse plants.

Soil Chemical Analysis

We assessed the relative importance of the abiotic filter (in the form of availability of nutrients and presence of toxic elements) by analyzing soil samples from the experimental site. Soil samples were taken at the Steudnitz site in July 2001 with a soil corer 2 cm in diameter and 10 cm in length. Ten soil cores of 10-cm depth were taken in three plots per block (for experimental design, see the section "Nurse Plants and Disturbance" later in this chapter) and mixed to form one sample per block, giving a total of six mixed samples for the whole experiment. The soil samples were sieved to 2 mm and any plant material removed. In the lab, each sample was ground and analyzed for total soil N (percentage of dry soil) (Hendershot method), P, K, Ca, Mg, and Cd, as well as for plant-available P and K (mg/100 g soil). Cd, P, Ca, Mg, Na, and K were extracted using aqua regia (HCl and HNO_3) and measured for cations using inductively coupled plasma–optical emission spectroscopy (ICP-OES). Plant-available P and K were extracted using a mixed solution of calcium lactate and calcium acetate (CAL method, Schüller 1969).

We analyzed soil-chemistry data with a one-way ANOVA using the three main blocks as treatment factors. These and all other analyses reported here were performed using SPSS.

Although the elements P, Na, N, Cd, K, and plant-available K were found at varying concentrations across the gradient, no significant block (gradient) effect was found (Table 14.1). There was a gradient effect for the elements Mg ($p = 0.0006$) and Ca ($p = 0.027$), however, with Mg increasing and Ca decreasing farther away from the emission source.

Nurse Plants and Disturbance

Degraded ecosystems such as the Steudnitz grassland can often resemble the extreme abiotic conditions of arid desert or alpine habitats. In such systems, facilitation between different plant species in the form of nurse plant effects may play a crucial role in determining which species can establish themselves in the community. We used a multifactorial block design to study the effects of nurse plants and aboveground disturbance on plant assembly

TABLE 14.1

Mean and Standard Error of Total P, K, Na, Mg, Ca, Cd (mg/100 g soil) and Percent N Content (g · kg⁻¹) and Plant-Available P and K (mg/100 g soil) in Soil Taken from Three Different Sites along the Experimental Gradient of Three Blocks[1,2]

Element	Block number	Mean[3]	Standard error
P (total)	1	120.5	5.5
	2	112.5	1.5
	3	112.0	1.0
K (total)	1	3.0	0.4
	2	3.8	0.01
	3	4.5	0.2
Na	1	15.0	1.0
	2	12.5	0.5
	3	12.5	0.5
Mg	1	2.9*	0.2
	2	3.6*	0.005
	3	4.2	0.005
Ca	1	325.0*	6.0
	2	320.0	1.5
	3	299.0*	0.5
Cd	1	0.01	0.0002
	2	0.013	0.002
	3	0.012	0.001
N (total %)	1	0.15	0.03
	2	0.19	0.005
	3	0.23	0.02
P (plant available)	1	1.60	0.23
	2	1.38	0.008
	3	1.60	0.2
K (plant available)	1	0.23	0.001
	2	0.25	0.002
	3	0.37	0.06

[1]One standard error.
[2]Block 1 was closest to the factory (50-m distance), block 2 was at 150-m distance, and block 3 was situated 300 m from the factory.
[3]* = Statistical difference between main blocks at $p < .05$; $n = 2$ for each main block per element, $n = 6$ for each element.

at the Steudnitz field site. Thus, six blocks were established across a pollution gradient ranging from 50 to 300 m away from the former fertilizer factory. Each block consisted of three 1.5-m by 1.5-m plots, each of which was assigned one of the following treatments: A: no vegetation removal; B: aboveground vegetation removal; and C: aboveground vegetation removal plus nurse plant. Within each 1.5-m by 1.5-m plot, there were six subplots of 15 cm by 15 cm, each of which was randomly assigned one plant species for sowing. Before analysis, the six blocks were pooled into three main blocks across a gradient of pollution, at 50-m, 100-m, and 300-m distance from the factory; thus, each combination of disturbance with nurse plant and species had two replicates per block.

In December 2000, we transplanted adult rosettes of *P. hieracoides*, a dominant plant species on the lower slope of the field site whose abundance is increasing, from their natural positions to the experimental subplots designated as nurse plant treatments. We removed aboveground vegetation (using garden scissors) from appropriate plots once at the beginning of the growing season in 2001 (4 months after nurse plant transplanting) to simulate a natural disturbance. Seeds were sown following vegetation removal (both under nurse plants and on plots without nurse plants). Six native plant species—four forbs (A. *millefolium*, T. *japonica*, P. *sativa*, I. *conyzae*) and two grasses (E. *repens*, *Dactylis glomerata*)—were sown at a density of 0.4 g/m^2 (recommended by Stevenson et al. 1995 for calcareous grassland; number of seeds per gram was noted). In the nurse plant treatment, we lifted the leaves of the rosette, deposited the seeds on the soil under the leaves, and pressed the seeds lightly onto the soil. The species chosen for sowing were relatively abundant on the central and upper sections of the slope but did not occur on the lower slope (or occurred only in patches of very few individuals; for example, A. *millefolium* and P. *sativa*; Heinrich et al. 2001).

We made a census of the emergence of seedlings at regular intervals—initially every 2 weeks, followed by monthly recordings—by noting the spatial position of the seedlings and the number of individual stems. The highest number of seedlings on a given subplot during the growing season was divided by the number of seeds sown to obtain percent emergence. Timing of peak emergence varied among species but was generally about 1 month after the first emergence was recorded. Recording started on May 3, 2002, and ended on September 18, 2002.

We analyzed the number of seedlings and percent emergence of seedlings separately, because number of seedlings provided a continuous census over the growing season whereas percent emergence was used to assess the peak emergence rate of the growing season. We used a repeated-measures ANOVA to test the effect of nurse plant presence and vegetation removal on the number of emerged seedlings over time. We analyzed percent emergence with a one-way ANOVA using the time of emergence of the highest number of individual ramets or seedlings (that is, we calculated percent emergence only once, at the peak individual seedling numbers for the growing season). We used post hoc tests to identify which treatments were significantly different. Whenever data were nonnormally distributed, the data were log (for absolute numbers) or arcsine (for percentages) transformed before ANOVA analysis. We also analyzed the effect of blocks on emergence of individual seedlings using repeated-measures (split-plot) ANOVA.

Emergence rates were relatively low in all treatments, with $D.$ $glomerata$ showing the highest rate (up to 22 percent) and $I.$ $conyzae$ the lowest rate (< 1 percent; Figure 14.1). There were significant treatment effects on number of individual seedlings per subplot for the following species: $E.$ $repens$ ($p = 0.017$), $T.$ $japonica$ ($p = 0.015$), and $A.$ $millefolium$ ($p = 0.013$). Emergence rates (based on sowing density) gave slightly different results compared to the absolute number of individual seedlings counted. In general, either the presence of a nurse plant or vegetation removal had a positive effect on percent seedling emergence. For two species ($A.$ $millefolium$ and $D.$ $glomerata$), the removal of vegetation alone had the biggest effect on percent seedling emergence ($p = 0.001$ and $p = 0.067$, respectively; see Figure 14.1).

We also found nurse plant and vegetation effects on the absolute number of seedlings emerged (as opposed to the percent seedling emergence, which depended on how many seeds per species were sown), although these effects were less significant than they were for the percent emergence data. Time had a significant effect on the number of emerged seedlings in two of the six species sown: $A.$ $millefolium$ and $D.$ $glomerata$ (RMANOVA: $p = 0.03$ and $p = 0.003$, respectively), with peak emergence occurring at the end of May.

The presence of a nurse plant significantly increased percent emergence compared to the other treatments in $T.$ $japonica$ ($p = 0.001$), and this trend was nearly significant for $E.$ $repens$ ($p = 0.061$). There were no significant differences between treatments for either $P.$ $sativa$ or $I.$ $conyzae$ ($p > 0.05$). Emergence rates for $P.$ $sativa$ were generally so low that no real conclusions could be drawn from the data, apart from the lack of success of this species in emerging at this site.

Water Stress

Lack of water has been and still remains one of the dominant stresses characteristic of the abiotic filter in the Steudnitz grassland. Although quack grass ($E.$ $repens$) still dominates the vegetation on the central part of the slope, it has started to disappear slowly, whereas false oat grass ($A.$ $elatius$ L.) is expanding (Heinrich et al. 2001). It could be that $A.$ $elatius$ is less water stress–tolerant than $E.$ $repens$, so that the abiotic conditions changed during regeneration of the grassland (that is, topsoil formed slowly where sandy substrate initially existed during the degradation). A consequence would be that the relative dominance of these two grass species could switch as regeneration of the site proceeded. To assess the effects of water stress on plant growth of these two

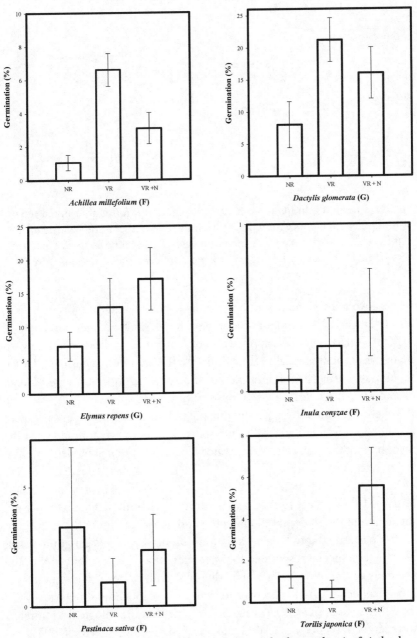

FIGURE 14.1 Mean percent emergence (with standard error bars) of six herb species (two grasses and four forbs, labeled G and F, respectively) sown in different disturbance regimes. NR: sown in plots with no nurse plant and no vegetation removal; VR: sown in plots with a single aboveground vegetation removal; VR+N: sown in plots under the leaves of the nurse plant (*Picris hiera-coides*).

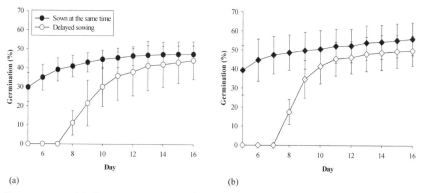

(a)   (b)

FIGURE 14.2 (a) Germination rate of *Arrhenatherum elatius* after initial sowing of E. repens. *(b) Germination rate of* Elymus repens *after initial sowing of A. elatius. Values are the means (± standard deviation) of nine replicates for each treatment. There was no significant difference between the treatments (*p < 0.05*).*

species, we grew E. *repens* and A. *elatius* in the greenhouse for 5 weeks at 18°C (day) and 15 °C (night) under natural light in Neubauer pots containing 600 g quartz sand. We applied an optimal fertilizer, developed by a German government agricultural institute (LUFA, *Landwirtschaftliche Untersuchungs- und Forschungsanstalt*), to the substrate. Species were planted according to three treatments: A: pure culture, with only one species per pot; B: "quarters," with both species in one pot but only one species in each quarter; and C: mixed culture, with both species per pot sown at random. At the beginning of the experiment, the water content was kept at 60 percent maximum water capacity (MWC); after the second week, it was decreased in half of the pots in each treatment to 45 percent MWC. The rest of the pots remained at 60 percent MWC as a control. We determined the biomass and the length of the roots and shoots at the end of the experiment.

We performed a two-way ANOVA on biomass and root length to assess how the two species responded to water stress when grown alone and in the presence of the other species. The different culture treatments had no significant effect on length of roots and fresh mass of roots and shoots ($p > 0.05$) for either species (E. *repens* or A. *elatius*), but all morphological parameters were larger at 60 percent MWC than at 45 percent MWC ($p < 0.003$).

Order of Arrival: Germination

As a further exploration of the competitive interaction between E. *repens* and A. *elatius*, we tested the effect of order of arrival by altering the timing of seed availability (delayed-sowing experiment). Seeds of *E. repens* and *A. elatius* were germinated in Neubauer pots containing 200 g quartz sand (40 percent MWC) at 20°–25°C for 16 days, under synthetic light (16 hours/day, 3500 lux). The seedling density amounted to 0.6 seedling/cm^2. Four different factors were applied in the experimental design: A: with one grass species sown at one time; B: with both grass species sown at the same time in one pot; C: with one species sown at two time intervals, one 5 days after the other (delayed sowing) in the same pot; and D: with the second species sown 5 days after the first (delayed sowing) in the same pot. We considered seeds to be germinated when the radix penetrated the testa. We counted seedlings every day to look for differences in the speed of germination. Within each experiment, there were three replicates per treatment, and the whole experiment was repeated three times at consecutive dates, giving a total of nine replicates per treatment factor.

We tested differences in germination rates between treatments using a one-way ANOVA with four factors (A, B, C, D; see above) and seven treatments: (1) *E. repens* (Er) sown alone; (2) *A. elatius* (Ae) sown alone; (3) both species grown together at the same time; (4) Er sown after Er; (5) Ae sown after Ae; (6) Er sown after Ae; and (7) Ae sown after Er. The germination rate of *A. elatius* reached 50 percent, compared to 60 percent in *E. repens*. There were no significant differences between any of the seven treatments (Figure 14.2; *E. repens* $p = 0.057$, *A. elatius* $p = 0.463$).

Order of Arrival: Preculture

Even when a particular species is no longer present in the extant vegetation, its presence at a particular point in the past could have an effect on how well conspecific individuals or plants of other species perform in the future. To test this hypothesis of the competitive interaction, we grew *E. repens* and *A. elatius*, in substrate in which the other species had been growing at an earlier time. *E. repens* and *A. elatius* were grown under synthetic light (16 hours/day, 3500 lux) in Neubauer pots containing 600 g quartz sand at 20°–25°C (60 percent MWC) for 6 weeks. We used the same optimal fertilizer as in the water stress experiment. As a preculture, we used sand as a substrate for growing *E. repens* or *A. elatius* for 4 weeks before the start of the experiment. The sand was not washed, but all plant particles were removed. The pots—including control pots without a preculture—were refertilized, so that all pots contained nutrients in saturation. This allowed us to distinguish the effects of preculture treatment from those of fertilization treatment. In addition to the control, we then applied four different treatments:

A = A. *elatius* with the same species in preculture; B = A. *elatius* with E. *repens* in preculture; C = E. *repens* with the same species in preculture; and D = E. *repens* with A. *elatius* in preculture. We determined the biomass and the mean length of all roots and shoots at 6 weeks.

We used a one-way ANOVA to see if the historical presence of a plant species in a substrate affected the performance of a subsequent species growing in that substrate. Specifically, we tested whether there were differences in biomass and in root and shoot length between the two preculture treatments. For A. *elatius*, the fresh mass of shoots was smaller without preculture than with preculture (of either species) ($p = 0.002$) (Figure 14.3). However, the two preculture treatments did not produce differences in either shoot length ($p = 0.655$) or root fresh mass ($p = 0.402$)—although length of the shoots ($p = 0.100$) and fresh mass of roots ($p = 0.362$) tended to be larger when the plants were grown after A. *elatius* than when grown after E. *repens*. Roots were significantly longer when the plants were grown with A. *elatius* rather than with *Elymus* as a preculture ($p = 0.021$).

For E. *repens*, the roots ($p = 0.002$) and shoots ($p = 0.001$) were longer and the fresh masses of roots ($p = 0.015$) and shoots ($p = 0.003$) were larger with preculture (of either species) than without. The length of the shoots ($p = 0.001$) and the fresh mass ($p = 0.005$) of the roots were significantly greater for plants grown after E. *repens* than for those grown after A. *elatius*.

Discussion

There is a growing consensus about the importance of natural disturbance for plant establishment (Canham and Marks 1985, Hobbs 1989, Chesson and Huntly 1997, Sakio 1997), which will have important consequences for the regeneration of communities in ecological restoration. In the disturbance experiment, all species except P. *sativa* were aided in their emergence by gap-creation disturbance (aboveground vegetation removal). Although Cavers and Harper (1967) did not create gaps in an experiment with introduced *Rumex* spp., they found a safe-site limitation for establishment but not for germination. They postulated a closed community that resisted invasion even though there were mature individuals of these species on-site and suggested that *Rumex* spp. were therefore probably remnants from a previous vegetation. It is possible that some form of disturbance at the *Rumex* site might open up the community to invasion. Eriksson and Ehrlen (1992) found that recruitment may be promoted by increasing microsite availability for germinating seeds (that is, safe sites), which usually was brought about by a disturbance of some kind. They also suggested that seed limitation in

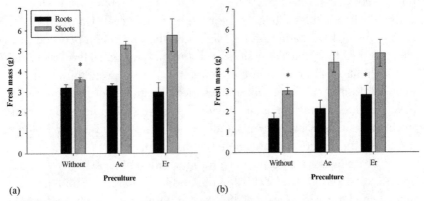

Figure 14.3 Fresh mass of roots and shoots of (a) *Arrhenatherum elatius* and (b) *Elymus repens* on different precultured sand. Treatments were: without preculture, with *A. elatius* (Ae) as preculture, or with *E. repens* (Er) as preculture. Values are the means (± standard deviation) of four replicates for each treatment. * indicates a statistically significant difference between treatments ($p < 0.05$).

plant communities is often more important than previously thought and that probably a combination of seed and microsite limitations acting simultaneously is a common reality. When seeds were artificially introduced with nurse plants, creating appropriate microsites, emergence occurred more often than with gap disturbance alone. Thus, we suggest that the current general lack of immigration from the top of the slope to the bottom of the slope of the species used in our study is related primarily to microsite limitation and secondarily to seed limitation.

Water availability is an important characteristic of a microsite. The current work showed that maximum water capacity had a larger effect on plant performance than whether a plant was grown alone or with other species. Very stressed plants often show visible changes in growth compared to unstressed plants (measurable as biomass or as shoot or root length; Swindale and Bidinger 1981, Fernández and Reynolds 2000). Jin et al. (2000) found that drought stress mostly affected plant shoots, which agrees with our findings. It is known that competition can also lead to a reduction in plant biomass (Fenner 1987, Gibson and Skeel 1996, Reader and Bonser 1998, Clausnitzer et al. 1999), although this was not shown in the current experiment. Even though we observed no clear morphological signs of stress through competitive interaction, it could be that the grass species in our study reacted at the biochemical level to such stress (see Sinnott 1960, Alscher and Cumming 1990, Bergmann et al. 1999). Plants are usually more sensitive to stress at the biochemical level than at the morphological level (Hauschild 1993, Lichtenthaler 1996), and further insight into the interaction between stress effects at the biochemical, physiological, and morphological level is needed.

Gradients in mineral pollutant content in the soil were found only for magnesium and calcium. Differences in soil texture (not measured) may have influenced the chances of individual plants establishing, but we found treatment effects of nurse plants and vegetation removal across all blocks, indicating their overriding effect on establishment.

Priority effects have been shown in quite different ecosystems (Connell and Slatyer 1977, Drake 1991, Ehmann and MacMahon 1996). It does not matter which species arrives first, however, if the abiotic conditions in a habitat are so extreme that only specialist plant groups can germinate and establish themselves there. Wagner (Chapter 13) found that for successful establishment, salt tolerance of incoming plant species at Steudnitz was a key required characteristic during the first few years of regeneration.

Biotic interactions (both negative and positive) between grass species could also have played a role in their changing abundance at the site over time. The influence of E. repens on the germination and growth of other species (especially dicotyledons) has been known for a long time (Bandeen and Buchholtz 1967, Kädling et al. 1990, Schulz et al. 1994). The influence of A. elatius on other species is not as well documented. Its competitive strength seems to depend on different abiotic and biotic factors (Campbell et al. 1991, Berendse et al. 1992).

In the preculture experiment, both grass species (E. repens and A. elatius) benefited more from growing in soil in which individuals of their own species had been grown previously than in soil without preculture or with preculture with the other species. Experiments of Eriksson and Eriksson (1998) described a positive effect on germination of sowing seeds after another species. They attributed the improved germination to an influence of the first sown species on the microenvironment, among other things. In the current study, preculture improved the growth of both species, perhaps due to improved nutrient availability via chemical effects in the microenvironment. On the other hand, emergence after delayed sowing showed that it was not important whether a grass species (E. repens or A. elatius) was sown at the same time or after a 6-day delay. In addition, the identity of the species first sown did not affect the germination rate (in other words, there was no nurse effect in the sense of Fenner 1987). It appears that the 6-day interval between sowing events was too short to engender any significant effects of order of arrival. Older plants growing on-site in the field may have a stronger effect on the growth of other neighboring species. Therefore, considering preculture and germination experiments together, nurse effects appear to be more important for early growth than for germination.

It seems that the current gradual decrease of *E. repens* and concomitant gradual increase of *A. elatius* at the Steudnitz site result from the different reactions of each species to pollutants in the soil (such as disproportions of macronutrients, high concentrations of heavy metals, high salinity) as well as from other environmental conditions such as drought (see Chapter 13). While the factory was working, *Elymus* could spread widely, because it could tolerate the existing soil conditions and other species could not (Heinrich 1984). *A. elatius* seems to be one of the species that could not establish while the soil was rich in pollutants (see Chapter 13). When pollutants in the soil gradually washed out (the abiotic filter changed; see Chapter 6), its establishment became possible. The current experiment suggests that *A. elatius* grows better on sand precultured with another grass (either of its own species or of *E. repens*). It could be that in Steudnitz, *E. repens* had a nurse effect on *A. elatius*, helping its establishment. Recently, Jones et al. (1994, 1997) made an important distinction between trophic effects and "physical ecosystem engineering" effects of one species on another. Trophic effects involve the consumption of resources, whereas nurse plants are physical ecosystem engineers in that they alter the environment by changing the availability and properties of resources, such as moisture and light (Corre 1983). Rousset and Lepart (2000) found that differences in microclimate under *Buxus sempervirens* and *Juniperus communis* shrubs in southern France, not differences in soil properties under the bushes, were responsible for improved germination of *Quercus humilis*. One of the most important findings from such studies is that the balance between positive and negative interactions can vary with the life-history stage of a plant (Sans et al. 1998). Nurse plants can provide safe sites for another plant's emergence (positive interaction) but then start to compete with the other plant once its needs change to having a safe site for establishment and persistence (negative interaction; Jones et al. 1997, Rousset and Lepart 2000). In the current study, the presence of a nurse plant initially had a positive effect on seedling emergence, but seedling establishment was not very successful once the seedlings had emerged (although we did not analyze this effect statistically). Flores-Martines et al. (1994) investigated the effect of cactus on the growth and fecundity of its nurse plant and found that not only did the cactus exert a strong negative effect on the nurse plant, but it could even replace the nurse plant in the long run. We observed a similar effect here: most of the nurse plants did not survive after a season of growth with invading seedlings.

The availability of safe sites for emergence and establishment is an important factor in the assembly of the Steudnitz grassland system. It seems that the order of arrival of plant species does affect the further development of the community, although the exact identity of the plants that arrive first might not play as important a role as initially hypothesized, given the results shown. However, the timing of arrival of individuals was important: later individuals did better than earlier ones in the preculture experiment. It remains to be seen whether the exact identity of the forerunner species plays a defining role in assembly if different nurse plant species are used in the field.

Results from such experiments would have implications for restoration of highly disturbed ecosystems, since nonspecificity of the nurse plant role would allow for the use of a range of different nurse plants to aid the establishment of selected species. On the other hand, should certain selected species require specific nurse plants to help them establish, the identity of these nurse plants will have to be determined by screening a variety of plants to find the one most suitable for encouraging further assembly of the community along the desired trajectory.

It is possible that restoration could be considerably facilitated in arid and extreme environments such as mining-spoil or dust-polluted sites, both by improving nutrient imbalances in the soil (see Bradshaw 1997 and Chapter 16 for examples) and by introducing suitable nurse plants for the establishment of desirable species. Callaway (1998) found a changeover from the relatively larger role of facilitation in more extreme abiotic environments to competition in more benign environments across an elevation gradient in the Rocky Mountains. Such a change in biotic filters forms one of the main tenets of Fattorini and Halle's dynamic filter model (see Chapter 6) for invasion of species into a regenerating system. As Hacker and Gaines (1997) point out, positive interactions play an important role in physically and biologically stressed systems and can have considerable effects on local species diversity through bottom-up effects (that is, effects at the local plant species–specific level that produce feedback effects on other plant species and also on other trophic levels).

Acknowledgments

We thank the various people who have helped in our research: Blandine Massonnet and Marzio Fattorini for help in the field; Annett Ruppe and the laboratory technicians at Agrar- und Umweltanalytik Jena for chemical analysis of soil samples; Tamara Gigolashvili and Hans Bergmann for discussions; and Markus Wagner, Tony Bradshaw, Ove Eriksson, and Tim Nuttle for constructive review of the manuscript. This research was funded by the *Deutsche Forschungsgemeinschaft* (DFG, German Research Council) as part of the Graduate Research Group on Functioning and Regeneration of Degraded Ecosystems (Project GRK 266).

REFERENCES

Alscher, R. G.; and Cumming, J. 1990. Stress responses in plants: adaptation and acclimation mechanisms. New York: Wiley-Liss.

Bakker, J. P. 1987. Restoration of species-rich grassland after a period of fertiliser application. In *Disturbance in Grasslands*, ed. J. Van Andel, J. P. Bakker, and R. W. Snaydon, 185–200. Dordrecht, The Netherlands: Junk.

Bandeen, J. D.; and Buchholtz, K. P. 1967. Competitive effects of quackgrass upon corn as modified by fertilization. *Weeds* 15:220–224.

Berendse, F.; Elberse, W. T.; and Geerts, R. H. M. E. 1992. Competition and nitrogen loss from plants in grassland ecosystems. *Ecology* 73:46–53.

Bergmann, H.; Lippmann, B.; Leinhos, V.; Tiroke, S.; and Machelett, B. 1999. Activation of stress resistance in plants and consequences for product quality. *Journal of Applied Botany* 73:153–161.

Bradshaw, A. D. 1983. The reconstruction of ecosystems. *Journal of Applied Ecology* 20:1–17.

Bradshaw, A. D. 1997. The importance of soil ecology in restoration science. In *Restoration Ecology and Sustainable Development*, ed. K. M. Urbanska, N. G. Webb, and P. J. Edwards, 33–64. Cambridge: Cambridge University Press.

Buckland, S. M.; Thompson, K.; Hodgson, J. G.; and Grime, J. P. 2001. Grassland invasions: effects of manipulations of climate and management. *Journal of Applied Ecology* 38:301–309.

Burke, M. J.; and Grime, J. P. 1996. An experimental study of plant community invasibility. *Ecology*, 77(3): 776–790.

Callaway, R. M. 1998. Competition and facilitation on elevation gradients in subalpine forests of the northern Rocky Mountains, USA. *Oikos* 82:561–573.

Campbell, B. D.; Grime, J. P.; Mackey, J. M. L.; and Jalili, A. 1991. The quest for a mechanistic understanding of resource competition in plant communities: the role of experiments. *Functional Ecology* 5:241–253.

Canham, C. D.; and Marks, P. L. 1985. The response of woody plants to disturbance: patterns of establishment and growth. In *The Ecology of Natural Disturbance and Patch Dynamics*, ed. S. T. A. Pickett and P. S. White, 197–216. New York: Academic Press.

Cavers, P. B.; and Harper, J. L. 1967. Studies in the dynamics of plant populations. I. The fate of seed and transplants introduced into various habitats. *Journal of Ecology* 55:59–71.

Cavieres, L. A.; Penaloza, A.; Papic, C.; and Tambutti, M. 1998. Nurse effect of *Laretia acaulis* (Umbelliferae) in the high Andes of central Chile. *Revista Chilena de Historia Natural* 71(3): 337–347.

Chambers, J. 1997. Restoring alpine ecosystems in the western United States: environmental constraints, disturbance characteristics, and restoration success. In *Restoration Ecology and Sustainable Development*, ed. K. M. Urbanska, N. R. Nigel, and P. J. Edwards, 161–187. Cambridge: Cambridge University Press.

Chesson, P.; and Huntly, N. 1997. The roles of harsh and fluctuating conditions in the dynamics of ecological communities. *American Naturalist* 150(5): 519–535.

Clausnitzer, D. W.; Borman, M. M.; and Johnson, D. E. 1999. Competition between *Elymus elymoides* and *Taeniatherum caput-medusae*. *Weed Science* 47:720–728.

Connell, J. H. 1980. Diversity and the coevolution of competitors, or the ghost of competition past. *Oikos* 35:131–138.

Connell, J. H.; and Slatyer, R. O. 1977. Mechanisms of succession in natural communities and their role in community stability and organization. *American Naturalist* 111:1119–1144.

Corre, W. J. 1983. Growth and morphogenesis of sun and shade plants. II: The influence of light quality. *Acta Botanica Neerlandica* 32(3): 185–202.

Denslow, J. S. 1985. Disturbance-mediated coexistence of species. In *The Ecology of Natural Disturbance and Patch Dynamics*, ed. S. T. A. Pickett and P. S. White, 307–324. New York: Academic Press.

Drake, J. A. 1991. Community assembly mechanics and the structure of an experimental species ensemble. *American Naturalist* 137(1): 1–26.

Ehmann, W.; and MacMahon, J. A. 1996. Initial tests for priority effects among spiders that co-occur on sagebrush shrubs. *Journal of Arachnology* 24:173–185.

Eriksson, O.; and Ehrlen, J. 1992. Seed and microsite limitation of recruitment in plant populations. *Oecologia* 91:360–364.

Eriksson, O.; and Eriksson, A. 1998. Effects of arrival order and seed size on germination of land plants: are there assembly rules during recruitment? *Ecological Research* 13:229–239.

Fenner, M. 1987. Seedlings. *New Phytologist* 106 (suppl.): 35–47.

Fernández, R. J.; and Reynolds, J. F. 2000. Potential growth and drought tolerance of eight desert grasses: lack of trade-off? *Oecologia* 123:90–98.

Flores-Martines, A.; Ezcurra, E.; and S. Sanchez-Colon. 1994. Effect of *Neobuxbaumia tetetzo* on growth and fecundity of its nurse plant *Mimosa luisana*. *Journal of Ecology* 82:325–330.

Franco, A. C.; and Nobel, P. S. 1989. Effect of nurse plants on the microhabitat and growth of cacti. *Journal of Ecology* 77:870–886.

Gibson, D. J.; and Skeel, V. A. 1996. Effects of competition on photosynthetic rate and stomatal conductance of *Sorghastrum nutans*. *Photosynthetica* 32:503–512.

Grime, J. P. 1979. *Plant Strategies and Vegetation Processes*. Chichester, England: John Wiley and Sons.

Gutterman, Y. 2000. Seed dormancy as one of the survival strategies in annual plant species occurring in deserts. In *Dormancy in Plants: From Whole Plant Behaviour to Cellular Control*, ed. M. Fenner, 139–159.Wallingford, UK: CAB International.

Hacker, S. D.; and Gaines, S. D. 1997. Some implications of direct positive interactions for community species diversity. *Ecology* 78(7): 1990–2003.

Hauschild, M. Z. 1993. Putrescine (1,4-diaminobutane) as an indicator of pollution-induced stress in higher plants: barley and rape stressed with Cr(III) or Cr(VI). *Ecotoxicology and Environmental Safety* 26:228–247.

Heinrich, W. 1984. Über den Einfluß von Luftverunreinigungen auf Ökosysteme. *Wissenschaftliche Zeitschriften der Friedrich-Schiller-Universität Naturwissenschaftliche Reihe* 33:251–289.

Heinrich, W.; Perner, J.; and Marstaller, R. 2001. Regeneration und Sekundärsukzession: 10 Jahre Dauerflächenuntersuchungen im Immissionsgebiet eines ehemaligen Düngemittelwerkes. *Zeitschrift für Ökologie und Naturschutz* 9:237–253.

Hobbs, R. J. 1989. The nature and effects of disturbance relative to invasions. In *Biological Invasions: A Global Perspective*, ed. J. A. Drake, H. A. Mooney, F. di Castri, R. H.

Groves, F. J. Kruger, M. Rejmanek, and M. Williamson, 389–405. New York: John Wiley and Sons.

Hobbs, R. J.; and Huenneke, L. F. 1992. Disturbance, diversity and invasion: implications for conservation. *Conservation Biology* 6:324–337.

Hobbs, R. J.; and Norton, D. A. 1996. Towards a conceptual framework for restoration ecology. *Restoration Ecology* 4:93–110.

Holl, K. D. 1998. Effects of above- and below-ground competition of shrubs and grass on *Calophyllum brasiliense* (Camb.) seedling growth in abandoned tropical pasture. *Forest Ecology and Management* 109:187–195.

Jin, S.; Chen, C. C. S.; and Plant, A. L. 2000. Regulation by ABA of osmotic-stress-induced changes in protein synthesis in tomato roots. *Plant Cell and Environment* 23:51–60.

Jones, C. G.; Lawton, J. H.; and Shachak, M. 1994. Organisms as ecosystem engineers. *Oikos* 69:373–386.

Jones, C. G.; Lawton, J. H.; and Shachak, M. 1997. Positive and negative effects of organisms as physical ecosystem engineers. *Ecology* 78(7): 1946–1957.

Kädling, H.; Weise, G.; Kreil, W.; Knabe, O.; Robowski, K.-D.; and Schuppens, R. 1990. Wert der Quecke (*Agropyron repens* L.) auf Graslandstandorten. *Archiv für Acker- und Pflanzenbau und Bodenkunde* 34:723–728.

Keddy, P. A. 1992. Assembly and response rules: two goals for predictive community ecology. *Journal of Vegetation Science* 3:157–164.

Kelt, D. A.; Taper, M. L.; and Meserve, P. L. 1995. Assessing the impact of competition on community assembly: a case study using small mammals. *Ecology* 76(4): 1283–1296.

Lichtenthaler, H. K. 1996. Vegetation stress: an introduction to the stress concept in plants. *Journal of Plant Physiology* 148:4–14.

Palmer, T. M.; Young, T. P.; and Stanton, M. L. 2002. Burning bridges: priority effects and the persistence of a competitively subordinate acacia-ant in Laikipia, Kenya. *Oecologia* 133:372–379.

Pywell, R. F.; Webb, N. R.; and Putwain, P. D. 1995. A comparison of techniques for restoring heathland on abandoned farmland. *Journal of Applied Ecology* 32(2): 400–411.

Reader, R. J.; and Bonser, S. P. 1998. Predicting the combined effect of herbivory and competition on a plant's shoot mass. *Canadian Journal of Botany* 76(2): 316–320.

Rousset, O.; and Lepart, J. 2000. Positive and negative interactions at different life stages of a colonizing species (*Quercus humilis*). *Journal of Ecology* 88:401–412.

Sakio, H. 1997. Effects of natural disturbance on the regeneration of riparian forests in the Chichibu Mountains, central Japan. *Plant Ecology* 132:181–195.

Sans, F. X.; Escarre, J.; Gorse, V.; and Lepart, J. 1998. Persistence of *Picris hieracoides* population in old fields: an example of facilitation. *Oikos* 83:283–292.

Schlag, R. N.; and Erschbamer, B. 2000. Germination and establishment of seedlings on a glacier foreland in the central Alps. *Arctic, Antarctic and Alpine Research* 32(3): 270–277.

Schüller, H. 1969. Die CAL-Methode, eine neue Methode zur Bestimmung des pflanzenverfügbaren Phosphates im Boden. Phosphates im Boden. *Zeitschrift für Bodenkunde* 123:48–63.

Schulz, M.; Friebe, A.; Kück, P.; Seipel, M.; and Schnabl, H. 1994. Allelopathic effects of living quackgrass (*Agropyron repens* L.): identification of inhibitory allelochemicals exuded from rhizome borne roots. *Journal of Applied Botany* 68:195–200.

Sinnott, E. W. 1960. *Plant Morphogenesis.* New York: McGraw-Hill.

Stevenson, M. J.; Bullock, J. M.; and Ward, L. K. 1995. Re-creating semi-natural communities: effect of sowing rate on establishment of calcareous grassland. *Restoration Ecology* 3(4): 279–289.

Strykstra, R. J.; Bekker, R. M.; and Bakker, J. P. 1998. Assessment of dispersule availability: its practical use in restoration management. *Acta Botanica Neerlandica* 47(1): 57–70.

Swindale, L. D.; and Bidinger, F. R. 1981. Introduction: the human consequences of drought and crop research priorities for their alleviation. In *Physiology and Biochemistry of Drought Resistance in Plants*, ed. L. G. Paleg and D. Aspinall, 1–13. Sydney: Academic Press.

Thompson, K.; and Grime, P. J. 1979. Seasonal variation in the seed banks of herbaceous species in ten contrasting habitats. *Journal of Ecology* 67(3): 893–921.

Tilman, D. 1997. Mechanisms of plant competition. In *Plant Ecology*, ed. M. Crawley, 239–261. Oxford: Blackwell Scientific.

Urbanska, K. M. 1997. Safe sites: interface of plant population ecology and restoration ecology. In *Restoration and Ecology of Sustainable Development*, ed. K. M. Urbanska, N. R. Webb, and P. J. Edwards, 81–110. Cambridge: Cambridge University Press.

Van der Heijden, M. A. G.; Klironomos, J. N.; Ursic, M.; Moutoglis, P.; Streitwolf-Engel, R.; Boller, T.; Wiemken, A.; and Sanders, I. R. 1998. Mycorrhizal fungal diversity determines plant diversity, ecosystem variability and productivity. *Nature* 396:69–72.

Weiher, E.; and Keddy, P. A. 1995. The assembly of experimental wetland communities. *Oikos* 73:323–335.

White, P. S.; and Jentsch, A. 2001. The search for generality in studies of disturbance and ecosystem dynamics. *Annals of Botany* 62:399–450.

Wisskirchen, R.; and Haeupler, H. 1998. *Standardliste der Farn- und Blütenpflanzen Deutschlands.* Stuttgart: Ulmer.

Zobel, M. 1992. Plant species coexistence: the role of historical, evolutionary and ecological factors. *Oikos* 65:314–320.

Chapter 15

Are Assembly Rules Apparent in the Regeneration of a Former Uranium Mining Site?

HARTMUT SÄNGER AND GOTTFRIED JETSCHKE

Mining dumps are often an inevitable consequence when aboveground or belowground resources are heavily exploited. After mining ceases, it is often necessary to revitalize the site and convert it back to natural or seminatural conditions. Hence, practical steps must be taken to restore the site to its previous conditions or to create comparable habitats. Often the first goal of restoration is simply to stabilize the substrate or develop a vegetation cover. Still, one must create or re-create not only a certain structure but also the corresponding function of an ecosystem (Bradshaw 1984, also see Chapter 16). If the time frame for restoration is short, special techniques must be applied. If the time frame is longer (on the order of 5–10 years or more), however, it may work better to rely on the processes of natural (that is, spontaneous) succession.

Mining dumps usually comprise extreme environmental conditions of microclimate, soil, and other properties. Only specially adapted plant species can survive and start vegetation development. The observation and study of the varied successional processes occurring on mining dumps can provide insights into the mechanisms that control the establishment of plant communities on abandoned and previously degraded sites in general. Unassisted succession is probably the simplest way to restore such sites, but restoration ecologists very often require (and desire) smaller or larger operational steps in order to reach a specified goal more quickly. To succeed in doing this, we need a better understanding of the driving forces of plant community succession.

Most authors agree that repeatable patterns of community development imply the existence of certain assembly rules (Lawton 1991, Keddy 1992, Wilson and Gitay 1995, Belyea and Lancaster 1999, Weiher and Keddy 1999, Wilson 1999). According to Wilson and Gitay (1995:374), such assembly rules are "ecological restrictions on the observed patterns of species presence or abundance that are based on the presence or abundance of one or more other . . . species." Weiher and Keddy (1999:267) see assembly rules as "explicit constraints that limit how assemblages are selected from a larger species pool." There are at least two schools of thought: one being that assembly rules should be based on interactions of the organisms alone, the other being that interactions of the organisms with the environment should be included (see Chapter 3 for a detailed discussion). We adopt the point of view that abiotic factors should be included because the filtering effect of abiotic conditions on the composition of a developing community seems to be very important for a plausible (and causal) explanation of the observed vegetation processes on our mining dumps (see Chapter 6). But we also consider any influence of early colonists on later ones as an integral part of possible assembly rules. Hence, we consider the existence of assembly rules as both the consequence of all interactions between the species in a real or potential community and the selective influence of the abiotic conditions at a given site.

In this chapter, we interpret our long-term data on natural successional changes occurring on dumps from uranium mining to assess whether the observed changes in the vegetation result from assembly rules or are merely caused by random events and influences. If assembly rules are evident, we may also ask how they are shaped under the extreme conditions of such sites. The answer, finally, will also contribute to the discussion of conceptual models of assembly (see Chapter 3).

Site Description: History and Current Vegetation Processes

Uranium was mined extensively in East Germany by SDAG Wismut (under strict control of the Soviet Union) between 1951 and 1990. The center of activity in present-day Thuringia was near the town of Ronneburg; the site has an overall area of 1322 ha, including 29 shafts, an opencast mining pit of 160 ha (with a volume of 84 million cubic meters), and 13 dumps with an overall area of 552 ha (and a volume of 179 million cubic meters). Although nearly all dumps were more or less regularly monitored, the data presented in this study come mainly from the special dump complex Paitzdorf; it is 25 ha in size (and 7.6 million cubic meters in volume). It was first constructed in 1966 as two adjacent conelike dumps (rising 100 m above the surroundings); after 1978, it was extended with some plateaulike parts of different heights, up to 35 m. The nearest settlements are between 600 and 1000 m from the dump. An aerial photograph of that region taken in 1992 is shown in Figure 15.1.

This dump complex is very interesting because it contains several dump types with different extents, slopes, and directions. A well-documented his-

FIGURE 15.1 The Paitzdorf uranium mining dump complex with surrounding area (in 1992). Photo: Wismut GmbH.

tory of construction allows correct dating of the observed vegetation. The surrounding landscape consists of habitats typical of the region, including agricultural fields, meadows, small woodlands, ruderal places, and settlements. Hence, the potential species pool for recolonization is rather large.

As a consequence of the mining activity, the dumps are more or less adverse biotopes that make high demands on pioneer species during the early phases of colonization: coarse-grained deposits, acidic raw soils, high surface temperatures in most places (because of the dark color of the substrate), and hard drainage water with high contents of sulfate, chloride, and uranium. The lithological composition of the deposited material is rather homogeneous, consisting mainly of schist, limestone, and diabase, with the main minerals being mica, quartz, chlorite, siderite, and pyrite.

The climate of the area is characterized by a mean annual temperature of 8°C and a mean annual precipitation of 630 mm (with a slight maximum during the summer months). The mean daily maximum temperature ranges from 8°C (in winter) to 24°C (in summer), the corresponding minimum from −10°C to 12°C. Hence, the macroclimate is moderate, slightly humid and warm, and typical for the temperate zone. The annual variation of average daily temperature and monthly rainfall is shown in Figure 15.2.

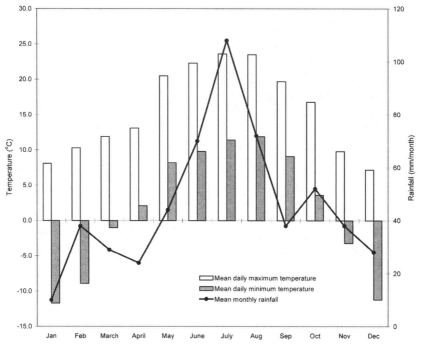

FIGURE 15.2 Mean daily temperature minima and maxima and mean monthly rainfall at the Paitzdorf mining dump site.

The area surrounding the dumps is primarily agricultural, being used either as fields for cereal production or as meadow or pastureland for cattle, with some woodland patches included. Hence, it forms a mosaic of different plant communities that serve as a source of diaspores. The potential natural vegetation of the region is a forest dominated by hornbeam, oak, and beech (Galio-Carpinetum; Scamoni 1964) with several layers of tree species, a well-developed shrub layer, and a species-rich herb layer with many diagnostically important species (*Galium sylvaticum, Stellaria holostea, Ranunculus auricomus, Campanula trachelium, Pulmonaria officinalis, Dactylis polygama, Brachypodium sylvaticum,* and *Festuca heterophylla*; plant names follow Wisskirchen and Haeupler 1998). Under these site conditions, the dumps are ideal sites for studying the pathways of primary succession and determining whether these changes are controlled by assembly rules or show a more random and nonregular pattern.

The Paitzdorf dump has been extensively monitored for many years. Vegetation and selected abiotic parameters (microclimate, plant-available nutrients) have been systematically recorded since 1992. The immigration and spread of diaspores, the germination ability of dominant species, the age structure of the trees, and successional changes have been recorded since 1996. As a result, many details of primary succession on uranium postmining dumps are known (Frank and Sánger 1994, 1995; Sänger and Vogel 1998).

In the last few years, the dump has been completely characterized with respect to species composition and abundance of vascular plants, mosses, and fungi. We measured 570 relevés according to the Braun-Blanquet method. Because the dump consists of several parts with different ages, slopes, and directions, we found a great variety of plant cover, including 255 vascular plant species altogether. We can distinguish three main vegetation types on the dump: (1) adverse sites only sparsely covered or nearly free of vegetation, (2) sites dominated by a dense layer of grasses and herbs, and (3) sites dominated by a pioneer birch woodland interspersed with oaks and pines. Photographs of the latter two types are shown in Figure 15.3. These observations allow a rough description of possible successional pathways (which are discussed in more detail later in this chapter).

About 20 plant associations have been described and typified; most of them are known in plant sociology, but a few do not fit into existing classifications and seem to be typical for certain mining sites. The wooded regions are similar to natural birch-oak associations on acidic soils (Betulo–Quercetum roboris). The early stages of herb grass communities are characterized as Echio-Melilotetum, Lactuco-Sisymbrietum, or similar pioneer communities; later, herbs such as *Tanacetum vulgare*, *Artemisia vulgaris*, and *Solidago canadensis*, as well the grass *Calamagrostis epigejos*, become dominant, together with small, woody *Rubus* spp.

Immigration and Spread of Species

The establishment of a plant community on a previously empty site can be initiated either from the seed bank (if there is already some soil containing seeds) or by immigration of species from outside through dispersal of diaspores. Before site conditions—such as water, nutrient, and light availability—and possible assembly rules can control the composition of the developing plant community, the actual arrival of viable diaspores determines the pool of applicant plant species (see Gleason 1926, Ridley 1930, Müller-Schneider 1977, Schupp 1995, Zobel 1997). Therefore, we consider availability of species an essential part of any assembly rule (as also suggested by Lockwood and Samuels in Chapter 4).

FIGURE 15.3 The two typical alternative vegetation patterns in early and intermediate succession: areas dominated by certain herbs and grasses (*left*) and areas dominated by a few pioneer tree species, mainly birch, willow, and poplar (*right*).

Immigration of Species from Outside

The 255 plant species found on the Paitzdorf dump contain 38 percent of the flora of the surrounding area up to a distance of 2000 m (675 species). This number is of the same order as for most of the other dumps nearby, although some larger deviations can be found. For example, the Drosen dump, with nearly the same size and age, carries only 147 species, but its surroundings are entirely farmland, whereas the Paitzdorf dump is embedded in a mosaic of different kinds of land use. This difference underlines the importance of the composition of the local species pool as the set of potential colonizer species.

What minimum distance must the various plant species have traveled to arrive at the dump? To answer this question, we split the surrounding area into concentric rings with boundaries at 50, 200, 500, 1000, and 2000 m distance from the borderline of the dump and numbered these rings 1 through 5, with 1 being closest. It is interesting that almost all species (there are three exceptions) show a monotonic pattern of occurrence in the sense that a species either occurs in all rings or does not occur in some inner rings but occurs in all other, farther rings. This pattern probably reflects the increase in ring area with distance from the dump.

Comparing the list of species found on the dump with those found in the five rings, we discovered that 122 of the dump species (48 percent) also occur from ring 1 on; hence, they most likely invaded from the nearest ring (< 50 m). Forty-three of the dump species (17 percent) occur in rings 2 to 5; hence, they had to cross a distance of 50 m or more. Another 43 species (17 percent) occur only in rings 3 to 5 and so traveled at least 200 m. Twenty-four species (9 percent) are found only in rings 4 and 5 and so must have traveled at least 500 m. The remaining 23 species (9 percent) occurring on the dump are found solely in ring 5, so they must have traveled ≥ at least 1000 m to arrive at the dump. There are no species on the dump that cannot be found within the surrounding 2000 m (see Figure 15.4a).

To look at the distance effect in another way, we asked what proportion of species that occur at different distances actually became established at the dump. These results are shown in Figure 15.4b. Even though the outer rings contained more species than the inner rings, and probably included more individuals of each species (producing more diaspores collectively) because of their greater area, these species were less likely to become established on the dump because of the greater distance they had to travel.

We can see clearly that the longer the distance to the dump, the lower the chance that a species will be found on the dump. Species that must cross a longer distance are therefore less likely to be included in the set of pioneer species, which shows the importance of dispersal. Therefore, mainly the species from the immediate neighborhood of the dump (which are also those found at almost all distances) need to be considered for the further action of possible assembly rules. We will see later that only a small portion of these highly frequent species will control the further vegetation development of the site.

Dispersal agent as well as distance affected which species colonized the dump. We classified the species according to their main agent of dispersal (although, obviously, many species will use more than one agent) and compared the frequencies of these dispersal types with those of the typical flora of East Germany (Frank and Klotz 1990). Seventy percent of the species found on the dump are dispersed mainly by wind (Figure 15.5). This percentage is substantially higher than the average percentage of wind-dispersed species in East Germany (59 percent) and shows the importance of wind as the primary dispersal agent. In relative terms, the difference is even more remarkable with respect to dispersal by animals: adhesive dispersal is much less common, but digestive dispersal (mainly by birds) plays a more important role than in the East German flora at large.

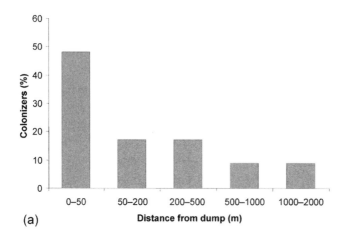

(a)

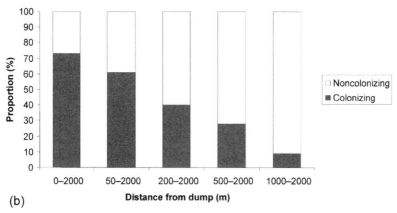

(b)

FIGURE 15.4 Relationship between the dump flora and the flora of the sur-rounding area at different distances. (a) Of the 122 species present on the dump, bars show how far they must have traveled by the minimum distance at which they are found in the surrounding area (origin of species). (b) Of the species present in each distance category, the proportion that successfully colonized the dump (establishment of species).

Spread of Species via Dispersal on the Dump

After the first species have reached the site, subsequent colonization can pro-ceed both from a further immigration of diaspores and from the local spread of diaspores of already established species. To obtain more information about

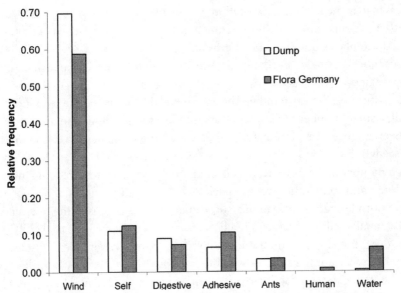

FIGURE 15.5 Relative frequency of the main different types of dispersal for plant species found on the Paitzdorf mining dump compared to all plant species of central Germany.

these sources, we put about 30 seed traps (each consisting of four collecting funnels with a diameter of 12 cm) across the site. They were emptied on a monthly basis over a 3-year period. The diaspores were determined to species level, and the number of diaspores of each species was counted.

At places nearly bare of vegetation, we still found a large number of diaspores (1000 to $10,000/m^2/year$); hence, there is no lack of potential colonists. The majority of species occurred in plant communities in the immediate neighborhood; they are mainly wind-dispersed and show C strategies ("competitors" in the sense of Grime 1979; species classified by Frank and Klotz 1990). But about 26 species were found only at distances of more than 10 m, so longer-distance dispersal from other areas of the dump or from outside is still important. Only for 9 species was short-distance dispersal across a few meters the most likely source; 4 of these species are in the top 10 most frequent species on the dump. Although many authors claim that trapped seeds essentially reflect the vegetation of the immediate surroundings (Robinson and Quinn 1988, Peart 1989, Bauer and Poschlod 1994, Bonn and Poschlod 1998), we can see that this does not hold true on the dump, at least for the areas that are nearly free of vegetation. These areas have a large input of seeds but act as a sink, since establishment is a rare event due to a lack of safe sites.

At the places dominated by herbs and grasses, about 10,000 to 100,000 seeds/m^2/year were recorded, mainly Asteraceae, Betulaceae, Caryophyllaceae, Papaveraceae, and Poaceae. Again, wind was the dominant agent of dispersal, but in this case the surrounding vegetation greatly dominated the seed input. Still, a small number of diaspores dispersed by animals (birds and ants) from longer distances were found. Overall, however, these patches serve as sources of seeds and tend to grow in size. In areas covered by birch trees, a similarly high number of diaspores were collected, but these consisted predominantly of birch and some other woody species. Rather small numbers of diaspores from outside species were found, suggesting that the crowns of the trees can substantially reduce the incoming flow of seeds (see also Nathan et al. 2002).

Along the slopes, which are mostly only sparsely vegetated, we found much smaller numbers of seeds in the traps, probably caused by turbulence effects. Since the slope soils are usually more coarse grained and have a more extreme microclimate, they are less easy to colonize and can therefore slow down the spatial spread of species.

Influence of Microclimatic Conditions

Throughout the literature, mining dumps in general have usually been described as extreme sites with respect to their microclimatic conditions (water balance, wind exposure, global solar radiation, temperature regime, and so forth). There are also unfavorable edaphic conditions (no soil or less developed soil, extreme pH, adverse chemical properties of the substrate). These conditions also exist at uranium mining dumps, but in addition, the dark color of the deposited slate material creates high temperatures in summer when, due to the sparse or missing vegetation, surface temperatures can reach more than 50°C. Such high temperatures create very unfavorable conditions for germination and establishment of plants. There are, of course, differences between north- and south-facing slopes; but generally, only species with a special adaptation can survive. Hence, high soil temperatures together with very low soil moisture and drying by wind act as harsh selection factors during the colonization process.

Influence of Soil Formation and Soil Chemistry

The pH of raw soils ranges from 7.5 near the surface to 6.5 below 20 cm depth, reflecting the slightly acidic properties of the deposited minerals. The

C:N ratio ranges from about 20 near the surface to 30 in lower soil layers—much below the extreme values known from brown coal and ore deposits. The deposited material of uranium dumps has a slightly higher average salt content than the surrounding area, but there is not much excess in chloride; hence, it is mainly the higher sulfate content (resulting from pyrite oxidation) that inhibits the development of vegetation on the dump.

A detailed analysis of soil samples shows that there is no lack of nutrients on the dumps. The phosphorus concentration (0.14–0.19 percent) is remarkably high compared to that of natural soils (although it is partially immobilized in calcium phosphate). Magnesium is also enriched, whereas potassium shows a rather natural value (0.20–0.29 percent). Therefore, uranium mining dumps clearly contrast with the common view of mining dumps as sites with extreme conditions for plant growth, at least with respect to nutrient availability. This contrast is supported by the fact that the deposited material weathers rather well and can form a near-surface layer of raw soil in less than 20 years. The absence of vegetation observed at some parts of the dump over many years can be explained only by factors other than nutrient availability: mainly, lack of diaspores and water and higher soil temperatures, salt concentrations, and pH at some parts of the dump. These factors can determine which species can colonize early and will be involved in the subsequent action of any assembly rules.

From the vegetation relevés, we can calculate the average Ellenberg indicator values and their variation along the dump. Ellenberg values are empirical integer numbers (ranging from 1 to 9) that characterize the requirements of a plant species for light, temperature, soil moisture, nutrients, pH, and other factors as they are observed in their natural communities in central Europe (Ellenberg 1979). These indicator values can be used to characterize the abiotic conditions at a given site quickly and to compare different sites or the same site at different times. Although these indicators are defined on an ordinal scale, often averages are calculated as for metric values. The averages of Ellenberg indicator values derived from the observed plant communities on the dump are rather close to the values characterizing the surrounding region, but variation is highest for reaction value (5.8–7.1), followed by nutrient value (5.3–6.1) and moisture, light, and temperature value. Because adequate nutrient availability is essential for ecosystem growth (Bradshaw 1984; also see Chapter 16), small-scale differences in soil properties and in microclimate are the most probable reasons for the early vegetation pattern observed on the dump, but average conditions do not differ substantially from those of the surrounding region.

Influence of Texture of the Substrate

A very important factor controlling the direction of vegetation development is provided by the grain size of the deposited and weathered material: if the grain size is large enough (greater than 5 cm), birches have a distinct advantage because they can germinate and establish better under these conditions. Conversely, many grasses and herbs colonize better at sites with fine-grained material (created as a result of weathering of larger particles). Consequently, variations in composition and early stages of weathering, combined with the randomness of diaspore inflow, may substantially determine whether a particular part of the dump gets colonized quickly by herbs and grasses or a pioneer birch woodland will develop. Hence, substrate grain size not only may influence the speed of revegetation (Martínez-Ruiz et al 2001) but also may determine the direction of succession. We will discuss this issue in more detail later in this chapter.

Biotic Interactions and Plant Strategy Types

We have seen in the previous two sections that the composition of the emerging plant communities on the dump is determined mainly by the availability of diaspores entering the site from the surrounding area and by the abiotic conditions of the site. Undoubtedly, sooner or later, biotic interactions also become relevant. Interactions between existing and arriving plant species can be competitive or facilitative and will further shape the community (Keddy 1992). The successful establishment and persistence of plant species out of the pool of potential colonizers, given existing conditions, is reflected in the prevailing plant strategies (CSR according to Grime 1979, split into the seven basic types). The results are shown in Figure 15.6: the most eye-catching difference between the mining dump and the surrounding region is the dominance of C and CSR species on the dump. Surprisingly, there are far fewer ruderals (R) and almost no stress tolerators (S) among the 255 dump species. This pattern indicates that after a few years, the conditions on the dump are favorable for competitive species to establish. Because public access to the dump is strictly prohibited, the site is not disturbed, and soon a shallow raw soil with high nutrient availability forms. Therefore, both ruderals and stress tolerators are outcompeted or displaced to some remaining areas with more extreme conditions.

If we sort the species by families, Asteraceae are the most dominant group, combining good wind-dispersal ability and strong competitiveness. Other families with a higher occurrence than in the regional pool (represented by

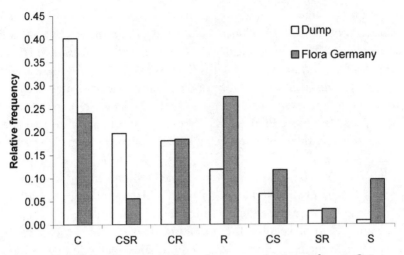

FIGURE 15.6 Relative frequency of the plant strategies (according to Grime 1979, as classified by Frank and Klotz 1990) of species found on the dump and of all species of central Germany.

the flora of East Germany) are Caryophyllaceae, Apiaceae, Polygonaceae, and Rosaceae, all of which have similar advantageous properties. Obviously, some tree species, especially birch, are well adapted to cope with the site conditions and, once established, to persist under competitive conditions. This persistence is supported by a mutualistic interaction with mycorrhizal fungi. Altogether, 14 tree species on the study site have been found to have mycorrhizal associations, and each species may have several fungi species forming mycorrhiza on it. The species with the most fungal species in association is birch, with 13 species, followed by pine (*Pinus sylvestris*), with 10 species. Together with some other genera (*Quercus, Acer, Populus, Carpinus, Salix*), these are the most dominant tree species on the dump. A systematic survey for pine seedlings showed that about 84 percent had mycorrhizal associations. Nearly all fungi species found on the dump form mycorrhizae. Hence, we conclude that mycorrhizal fungi constitute a very important facilitating factor during primary succession on mining dumps. Without these fungi, which arrive via spores from the surrounding fertile area, most of the tree species would probably not be able to colonize the dump successfully (Sheldon and Bradshaw 1975, Waaland and Allen 1987; see also Chapter 9).

Coming back to the observed sequences of succession, we find that there are about 10–15 species that are very frequent (and often locally abundant) on the dump. These include *Artemisia vulgaris* (C), *Betula pendula* (C), *Calamagrostis epigejos* (C), *Daucus carota* (CR), *Erigeron acris* (R), *Solidago canadensis* (C), *Tanacetum vulgare* (C), *Taraxacum officinale* (CSR), *Tripleurospermum perforatum* (CR), and *Tussilago farfara* (CSR). All these species are also very abundant in the surrounding landscape, have high seed production, and show a very effective method of long-distance dispersal by wind. These properties make them good colonizers, but they are also good competitors. These properties are reflected in their assigned plant strategy, given in parentheses.

Between 60 and 80 species arrived on the dump in the first 2 years, and many were observed to have germinated. However, only a small subset of these species will become dominant and suppress the other species, displacing them or keeping them at a low density. This process can be interpreted as the consequence of assembly rules related to plant strategy that control the composition of the developing community through biotic interactions (mainly competition and inhibition, but also mutualism and facilitation).

Alternative Stable States

We have already mentioned that the texture of soil or substrate in the early phases of colonization favors either herbs and grasses or birch. If by chance or local site conditions a particular plant community of grasses and herbs is formed first, it may, by means of clonal growth and local diaspore dispersal, establish a dense cover. This layer will prevent the further penetration of seeds from woody species and support a quasi-stable persistence of the grass-herb community over many years. Seeds from trees can establish only slowly at some sporadically created safe sites. If, on the other hand, the grain size of raw material and random events during colonization lead to a pioneer forest composed mainly of birch, the dense crown layer of the trees protects the site against many arriving wind-dispersed diaspores. Therefore, the spread of anemochorous species is reduced and only some zoochorous (that is, dispersed by animals) species are favored. As a consequence, the pioneer woodland, with a rather species-poor herb layer, can also persist over many years before a slow altering of species composition takes place. A simplified scheme of these alternative successional pathways is given in Figure 15.7.

These observations favor the view that some small and random differences both in substrate properties and in the early phases of diaspore immigration and settlement can essentially control the further development of the site over a long period. At least two alternative pathways of succession are

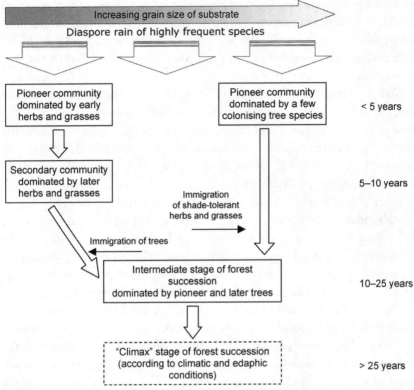

FIGURE 15.7 Pathways of primary succession on the Paitzdorf dump displaying current alternative quasi-stable states and probable eventual convergence.

possible, demonstrating that there are strong assembly rules controlling the composition of a developing community. In each case, whichever group of species establishes first creates conditions that tend to perpetuate that assemblage by excluding subsequent colonization by other species (the inhibition model of Connell and Slatyer 1977). Which assemblage that is depends on site properties and random events. The possibility and consequences of such alternative stable states are discussed in more detail in Chapters 3 and 4.

Discussion and Conclusions

Our observations have shown that the colonization and natural revegetation of uranium mining dumps can occur in a rather short period of less than 20 years. This result is in accordance with other observations (for example, Martínez-Ruiz et al. 2001). In spite of the rather harsh conditions on these dumps, many species are able to reach the dump and establish successfully.

During the early phases, abiotic factors may be very important as selection forces, but the availability of immigrating diaspores is also an essential element that affects the composition of the early communities. Later, biotic interactions become more pronounced and will strongly rule the next steps of succession.

We showed the possible origins of the species found on the dump. We can also ask why particular species from the surrounding area do not occur on the dump. We checked all 75 species occurring within the 200-m distance ring for their properties, requirements, and abundances and found that about 40 percent lack suitable habitat on the dump due to the harsh abiotic conditions. Twenty-five percent do not have a proper dispersal strategy to reach the dump soon enough (relative to other species), and 20 percent have very low abundance and so do not produce enough diaspores. Some species may be absent for several of these reasons, and others are just cultivated by humans under special conditions or may have other limitations. We can see again that an appropriate dispersal type, mainly wind dispersal, is very important for a species to be among the early colonizers of a nearby, empty site.

Because the surrounding area (up to 2000-m distance) has a high species richness (675 species) and a variety of biotopes, it provides a large regional species pool for the colonization of the dump. From the very beginning of the dump's existence, there has been an enormous input of diaspores, from which site conditions select suitable colonizers. Therefore, it is not very surprising that on the dump a certain set of about 10–15 species are highly frequent and often locally abundant. All these species combine high seed production with a very effective method of long-distance dispersal to colonize early, but they are also superior in competition with other species. Hence, these species mainly determine the subsequent steps of succession, running either through a community dominated by grasses and herbs or through a pioneer woodland stage. All other species are more or less interchangeable within a certain set of properties and interactions; they are subordinate actors in a game that is almost exclusively dominated by a few highly frequent species and can run only along the two main pathways displayed in Figure 15.7.

To answer our main question, there is no doubt about the action of assembly rules in the control of vegetation development on postmining dumps. Primarily, the availability of diaspores is essential to start the colonization of the dump. This availability is essentially determined by the local and regional species pool and by the subset of species with the capability for long-distance dispersal. After their arrival, abiotic site conditions further control which species actually establish on the dump. This combination of dispersal and abiotic factors is very important in determining the composition of early communities on the dump. Because there is no disturbance by humans or large animals and because of the rather quick weathering and raw soil formation, biotic interactions will become more and more important after a few years. We have mentioned reasons that uranium postmining dumps are a very special type of site, but they are not as extreme as many other postmining sites. The dump material, which was totally sterile upon deposition, rather soon becomes a fertile habitat for a particular set of species.

If we consider the rules for species composition and community development, so-called priority effects—in which the early arrival of organisms affects the subsequent assembly of the community—are clearly operating at the site. Small differences of substrate, diaspore input, and germination may determine how the communities emerging on the dump are built up. This may cause a certain switching effect between alternative pathways of succession that can persist over a relatively long period. Once a certain community (birch woodland or grass-herb) is established, the successful colonization of many other species will be inhibited or substantially delayed. Nevertheless, both pathways will eventually produce some natural forest community typical for the region, unless the area is managed to prevent it. Therefore, following the discussion in Chapter 3, although alternative stable states are evident in the relative short-term dynamics of uranium mining dumps, a deterministic endpoint can be predicted (also supporting Clements's 1916 and 1936 "organismic" concept of ecosystems).

In many cases, especially in temperate forest biomes, woodland or forest is the desired final stage of restoration, and its achievement is desired as soon as possible. Currently, controlled revegetation was started on a nearby postmining site after the dump material was leveled. Trees were planted at a very early stage of the active restoration effort (during the first year) into an almost sterile substrate to accelerate succession to a forest, but mortality rates after the first year were (predictably?) very high. On the other hand, about 60 species immigrated spontaneously during the first year; about 20 of them became rather frequent and partially formed a herbal layer. Moreover, some patches consisting mainly of birch with some oak and alder could be found after 3 years. Our observations emphasize that there is almost no better mechanism than spontaneous succession for revegetating and rehabilitating areas that have been degraded by mining activity. At least under the conditions of a moderate, humid climate in the temperate zone, natural succession may start and pass through its early stages within a couple of years. This contrasts with well-known observations from semiarid or arid regions where the wounds of human landscape change—for example, road construction— can be seen over decades. However, where conditions allow rapid weathering of dump material to provide sufficient nutrients, and where the surrounding area has sufficiently high species richness, almost nothing additional has to be done. Hence, the best course of action for a restorationist is to guarantee good conditions for natural succession (for example, covering the area with some raw soil if possible and necessary, avoiding erosion along slopes, and so forth), because it is a very powerful tool provided the conditions are not too extreme and the site is located in a diverse and intact natural landscape. Some moderate actions may facilitate restoration, but usually no major reconstruction is necessary. Knowing more about the assembly rules governing natural revegetation will not only give us better insights into the underlying ecological processes but also save a lot of money.

REFERENCES

Bauer, U.; and Poschlod, P. 1994. Ökologie und Management periodisch aufgelassener und trockengefallender kleinerer Stehgewässer im oberschwäbischen Voralpengebiet. Veränderung der Phytozönose durch Sömmerung am Beispiel von Gloggere- und Tiefweiher. *Veröffentlichungen PAÖ* 8:337–351.
Belya, L.; and Lancaster, J. 1999. Assembly rules within a contingent ecology. *Oikos* 86:402–416.
Bonn, S.; and Poschlod, P. 1998. *Ausbreitungsbiologie der Pflanzen Mitteleuropas.* Wiesbaden: Quelle and Meyer.
Bradshaw, A. D. 1983. The reconstruction of ecosystems: presidential address to the British Ecological Society, December 1982. *Journal of Applied Ecology* 20:1–17.
Bradshaw, A. D. 1984. Ecological principles and land reclamation practice. *Landscape Planning* 11:35–48.
Clements, F. E. 1916. *Plant Succession.* Publication No. 242. Washington, D.C.: Carnegie Institution.
Clements, F. E. 1936. The nature and structure of the climax. *Journal of Ecology* 22:9–68.
Connell, J. H.; and Slatyer, R. O. 1977. Mechanisms of succession in natural communities and their role in community stability and organization. *American Naturalist* 111:1119–1144.
Ellenberg, H. 1979. Zeigerwerte der Gefäßpflanzen Mitteleuropas. *Scripta Geobotanica* 9:1–106.
Frank, D.; and Klotz, S. 1990. Biologisch-ökologische Daten zur Flora der DDR. 2. Aufl. *Wissenschaftliche Beiträge Martin-Luther Universität Halle-Wittenberg* 32(P41): 1–167.
Gleason, H. A. 1917. The structure and development of the plant association. *Bulletin of the Torrey Botany Club* 44:463–481.
Gleason, H. A. 1926. The individualistic concept of the plant association. *Bulletin of the Torrey Botany Club* 53:7–26.
Grime, J. P. 1979. *Plant Strategies and Vegetation Processes.* Chichester, England: John Wiley and Sons.
Keddy, P. 1992. Assembly and response rules: two goals for predictive community ecology. *Journal of Vegetation Science* 3:157–164.

Lawton, J. H. 1991. Are there assembly rules for successional communities? In *Colonization, Succession and Stability*, ed. A. J. Grey, M. J. Crawley, and P. J. Edwards, 225–244. London: Blackwell Scientific.

Martínez-Ruiz, C.; Fernández-Santos, B.; and Gómez-Gutiérrez, J. M. 2001. Effects of substrate coarseness and exposure on plant succession in uranium-mining wastes. *Plant Ecology* 155:79–89.

Müller-Schneider, P. 1977. Verbreitungsbiologie (Diasporologie) der Blütenpflanzen. *Veröffentlichungen Geobotanisches Institut ETH, Stiftung Rübel* 61:1–226.

Nathan, R.; Horn, H. S.; Chave, J.; and Levin, S. A. 2002. Mechanistic models for tree seed dispersal by wind in dense forests and open landscapes. In *Seed Dispersal and Frugivory: Ecology, Evolution and Conservation*, ed. D. J. Levey, W. R. Silva, and M. Galetti, 69–82. Wallingford, UK: CAB International.

Olsson, E. G. 1987. Effects of dispersal mechanism on the initial pattern of field-forest succession. *Acta Oecologica* 8:379–390.

Peart, D. R. 1989. Species interactions in a successional grassland. I. Seed rain and seedling recruitment. *Journal of Ecology* 77:236–251.

Ridley, H. N. 1930. *The Dispersal of Plants Throughout the World*. Ashford, Germany: L. Reeve.

Robinson, G. R.; and Quinn, J. F. 1988. Extinction, turnover and species diversity in an experimentally fragmented California annual grassland. *Oecologia* 76:71–83.

Sänger, H. 1994. Flora and vegetation of uranium mining areas in the former GDR. In *Progress to Meet the Challenge of Environmental Change*, ed. J. H. Tallis, H. J. Norman, and R. Benton. Proceedings of the VI International Congress of Ecology, Manchester, UK.

Sänger, H. 1995. Flora and vegetation on dumps of uranium mining in the southern part of the former GDR. *Acta Societas Botanicorum Poloniae* 64:409–418.

Sänger, H.; and Vogel, D. 1998. Untersuchungen zur Flora und Vegetation in bergbaubedingt salzbelasteten Feuchtgebieten. *Hercynia N.F.* 31:201–227.

Scamoni, A. 1964. Karte der natürlichen Vegetation der Deutschen Demokratischen Republik (1: 500 000) mit Erläuterungen. Beiträge zur Vegetationskunde. *Feddes Repertorium* 6 (suppl.): 14.

Schupp, E. W. 1995. Seed-seedlings conflicts, habitat choice, and patterns of plant recruitment. *American Journal of Botany* 82:3:399–409.

Sheldon, J. C.; and Bradshaw, A. D. 1975. The reclamation of slate waste tips by tree planting. *Landscape Design, Journal of the Institute of Landscape Architecture* 113:31–33.

Waaland, M. E.; and Allen, E. B. 1987. Relationships between VA mycorrhizal fungi and plant cover following surface mining in Wyoming. *Journal of Range Management* 40:271–276.

Weiher, E.; and Keddy, P. 1999. Assembly rules as general constraints on community composition. In *Ecological Assembly Rules: Perspectives, Advances, Retreats*, ed. E. Weiher and P. Keddy, 251–271. Cambridge: Cambridge University Press.

Wilson, J. B. 1999. Assembly rules in plant communities. In *Ecological Assembly Rules: Perspectives, Advances, Retreats*, ed. E. Weiher and P. Keddy, 130–164. Cambridge: Cambridge University Press.

Wilson, J. B.; and Gitay, H. 1995. Limitations to species coexistence: evidence for competition from field observations, using a patch model. *Journal of Vegetation Science* 6:369–376.

Wisskirchen, R.; and Haeupler, H. 1998. *Standardliste der Farn- und Blütenpflanzen Deutschlands.* Stuttgart: Ulmer.

Zobel, M. 1997. The relative role of species pools in determining plant species richness: an alternative explanation of species coexistence? *Trends in Ecology and Evolution* 12:266–269.

Chapter 16

The Role of Nutrients and the Importance of Function in the Assembly of Ecosystems

ANTHONY D. BRADSHAW

Our thoughts are shaped by our experience. In the present-day world, our first experience of assembly is usually at home or nursery school, trying to fit shapes into holes and fitting bricks together to make shapes. Later, many of us progress to gluing pieces of plastic together to make more complex structures such as model aircraft or ships. But assembly remains a process of putting things together—just that and no more. Later, some of us go on to make models that work: cranes that lift objects or airplanes that fly. These working models introduce an altogether different aspect of assembly: the final assessment of success is not whether what we assemble looks good, but whether it actually does something. If you do not understand the difference, try talking to someone who has just completed his first flying model airplane and is trying to make it fly—if you can make contact between the bouts of fury.

Equally, we can consider car assembly lines. Considerable effort is expended in putting together many different pieces of material, until something emerges that looks like a car. But the final, important criterion is whether the car functions properly. Any manufacturer who tries to sell cars of which even 1 percent does not function is unlikely to prosper.

Function is a critical quality of the assembly process. Yet when we think about assembly, what usually comes to mind, prompted by our childhood experiences, is a simple process of putting pieces together. This image fails to appreciate what assembly is fundamentally about.

Here, arguments are presented that have been developed in relation to the restoration of ecosystems in rather extreme environments. Therefore, the chapter is included in Part IV, "Assembly Rules in Severely Disturbed Environments," yet the arguments are equally applicable to what can happen in more normal environments, because it will be appreciated (by plant ecologists in particular) that small variations in nutrients and other resources can have major effects on ecosystems and on the balance of species within them. It follows that nutrient resources and nutrient supply must be critical components of all assembly rules, although in some situations they will be less dominating than in others. A number of the publications referenced herein may seem rather old. But these citations reflect the fact that much of the work valid in the present context was of topical interest some time ago and has now reached the stage when it is often overlooked.

The Term *Assembly* Applied to Restoration

An ecosystem is an assemblage of species and individuals at a particular place. But when Tansley (1935) devised the term *ecosystem*, he also included climate and soil, with the implication that they, with species, formed an interacting whole. Therefore, we have realized that an ecosystem has many features, not only its plants and animals but also the functions taking place within it, such as growth, nutrient accumulation, and cycling. It is convenient to represent *structure* and *function* as two major attributes of an ecosystem, critical attributes that must be considered if we are to restore ecosystems (Bradshaw 1984). The two attributes can be the two axes of a graph on which different levels and types of ecosystem degradation and restoration can be represented (Figure 16.1).

If we apply these thoughts to the details of assembly of ecosystems, it becomes apparent that ecosystem function and the processes it involves are as important to successful assembly as the collecting together of species. Indeed, because an ecosystem is composed of living creatures, it is impossible for an ecosystem to exist without the processes that are part of the characteristic of function. In particular, an ecosystem cannot exist without growth, not least because many species arrive merely as small propagules.

Restoration, which is in effect a process of reassembly, must therefore take into account growth and other functional processes. In this respect, restoration copies the processes that occur in natural succession (Bradshaw 1983). Hence, current ideas about the processes occurring in natural succession may tend to dominate our thinking about the assembly rules we hope to apply in restoration. Specifically, our thinking tends to be dominated by the concepts originally enunciated by Connell and Slatyer (1977) about the mechanisms of succession. They identified three major processes—facilitation, tolerance, and inhibition—that determine the sequence of species, all of which are about interactions between species. This framework was innovative and excellent and has stimulated a great deal of thought about what

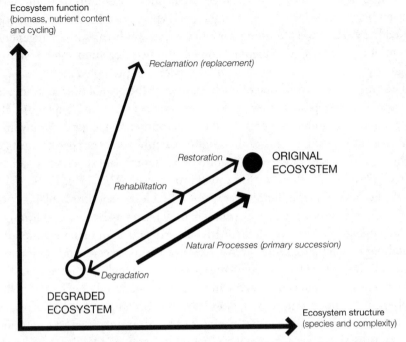

FIGURE 16.1 Diagrammatic representation of natural succession and restoration processes (after Bradshaw 1984). Ecosystem development can be quantified in the two dimensions of structure and function. Mineral nutrition is an important part of ecosystem function. In a degraded ecosystem, it is at a low level and will exercise considerable filtering effects on the incoming species. The mineral nutrition of the ecosystem will improve from natural physico-chemical and biological processes, strongly assisted by the activities of the incoming species. It will always exercise a strong control (filtering process) over the progress of ecosystem development, but the nature of this control will change markedly over time.

takes place in succession. But it has made it easy to overlook the actual underlying functions and resources that must be involved (for example, see Young et al. 2001). Although Connell and Slatyer pay some attention to resource acquisition, the general impression is that succession is dominated by the different ways space is made available for incoming species. So it has been all too easy to derive an image that ecosystem assembly is merely plugging the right pieces into the right slots. This chapter will examine problems with this approach in the context of plants, which is justifiable because of their fundamental contributions to all ecosystems.

Assembly rules envisage that there are explicit constraints that limit how assemblages are selected from a larger species pool (Keddy 1992, Weiher and Keddy 1999). At the Dornburg workshop on assembly rules in the regeneration of ecosystems, a conceptual model, the two-step filter model, subsequently recognized as a dynamic filter model, was presented and discussed. This model has formed one of the backdrops to the goal of the workshop: to strive for a conceptual framework for restoration ecology (see Chapter 6). The dynamic filter model accepts the trait-environment concept of Keddy and Weiher (1999) and envisages that, in the assembly process, species must go through a series of filters. The historic responses of plant species to changes in climate seem to illustrate this sort of process well (Díaz et al. 1999).

The problem with the filter concept is that it tends to imply that assembly is a process in which species merely undergo a passive "passing through slots" operation. Yet ecosystems are dynamic organizations having functions without which they cannot survive. As stated earlier, species and ecosystems cannot develop or be restored without growth—which, particularly in the plant component of ecosystems, can be limited by a number of factors of the physical environment; these factors must be corrected if growth is to take place. Assembly rules cannot therefore be constraints imposed between organisms alone (see Chapter 1). Abiotic filtering is a critical process, even if Wilson (1999) would prefer to exclude it from assembly rules. This exclusion seems to be a very unrealistic suggestion because assembly cannot take place without some sort of filtering. But these filters are complex; in particular, there are the interactions at any one time between the community of organisms and ecosystem processes (see Chapter 7). So filtering is dynamic.

Limiting factors are likely to be manifest particularly in severely degraded and derelict land, although they will also be important in less damaged ecosystems. It has been recognized for many years that restoration of all sorts depends on the successful recognition and treatment of limiting factors; as a result, these factors are well listed and studied in restoration literature (Table 16.1), although their effects on individual species are not always well documented. It is not improper to say that limiting factors represent a basis of assembly rules for ecosystem restoration, since ecosystems can be assembled only if the problems have been addressed in the ways suggested. They are all part of the complex of factors contributing to both abiotic and biotic filters, each of which can be manipulated to achieve restoration (see Chapter 4).

For each problem identified in Table 16.1, the real situation is complex, because of the nature of plants and their growth and adaptation to external factors—as any consultation of relevant ecological textbooks will show. To

TABLE 16.1

The Physical and Chemical Problems That Can Be Found in Degraded
Terrestrial Ecosystems and Their Short- and Long-Term Treatments if
Ecological Restoration Is to Be Achieved[1]

Category	Problem	Immediate treatment	Long-term treatment
Physical			
Texture	Coarse	Add organic matter or fine material	Establish vegetation
	Fine	Add organic matter	Establish vegetation
Structure	Compact	Rip or scarify	Establish vegetation
	Loose	Compact	Establish vegetation
Stability	Unstable	Apply stabilizer or incorporate nurse species	Regrade or establish vegetation
Moisture	Wet	Drain	Drain
	Dry	Irrigate or apply mulch	Use tolerant species
Nutrition			
Macronutrients	Nitrogen	Add fertilizer	Incorporate legume
	Others	Add fertilizer and lime	Add fertilizer and lime
Micronutrients	Deficient	Add fertilizer	
Toxicity			
pH	Low	Add lime	Add lime or use tolerant species
	High	Add pyritic waste or organic matter	Allow weathering
Heavy metals	High	Add organic matter or use tolerant plants	Apply inert covering or use tolerant plants
Salinity	High	Allow weathering or irrigate	Use tolerant species

[1]Data from Bradshaw 1983.

illustrate these complexities, as well as the control that functional characters can have on the process of assembly, this chapter will examine just one factor, mineral nutrition, to explore the importance of incorporating functional characters and their effects into a set of assembly rules.

Mineral Nutrients as a Critical Resource

Since plant and ecosystem growth cannot occur without adequate resources, resource availability and acquisition must be fundamental. It has been suggested that nutrient relations can be considered ingredients of a central "black box" of the model in which interactions occur (see Chapter 3). But relegating nutrients to a black box hardly does them justice, not the least because of all we know about them and their importance, particularly in relation to restoration. For example, Chapter 17 discusses the specific effects of disturbance on nutrients and other resources in the context of restoration.

There are four major resources that are critical to the development of an ecosystem. They are of two general types:

1. Renewable: water, light, CO_2
2. Nonrenewable: nutrients

Renewable resources are those that can be supplied from outside the ecosystem and are therefore usually readily available, although they may sometimes be available only in limited amounts; for instance, water in an arid climate. But their supply is not limited in the long term. In contrast, supplies of nonrenewable resources, those that are normally supplied from within the ecosystem, such as soil phosphorus, can be completely limited and therefore can be easily exhausted unless they are recycled. This means that a recycling system must be in operation. Mineral nutrients, as nonrenewable resources, can therefore exercise considerable control on ecosystem development and must be an important part of assembly rules.

In particular, we must remember that concentrations, and therefore supplies, of mineral nutrients can vary considerably between different environments. This variation occurs initially as a result of chemical variations in rock formations and the soils and drainage waters derived from them (for example, Clarke 1924). But it is also due to the physical accumulation or leaching processes that are features of particular environments, which can radically alter original nutrient concentrations.

In general, nutrient concentrations are most likely to be deficient in terrestrial situations, where the mobility of nutrients is restricted, and least likely to be deficient in aquatic situations, where water movement enables nutrient flux and maintains supply (Ehrenfeld and Toth 1997; also see Chapter 2). Terrestrial soils can also show considerable differences in nutrient concentrations, not only because of historical and chemical phenomena but also because of natural biological accumulation at the soil surface occurring as a result of the growth and foraging activities of plants (for example, Knabe 1973). For this reason, nutrients are least likely to be limited in long-established topsoils in temperate climates. But such accumulation processes can be prevented in soils in leaching climates, so great variation is possible. Because of these specific complexities, further consideration will be limited to what occurs in terrestrial ecosystems.

Mineral Nutrients in Restoration Situations

In restoration situations, where serious damage to an ecosystem has occurred, levels of mineral-nutrient supply are of particular significance. If all that has occurred has been the loss of the original vegetation cover, then mineral supply is unlikely to have been altered greatly because the supply will have remained in the soil. But in situations where there has been disturbance by mining or industrial activity, in which original soils are lost and subsoils or industrial materials left behind, there will inevitably be deficiencies of particular nutrients, although others may be in excess. The disturbances that are considered here are therefore rather larger than those considered by Fattorini and Halle in Chapter 6. They do not involve the loss just of species but of fundamental resources without which an ecosystem cannot function.

Restorationists, not least those eager to achieve quick results, have understood these deficiencies well. Where original topsoil has been lost, it has usually been found necessary to apply at least 200 kg/ha of a 20-10-10 NPK fertilizer as an instant remedy for nutrient deficiency (for example, Mays and Bengtson 1978). Without fertilization, any plants that have been sown may fail completely. But the addition of fertilizer may meet the needs of the developing vegetation only over the first 2 years. Further additions—particularly of nitrogen, because it is the element that plants need most—are commonly required; otherwise, growth collapses (for example, Bloomfield et al. 1982). Nutrient addition is one of the major components of restoration aftercare. An extensive literature, going back many years, addresses the nutrient requirements of restoration (for example, Whyte and Sisam 1949, Berg 1973, Bradshaw et al. 1975).

Mineral Nutrients in Assembly

The first conclusion, and possible assembly rule, that we can arrive at is that:

- Adequate nutrients are important for assembly, because they allow ecosystem growth and development to occur.

The level of nutrients that is adequate depends on the type of ecosystem being assembled and its characteristics, particularly its rate of growth (Grime 1979). However, we must remember that there is a second rule:

- Different species can have very different adaptations and responses to different nutrients.

Ecologists have established this fact well. The magnitudes of the differences are exemplified by the extensive analysis of species distribution in the

Sheffield region of England, for example (Grime and Lloyd 1973). As a result, we cannot generalize the pattern of growth of species in relation to nutrient levels for all species because these patterns are essentially specific to individual species in relation to individual nutrients (for example, Bradshaw et al. 1958).

Then, there is species interaction to be considered. Although many factors affect interactions between species, an important rule must be that:

- The effects of nutrient level on species interaction can be drastic because different nutrient responses are manifested in different rates of growth.

The pioneering work of de Wit and his colleagues at Wageningen has investigated this effect by looking at relative replacement rates in two-species mixtures under different nutrient regimes (de Wit and van den Bergh 1965; van den Berg 1968). Thus, another rule is that:

- Different ecosystems will arise because of different levels of nutrient availability in complex and not easily predictable ways.

Perhaps the most outstanding evidence of this phenomenon is the 150-year-old Park Grass Experiment at Rothamsted Experimental Station in England. Different fertilizer treatments applied annually to different sections of an originally uniform area of grassland have brought about startling changes in species composition and growth (Brenchley and Warrington 1969). This remarkable experiment is the forerunner of the experiments by Weiher and Keddy (1995), which show that different nutrient conditions can give rise to different communities in sites with identical species pools. It may also explain Lockwood and Pimm's (1999) concern about the difficulty of achieving a specific set of species in restoration, which is almost certainly often a result of unappreciated subtle soil variation. Such variation can act in a meticulous and restricted manner, as shown by the work of Snaydon (1962) on the microdistribution of white clover (*Trifolium repens*) in hill grasslands in Wales.

From all this we must conclude that there is an assembly rule of general application:

- The outcome of any ecosystem assembly, in terms of both general performance and species composition, depends substantially on nutrient supply and availability.

Complications

Although the importance of nutrients is well understood by modern, ecologically experienced restorationists, it was not so well understood by earlier workers, who were concerned merely with achieving a vegetation cover of some sort that would provide stabilization and simple visual improvement but who were not too concerned with its species composition (Bradshaw 1998). After early errors, however, they came to understand the importance of providing an adequate nutrient supply to enable even a simple ecosystem to be successfully established.

In any restoration process where a specific ecosystem containing particular species is required, the application and management of nutrients will be a complicated matter. It must take into account the subtleties of soil conditions, particularly because soil materials may differ markedly not only in the total amounts of a particular nutrient they contain but also in their ability to bind that nutrient and render it unavailable. Such properties can be very significant in the case of phosphorus (for example, Fitter 1974).

Individual species can show complex, not necessarily linear, patterns of nutrient response. High levels of nutrients can have positive effects on the growth of some species and negative effects on others. This difference was observable in the early experiments of Jefferies and Willis (1964) and is obvious in any restoration involving ecosystems normally found on poor or acid soils. Nevertheless, the treatments required to obtain the most effective assembly of such species may not always follow expectation nor be easily predictable, as exemplified by Putwain's experiments (Environmental Advisory Unit 1988) on the restoration of heath species (*Calluna* and *Erica* spp.) on mined land in southwestern England. It is well known that these species are adapted to low-nutrient conditions. Yet under field conditions, their establishment in some sites was greatest at the highest level of fertilizer used (300 kg/ha NPK 17-17-17), a finding that is supported by the field experiments of other workers (for example, Gore 1975, Helsper et al. 1983). This phenomenon almost certainly results from the extreme conditions at the sites investigated, but it also reflects the possibility that nutrient requirements for successful seedling establishment are different from those for subsequent community maintenance, when competition from other species that are more responsive to nutrients may occur.

We must also consider the important fact that the activities of plants can alter the soil mineral nutrient composition dynamically (see Chapter 7). By means of their spreading root systems and associated mycorrhizae (see Chapter 10), plants can forage for nutrients over a large area. These nutrients are brought back to the main body of the plant, after which they are shed, normally in an organic form in litter. Nutrients are progressively accumulated in areas adjacent to plants, in a form that makes them more available than they were previously. In a reclamation situation, this process has been well illustrated for phosphorus and potassium (Knabe 1973). Thus, the interactions between abiotic and biotic factors (considered as filters) can be subtle and fundamental (see Chapter 5).

An important aspect of both natural succession and restoration is that the processes leading to ecosystem development are not completed suddenly, but over a period of time. Figure 16.1, in which both structure and function are represented as showing progressive increases, is a fair representation of what happens.

The Particular Problems of Nitrogen

For nitrogen, there are a number of additional considerations. The nutrient that plants require in the largest amounts is nitrogen. It is characteristically stored in the organic matter of soils and therefore occurs mainly in the topsoil, with negligible amounts in the subsoil and similar materials. In restoration situations where original topsoil has been lost, there is inevitably a severe shortage of nitrogen.

Collapse of the ecosystem under restoration is commonplace where nutrient deficiencies have been treated with fertilizer application. Analysis and experimentation show that this collapse is most often due to a recurrence of nitrogen deficiency (Bloomfield et al. 1982). Typically, because of leaching losses, the amount of nitrogen added is adequate only for a single year's growth, and little is left in the soil for subsequent years. At the same time, nitrogen holds a special position as a soil nutrient because it is stored only in organic matter, of which only a small proportion decomposes annually. So about 1000 kg/ha nitrogen needs to be stored in the soil organic matter as capital to provide for the annual needs of temperate ecosystems (Bradshaw 1999). A simple assembly rule is that:

- Nitrogen plays a critical role in ecosystem assembly, and the total amounts provided must be adjusted to suit the component species' annual needs for mineral nitrogen.

For a given amount of organic nitrogen capital in the soil, the amount of mineral nitrogen released will depend on the rate of mineralization, which itself depends on the rate of decay of organic matter (Swift et al. 1979). Species can have very specific adaptations to such variation in nitrogen supply. For example, Woldendorp (1983) demonstrated that *Plantago lanceolata* has a markedly greater flexibility in its nitrogen metabolism than *P. major*. This greater flexibility includes the relevant enzyme activity, and so *P. lanceolata* is able to grow in a wider range of habitats than *P. major*. Another assembly rule, therefore, is that:

• The rate of decay of organic matter is a significant factor because of its effects on the rate of nitrogen mineralization.

Nitrogen levels can be augmented in soils by the addition of nitrogen-rich organic matter, as well as by the application of fertilizer. In contrast to the nitrogen in fertilizers, which is readily leached especially in restoration situations (Dancer 1975), nitrogen in organic materials may be released slowly. Therefore, another rather practical rule must be that:

• The means of adding nitrogen must be well understood if an appropriate level of supply is to be achieved.

Although a small amount of nitrogen comes from atmospheric sources, substantial augmentation of nitrogen in soils can be brought about by natural biological fixation, notably by the rhizobium associated with legume species but also by actinomycetes in other species. Natural biological fixation is a powerful process, easily able to contribute more than 100 kg/ha/year of nitrogen without cost (Dancer et al. 1977). The process occurs commonly in natural succession, leading to a progressive improvement in nitrogen status and growth of the developing ecosystem, whether one considers soil, grass, or trees (Marrs et al.1983). Outstanding examples of the progressive buildup of nitrogen in developing ecosystems within 100 years are the buildups occurring on glacial moraines in Alaska (Crocker and Major 1955) and on mining wastes in Minnesota (Leisman 1957). Nitrogen-fixing species are widely used to provide nitrogen in ecosystem restoration. The nitrogen accumulated is available not only to the host species but also to other species in the ecosystem within a short time (for example, Kendle and Bradshaw 1992). Nitrogen-fixing species are powerful "ecosystem engineers." Surprisingly, these species were overlooked by Jones et al. (1994) in their review of ecosystem engineers. Nitrogen-fixing species provide excellent examples of the occurrence of facilitation and of the overlap between abiotic and biotic filters discussed in Chapter 6. A further rule is that:

• Nitrogen-fixing species, either found naturally in the ecosystem or purposefully introduced, can augment soil nitrogen and thus facilitate restoration.

Such nitrogen-fixing species, however, have their own nutrient requirements, particularly for elements such as calcium and phosphorus. These nutrients must be managed carefully to ensure that nitrogen-fixing species grow appropriately. It is possible for excessive nitrogen fixation to occur, through the wrong choice of species or excessive encouragement of nitrogen-fixing species by the presence of these other nutrients, upsetting the balance of species in the restored ecosystem. Therefore, the use and management of nitrogen-fixing species as a means of augmenting soil nitrogen levels, although an economical method, must be carried out with care.

Nitrogen is just one of the nutrients required in ecosystems, but it is important in all ecosystems and therefore in ecosystem restoration. Indeed, it is crucial in restoration. It is a partly renewable resource, and yet it is usually markedly limited. Its effects on different species are complex and multifarious. Complex plant-animal interactions can occur, which is well shown by the many different examples given by Lee et al. (1983) and other authors. Despite all the ecological research on nitrogen, we still know too little about its role in restoration situations; for instance, its behavior in stored soils. The changes that occur when fertile soil is stockpiled can be far-reaching (Davies et al. 1998). There are other equally important nutrients, however, each with its own special characteristics within soils and its own relationships with plants. For instance, despite the fact that removal and return of topsoils is a well-established restoration technique, the cycling of phosphorus in reinstated topsoils can be altered unwittingly in rehabilitation processes, leading to lasting effects on the sustainability of restored ecosystems (Grigg et al. 1998).

Conclusions

Mineral nutrients have pervasive and substantial effects on ecosystem development, even in relatively undisturbed situations. They contribute to narrow-sense assembly rules (see Chapter 3) and also to the rules derived from successional processes, as contrasted by White and Jentsch (Chapter 17). We can see their effects best by using a reductionist approach in which single effects are studied. They can be considered as individual abiotic filters or part of an abiotic filter complex. Their incorporation into conceptual filters and assembly rules focuses attention on their effects, allowing us to take action to mollify or overcome these effects in restoration situations. But nutrient effects are complex and interactive, making the dynamic filter concept (Chapter 6) especially helpful. Such interactions involve the plants themselves as well as other components of the ecosystem. It is difficult to simplify the attention we should pay to nutrients into a series of assembly rules. It is similarly difficult to derive a simple filter model to describe the ecological situations nutrient availability can bring about, since the filter can change markedly in front of us with time. We need more information on what is going on in the black box.

Bearing in mind that individual plant and ecosystem growth depends on nutrient availability, the following general guidelines (as distinct from assembly rules) can be enunciated:

- Any single nutrient can have controlling effects on the growth of a species and therefore on its status in an ecosystem.
- Different nutrients have different effects.
- Nutrients differ in the ways they are stored and released in soils.
- Individual plant species differ greatly in their patterns of response to nutrients.
- The activities of plants can alter the amounts and availabilities of nutrients in the soil.
- All these differing effects can interact, and the "filter system" is therefore dynamic.
- Although the effects of nutrients are most obvious where ecosystem degradation is severe, they can be just as important and controlling where there is only slight damage.

Wagner (Chapter 13) shows how complex this situation can be, since a range of abiotic factors, only partly related to nutrients, can operate. In other cases (for example, see Chapter 15), nutrient abiotic factors may hardly operate at all, and biotic factors may be more important. Actual assembly rules, therefore, will be complex, and their relative importance differs between individual sites and ecosystems.

Nutrient availability is critical in ecosystem restoration. Although some generalizations can be made about species adaptation to nutrients, the situation is complex. Nevertheless, mineral nutrients, as an abiotic filter, interact substantially with any biotic filter. Because of the powerful effects of nutrients in restoration work, the contribution of nutrients and the alleviation of any deficiencies must be considered carefully, based on the results of experimentation (see Chapter 2). We may wish to accept the existing mineral nutrients in an ecosystem without change, we may wish to change them by suitable treatments, or we may wish to allow the plants to do the job for us. Therefore, despite the importance of nutrients, it may be difficult to define specific rules that describe their effects and that apply to their management in ecosystem assembly.

Acknowledgments

I am grateful for the invitation and support of the University of Jena in allowing me to attend the Dornburg workshop, and to Philip Putwain, Markus

Wagner, Vicky Temperton, and Tim Nuttle for their thoughtful comments on this chapter.

REFERENCES

Berg, W. A. 1973. Evaluation of P and K soil fertility tests on coal-mine spoils. In *Ecology and Reclamation of Devastated Land*, vol. 1, ed. R. J. Hutnik and G. Davis, 93–104. New York: Gordon and Breach.

Bloomfield, H. E.; Handley, J. F.; and Bradshaw, A. D. 1982. Nutrient deficiencies and the aftercare of reclaimed derelict land. *Journal of Applied Ecology* 19:151–158.

Bradshaw, A. D. 1983. The reconstruction of ecosystems. *Journal of Applied Ecology* 20:1–17.

Bradshaw, A. D. 1984. Ecological principles and land reclamation practice. *Landscape Planning* 11:35–48.

Bradshaw, A. D. 1998. From green cover to something more: the development of principles and practice in land reclamation. In *Land Reclamation: Achieving Sustainable Benefits*, ed. H. R. Fox, H. M. Moore, and A. D. McIntosh, 339–350. Rotterdam: Balkema.

Bradshaw, A. D. 1999. The importance of nitrogen in the remediation of degraded land. In *Remediation and Management of Degraded Lands*, ed. M. H. Wong, J. W. C. Wong, and A. J. M. Baker, 153–173. Boca Raton, Fla.: Lewis Publishers.

Bradshaw, A. D.; Dancer, W. S.; Handley, J. F.; and Sheldon, J. C. 1975. The importance of land revegetation and the reclamation of the china wastes in Cornwall. In *The Ecology of Resource Degradation and Renewal*, ed. M. J. Chadwick and G. T. Goodman, 363–384. Oxford: Blackwell Scientific.

Bradshaw, A. D.; Lodge, R. W.; Chadwick, M. J.; and Jowett, D. 1958. Experimental studies into the mineral nutrition of several grass species. I. Calcium level. *Journal of Ecology* 46:749–757.

Brenchley, W. E.; and Warrington, K. 1969. *The Park Grass Plots at Rothamsted 1856–1949*. Harpenden, UK: Rothamsted Experimental Station.

Clarke, F. W. 1924. *The Data of Geochemistry*. 5th ed. Washington, D.C.: U.S. Geological Survey.

Connell, J. H.; and Slatyer, R. O. 1977. Mechanisms of succession in natural communities and their role in community stability and organization. *American Naturalist* 111:1119–1144.

Crocker, R. L.; and Major, J. 1955. Soil development in relation to vegetation and surface age at Glacier Bay, Alaska. *Journal of Ecology* 43:427–448.

Dancer, W. S. 1975. Leaching losses in the reclamation of sand spoils in Cornwall. *Journal of Environmental Quality* 4:499–504.

Dancer, W. S.; Handley, J. F.; and Bradshaw, A. D. 1977. Nitrogen accumulation in kaolin mining wastes in Cornwall. II Forage legumes. *Plant and Soil* 48:303–314.

Davies, R. S.; Younger, A.; Hodgkinson, R.; and Chapman, R. 1998. Nitrogen loss from a soil restored after surface mining. In *Land Reclamation: Achieving Sustainable Benefits*, ed. H. R. Fox, H. M. Moore, and A. D. McIntosh, 235–240. Rotterdam: Balkema.

de Wit, C. T.; and van den Bergh, J. P. 1965. Competition between herbage plants. *Netherlands Journal of Agricultural Science* 13:212–221.

Díaz, S.; Cabido, M.; and Casanoves, F. 1999. Functional implications of trait-environment linkages in plant communities. In *Ecological Assembly Rules: Perspectives,*

Advances, Retreats, ed. E. Weiher and P. Keddy, 338–362. Cambridge: Cambridge University Press.

Ehrenfeld, J. G.; and Toth, L. A. 1997. Restoration ecology and the ecosystem perspective. *Restoration Ecology* 5:307–317.

Environmental Advisory Unit. 1988. *Heathland Restoration: A Handbook of Techniques.* Southampton: British Gas.

Fitter, A. H. 1974. A relationship between phosphorus requirement, the immobilization of added phosphate and the phosphate buffering capacity of colliery shales. *Journal of Soil Science* 25:41–50.

Gore, A. J. P. 1975. An experimental modification of upland peat vegetation. *Journal of Ecology* 69:85–96.

Grigg, A. H.; Mulligan, D. R.; and Dahl, N. W. 1998. Topsoil management and long term reclamation success at Weipa, North Queensland, Australia. In *Land Reclamation: Achieving Sustainable Benefits,* ed. H. R. Fox, H. M. Moore, and A. D. McIntosh, 249–254. Rotterdam, The Netherlands: Balkema.

Grime, J. P. 1979. *Plant Strategies and Vegetation Processes.* Chichester, England: John Wiley and Sons.

Grime, J. P.; and Lloyd, P. S. 1973. *An Ecological Atlas of Grassland Plants.* London: Edward Arnold.

Helsper, H. P. G.; Glenn-Lewin, D.; and Werger, M. J. A. 1983. Early regeneration of *Calluna* heathland under various fertilisation treatments. *Oecologia* 58:208–214.

Jefferies, R. L.; and Willis, A. J. 1964. Studies on the calcicole-calcifuge habit. II. The influence of calcium on the growth and establishment of four species in soil and water culture. *Journal of Ecology* 52:691–707.

Jones, C. G.; Lawton, J. H.; and Shachak, M. 1994. Organisms as ecosystem engineers. *Oikos* 69:73–86.

Keddy, P. A. 1992. Assembly and response rules: two goals for predictive community ecology. *Journal of Vegetation Science* 5:157–164.

Keddy, P. A.; and Weiher, E. 1999. The scope and goals of research on assembly rules. In *Ecological Assembly Rules: Perspectives, Advances, Retreats,* ed. E. Weiher and P. Keddy, 1–20. Cambridge: Cambridge University Press.

Kendle, A. D.; and Bradshaw, A. D. 1992. The role of soil nitrogen in the growth of trees on derelict land. *Aboricultural Journal* 16:103–122.

Knabe, W. 1973. Investigations of soils and tree growth on five deep-mine refuse piles in the hard-coal region of the Ruhr. In *Ecology and Reclamation of Devastated Land,* vol. 2, ed. R. L. Hutnik and G. Davis, 307–324. New York: Gordon and Breach.

Lee, J. A.; McNeill, A. S.; and Rorison, I. H., eds. 1983. *Nitrogen as an Ecological Factor.* Oxford: Blackwell Scientific.

Leisman, G. A. 1957. A vegetation and soil chronosequence on the Mesabi Iron range spoil banks, Minnesota. *Ecological Monographs* 27:221–237.

Lockwood, J.; and Pimm, S. L. 1999. When does restoration succeed? In *Ecological Assembly Rules: Perspectives, Advances, Retreats,* ed. E. Weiher and P. Keddy, 363–392. Cambridge: Cambridge University Press.

Marrs, R. H.; Roberts, R. D.; Skeffington, R. A.; and Bradshaw, A. D. 1983. Nitrogen and the development of ecosystems. In *Nitrogen as an Ecological Factor,* ed. J. A. Lee, S. McNeill, and I. H. Rorison, 113–136. Oxford: Blackwell Scientific.

Mays, D. A.; and Bengtson, J. W. 1978. Lime and fertilizer use in land reclamation in humid regions. In *Reclamation of Drastically Disturbed Lands,* ed. F. W. Schaller and P. Sutton, 307–328. Madison, Wis.: American Society of Agronomy.

Snaydon, R. W. 1962. Microdistribution of *Trifolium repens* L. and its relation to soil factors. *Journal of Ecology* 50:133–143.

Swift, M. J.; Heal, O. W.; and Anderson, J. W. 1979. *Decomposition in Terrestrial Ecosystems.* Oxford: Blackwell Scientific Publications.

Tansley, A. G. 1935. The use and abuse of vegetational concepts and terms. *Ecology* 16:284–307.

van den Berg, J. P. 1968. Distribution of pasture plants in relation to chemical properties of the soil. In *Ecological Aspects of the Mineral Nutrition of Plants*, ed. I. H. Rorison, 11–24. Oxford: Blackwell Scientific.

Weiher, E.; and Keddy, P. A. 1995. The assembly of experimental wetland plant communities. *Oikos* 73:323–335.

Weiher, E.; and Keddy, P. A. 1999. Assembly rules as general constraints on community composition. In *Ecological Assembly Rules: Perspectives, Advances, Retreats*, ed. E. Weiher and P. Keddy, 251–271. Cambridge: Cambridge University Press.

Whyte, R. O.; and Sisam, J. W. B. 1949. *The Establishment of Vegetation on Industrial Waste Land.* Aberystwyth, UK: Commonwealth Bureau of Pasture and Field Crops.

Wilson, J. B. 1999. Assembly rules in plant communities. In *Ecological Assembly Rules: Perspectives, Advances, Retreats*, ed. E. Weiher and P. Keddy, 130–164. Cambridge: Cambridge University Press.

Woldendorp, J. W. 1983. The relationship between the nitrogen metabolism of Plantago species and the characteristics of the environment. In *Nitrogen as an Ecological Factor*, ed. J. A. Lee, S. McNeill, and I. H. Rorison, 137–166. Oxford: Blackwell Scientific.

Young, T. P.; Chase, J. M.; and Huddleston, R. T. 2001. Comparing, contrasting and combining paradigms in the context of ecological restoration. *Ecological Restoration* 19:5–18.

PART FIVE

Disturbance and Assembly

The next three chapters reflect the increasing recognition of the importance of disturbance in fundamentally shaping the nonequilibrium dynamics of ecological communities over long stretches of time. The importance of disturbance regimes for the development of communities is explored in both terrestrial and aquatic ecosystems. Chapter 17 proposes how disturbance regimes per se (rather than species composition) can indicate the status of an ecosystem in restoration. Chapter 18 reviews similar issues for running-water systems, focusing on how spatial and temporal patterns of disturbance affect ecosystem recovery following floods. Chapter 19 investigates how the interplay of antagonistic disturbances in the hyporheic zone of a river influences abiotic conditions, which in turn determine biotic community composition.

Chapter 17

Disturbance, Succession, and Community Assembly in Terrestrial Plant Communities

PETER S. WHITE AND ANKE JENTSCH

Although restoration ecologists may conceive of disturbance as limited to the period before restoration begins, disturbance actually plays two roles: it creates the initial conditions for change, and it produces the continuing dynamics that control the establishment and turnover of individuals and species (see Chapter 19). In the last several decades, a rich literature has developed on the ecological role of disturbance and its influence on succession (for example, White and Jentsch 2001). In this chapter, we bring insights from the disturbance literature to the subject of community assembly in restoration ecology.

Just as disturbance causes a wide variety of effects, restoration addresses a wide assortment of starting points and goals. The starting points range from nearly intact ecosystems requiring only the reintroduction of fire or a missing species, for example, to highly degraded ecosystems in which few original species are present and the substrate is low in nutrients or high in toxic substances (see, for example, the chapters in Part IV, "Assembly Rules in Severely Disturbed Environments"). Restoration goals also range widely— from simply reestablishing primary production (for example, limited function) to reestablishing an ecosystem with all of its native species and natural processes (for example, full structure and function). Although goals must be feasible for each site (White and Walker 1997, Holmes and Richardson 1999), a common denominator of all restoration projects is the goal of sustainability and minimum ongoing human effort. Understanding the role of disturbance and succession is essential to understanding the mechanisms of sustainability.

We focus on terrestrial vegetation (for aquatic ecosystems, see Chapters 18 and 19). Vegetation forms the structure of the physical habitat for heterotrophs, plays a critical role in soil formation and retention, and provides the energy available to other trophic levels. Further, restoration of degraded sites can be limited by primary production because of the scarcity of soil resources or the presence of toxins (see Chapters 13 and 16). When focusing on one trophic level, competition is often the dominant interaction assumed to underlie community characteristics. This focus also implies a resource- and functional-trait–based conceptual frame (Díaz et al. 1999). Nevertheless, the vegetation may not predict other trophic levels, and predators and herbivores can modify community assembly.

The focus on vegetation reveals a second bias. Restoration has two conceptual bases that emerged from different schools and from studies of different taxonomic groups (see Chapters 3 and 4; also see Young et al. 2001). The succession concept emerged from plant ecology. By contrast, the assembly rules concept emerged from studies of animal communities in the context of island biogeography. How are these two approaches related? After Young et al. (2001), we argue that filters, assembly rules, disturbance, and succession are part of a larger concept, community assembly. Regardless of terminology, what is most important is process-based restoration (see Chapter 4). Studies of disturbance and succession contribute to our understanding of the processes that shape communities and contribute to restoration.

The variation among restorations and disturbances makes generalization a challenge. Our goal is to find approaches to organizing this variation in ways useful to restoration ecology. We begin with a brief overview of disturbance concepts. We then address the relationship of filters, assembly rules, and succession. We further bridge the two issues by turning to the importance of disturbance in community assembly and restoration. Finally, we present a synthesis of the implications for restoration ecology.

Disturbance Concepts

Disturbance is a relatively discrete event in time that disrupts ecosystem, community, or population structure and changes the resources, substrate availability, or physical environment (White and Pickett 1985). Disturbance has a wide range of effects, occurs at a wide range of spatial and temporal scales, and affects all levels of organization. Disturbance descriptors are used to characterize individual disturbances and to describe disturbance regimes. These descriptors include temporal characteristics (such as frequency and seasonality), spatial characteristics (such as patch size, shape, and distribution), magnitude (the percentage of the biomass of an ecosystem affected), specificity (to species, size, or age classes), and synergisms (disturbance interactions and feedbacks) (White et al. 1999). It is important to conceptualize disturbance as a variable and to take a mechanistic approach to understanding its influence. For example, Johnson and Miyanishi (1995) have shown how to model and describe the wide range of fire effects quantitatively. In this section, we discuss several disturbance characteristics that have special relevance to restoration.

Disturbances vary widely in severity and thus leave variable amounts of influence from the predisturbance site. The term *ecological legacy* (Swanson and Franklin 1992) describes the living organisms, the dead organic matter, and the physical structures created by disturbance that remain after disturbance, including plants, fungi, bacteria, insects, and other species with large ecosystem effects. Because of variability in this legacy, disturbances produce a family of potential successions, ranging from primary to secondary successions but including a range of secondary successions that differ greatly in their biological and physical starting points (Figure 17.1). Structures created by the disturbance, and therefore part of the legacy, include nurse logs, pits and mounds, and exposures of mineral soil formerly covered by thick organic matter. Such effects create variation in the microsites available for regeneration (Grubb 1977). The natural rate of succession depends critically on legacy and its heterogeneity. Severely degraded sites are likely to have very little legacy and low microsite heterogeneity, thus limiting the species available at the site and restricting those that can establish.

Disturbance can lead to sustained or temporary changes in site resources. For example, soil nutrients may be leached from the site, as demonstrated for dry acidic grasslands with mechanical ground disturbance (Jentsch 2001). Repeated disturbances at close intervals may lead to the sustained export of nutrients. Disturbances can also increase site resources: flood and avalanche depositional zones receive nutrients and organic matter from elsewhere in the landscape. In contrast, many disturbances, although they may temporarily increase resources (space, light, nutrients, and water), leave the overall site potential unchanged and result in straightforward secondary successions that lead back to the predisturbance resource levels. When nutrients are available after disturbance, early colonists can be important in holding those resources on-site and may thus be important to the nutrient capital and eventual course of succession.

Disturbances that have historical precedent and produce conditions that are within the historical bounds of variation for an ecosystem may produce responses different from those produced by disturbances that are novel or create conditions that are outside those bounds. At evolutionary time scales, historical precedent would be ultimately responsible for the range of life his-

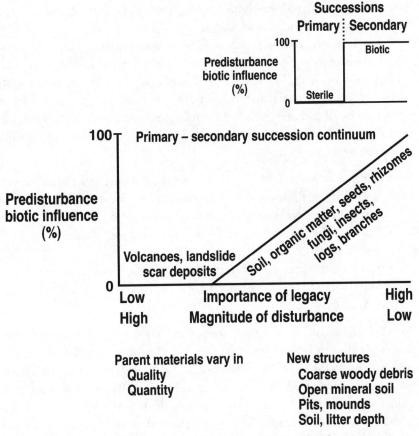

FIGURE 17.1 Ecological legacy. Disturbance creates a family of successions, ranging from primary to secondary, that differ in amount of legacy as a result of differences in disturbance severity (after White and Jentsch 2001). The insert shows the traditional view: namely, that all primary and secondary successions were distinct but homogeneous. The diagonal line shows the increase in the influence of the predisturbance ecosystem as diversity severity decreases. This influence or legacy of the predisturbance ecosystem consists of such entities as soil, organic matter, seeds, rhizomes, fungi, insects, logs, and branches.

tories present and the occurrence of species adapted to the disturbance. Thus, there is a twofold historical contingency. First, in ecological time, only those species with access to the site can participate in recovery (this presence of species—as living plants or seeds—is itself influenced by the history of disturbance). Second, in evolutionary time, species adaptations are functions of previous evolution. Both contingencies determine the diversity of functional responses within a restored site.

Disturbance specificity is critical to the role disturbance plays in succession and restoration. Some disturbances discriminate according to the size and age of individuals (White et al. 1999). For example, windstorms in forests primarily affect the oldest and largest trees. Depending on the successional age of the forest and the severity of the damage (for example, size of patches and amount of soil disturbance), these disturbances can have all possible effects on succession:

- Compositional stability (cyclic succession, called Class I disturbances by White and Pickett 1985, or random replacement on small patches fitting the carousel model of van der Maarel 1996)
- Resetting the system to an early-successional state (Class II disturbances)
- Accelerating succession through the death of early-successional dominants and the release of later successional stems from the understory

Whether disturbances accelerate succession, set succession back to earlier stages, or have no effect on succession, the common mechanism is the removal of established dominants—that is, the removal of inertia in the community. Because established dominants can inhibit change, restoration may sometimes require disturbance to accelerate the response of environmental conditions to restoration (as long as mineralized resources and soil are not made vulnerable to erosion by the disturbance itself).

When disturbances primarily affect dominant species, other species may increase after disturbance, even if their functional traits are similar to those of the previous dominants (the resilience hypothesis of Walker et al. 1999). This increase of subordinate species occurs if dominant and subordinate species within the same functional group differ in their ability to respond to disturbance. Thus, redundancy is important in ensuring persistence in ecosystem function at the restoration site in the face of disturbance. When disturbances are specific to the dominant species, the situation is similar to exploiter-mediated coexistence (herbivore, predator, or disease attack on the competitive dominants; Paine 1974). When disturbances are specific to larger individuals regardless of species, the situation is similar to compensatory mortality (the death of the largest individuals allows smaller individuals to persist and increase; Connell 1978).

By contrast, some disturbances have the opposite specificity: they affect nondominants or invading species. For example, ground fires in forests act primarily on the youngest and smallest individuals. Such fires can crop invading successional species of later successional stages, thereby preventing succession. In savannas and grasslands, frequent fires kill invading woody stems but leave perennial forbs and grasses intact belowground, perpetuating the community in a continually arrested successional state.

Disturbances usually create patchiness and heterogeneity whether at the scale of the individual, the community, or the landscape (Jentsch et al. 2002a). Although the disturbance process creates pattern, pattern can influence process, as when the distance between disturbed patches in grasslands limits species establishment to those species that can disperse over those distances and that rely on the open substrate created by disturbance (Jentsch 2001). The effect of pattern on process is an important issue in restoration because our goals must include establishing a pattern that can support a desired process.

Landscape ecology can help us find the appropriate dimensions for reintroduced disturbance regimes. Turner et al. (1993) proposed that dynamic equilibrium in landscapes was a function of two ratios: the ratio of disturbed area to the area of the entire site and the ratio of disturbance frequency to the time required for successional recovery to the predisturbance state. Equilibrium occurs when the sum of disturbed patches is small relative to the site area and disturbance frequency is low relative to the time needed for recovery. If a restored site is all in one age state (whether that is recently disturbed or long undisturbed), it will lose species that characterize the other age states (Pickett and Thompson 1978) and thus will also lose its ability to respond to disturbance. Thus, an appropriate disturbance regime needs to be reintroduced if composition and dynamics are to be fully restored. Radeloff et al. (2000) described management strategies to avoid species loss when landscapes pattern does not support equilibrium dynamics because of habitat isolation and fragmentation.

The scale of disturbance affects the balance of stochastic and deterministic processes within communities. For example, individual and random mortality within plant communities results in a fine-scale gap dynamics that is seen as a within-community phenomenon (White and Jentsch 2001). As noted earlier, fine-scale disturbance can result in a predictable cyclic succession or a series of random replacements among community dominants. Small-scale, random replacement has been termed the carousel model of vegetation dynamics, with shifting locations of species over time but no overall successional trend at the community scale (van der Maarel and Sykes 1993, van der Maarel 1996). On the other hand, large patches of disturbance produce successional change that is usually seen as a between-community phenomenon driven by deterministic processes (that is, species are sorted on a successional axis by adaptations; White and Jentsch 2001). Even in this case, disturbance may promote stochastic effects on composition if the order of arrival (a form of sequence effect, see Chapter 4) determines dominance among several possible colonists. In sum, disturbances produce a range of effects and responses; this range can be organized according to seven factors of variation (Table 17.1).

Community Assembly: The Relationship of Filters, Assembly Rules, and Succession

In examining the role of disturbance in community assembly, we first ask two questions: Does disturbance simply modify environmental filters or does it act as a filter in its own right? And what is the relationship of succession to assembly rules? We start with the conceptual background that will allow us to answer these questions.

Community assembly can be separated into (1) filters that determine which species are, in a sense, part of the game; and (2) species interactions that determine the success of species in that game. Some authors have referred to species interactions as the biotic filter or have used other terms such as biotic interactions, internal dynamics, or system models (see Chapters 6 and 7; also see Belyea and Lancaster 1999, Keddy and Weiher 1999). We include such biotic interactions as the domain of assembly rules rather than as filters (see Chapter 7). In this sense, we follow Wilson and Gitay (1995), who restricted assembly rules to species interactions. Filters can then be separated into environmental filters, which determine whether species can tolerate the physical conditions of the site (see Chapter 5); dispersal filters, which determine whether species can disperse to the site; and, as we explain shortly, disturbance filters.

Belyea and Lancaster (1999) treat succession as part of internal dynamics and thus fertile ground for finding assembly rules. We agree but propose that there are two very different cases of the assembly process, corresponding to the original or narrow-sense definition of assembly rules (for example, Diamond 1975) and autogenic succession (Figure 17.2). Historically, assembly rules were developed in the context of the theory of island biogeography (MacArthur and Wilson 1967, Keddy and Weiher 1999, Young et al. 2001). In its simplest form, the theory predicts that the species present on an island at any one time are a random subset of the species pool. The number of species, of course, depends on the size and isolation of the island, but there is continual turnover of species over time. Diamond (1975) argued that pat-

TABLE 17.1

*Seven Factors That Cause Variation in the Role of Disturbance
in Ecosystems*[1]

1. *Historical contingency:* Disturbance can be within or outside the bounds of
   variation of past disturbances and within or outside the range of physiological and
   dispersal abilities of the species present.
2. *Resource availability:* Ecosystems differ in absolute resource availability, and
   disturbances can leave resources unchanged, increase resources, or decrease
   resources relative to the predisturbance condition.
3. *Ecological legacy:* Disturbances vary in magnitude and thus leave behind a range
   of influences from the predisturbance ecosystem.
4. *Frequency:* Disturbances vary in recurrence, affecting which species can reach
   reproductive maturity.
5. *Patch size:* Disturbances vary in patch size; larger patches often have more
   drastically altered conditions, and size may affect colonization rate across the
   patch.
6. *Specificity:* The species and age or size classes affected may differ depending on
   disturbance type.
7. *Pattern and process:* Process creates pattern, but pattern influences process.

[1]See White and Jentsch 2001, White et al. 1999.

terns of co-occurrence among birds on oceanic islands were not random but
rather were determined by rules that set how niche space (fixed by resources
on the island) could be divided. Assembly rules subsequently were applied
to other situations (see Chapter 3), but often in the context of how subsets
of a potentially available species pool at a site could divide up or coexist on
a given resource base. In most cases, narrow-sense assembly rules assume
that environmental conditions and the resource base are givens.

In contrast to narrow-sense assembly rules, succession in terrestrial communities involves change in the physical environment and resource base
driven by the developing community, particularly an increase in biomass
(which modifies temperature, moisture, and light); the production of organic
detritus; the influence on soil; and the partitioning of soil resources among
living organisms, dead organic matter, and the physical substrate (for example, Jentsch 2001). Total niche space and environment, in this sense, change
with succession. Clements (1916) termed this organism-driven change in
environment "reaction" and considered it the driving variable of succession.

The study of narrow-sense assembly rules and terrestrial succession also
differs in other ways. In succession, species residence time (until late succession) is limited to one or at most a few generations. In assembly rules,
species residence is considered only if it occurs over multiple generations
(Young et al. 2001). This difference is a matter of focus: in succession, the
emphasis is on turnover of species; in assembly rules, the emphasis is on the

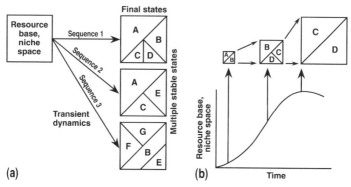

FIGURE 17.2 Succession and assembly rules. (a) In narrow-sense assembly rules, the focus is on final states and alternative ways of dividing total niche space under the assumption of stable niche space and resource base. (b) In succession, the focus is on directional change in species composition accompanied by an increase in total niche space and resource base. Capital letters represent species.

final community composition. In succession, species interactions cause environmental change, which causes compositional change, leading to only a few possible outcomes. In assembly rules, in contrast, species interactions cause change directly, not through the environment, thus leading to many possible outcomes (see Chapter 4). Finally, in terrestrial ecosystems, succession is usually studied relative to plants competing for the same resources, whereas assembly rules are usually studied relative to animals, where the strengths of species interactions depend on the degree of competition for similar resources (niche overlap or separation).

We digress here to make an observation about scale with reference to stochastic influences on composition, stability, and the notion of multiple stable states. Narrow-sense assembly rules and successional perspectives also differ in their findings on these phenomena. The original island biogeography literature (MacArthur and Wilson 1967) predicts that the community, because it is insular, can contain only a subset of all possible species (because it is smaller than and isolated from the mainland). A similar context underlies microcosm experiments: the final species number must be smaller than the total number of available species. The phenomenon of small-scale subsets of species composition is a prerequisite to the idea of multiple stable states, since we can ask whether different subsets can divide up the same resources in a stable way. It follows that none of the dominant species and none of the subset combinations is better—they are just alternative ways of using the same resource base. These principles must be taken into account in restoration efforts that focus on small, isolated sites. Which subset develops at such a site may depend on the sequence effect of a stepwise community assembly process (see Chapter 4) or on unknown stochastic effects. When restoration addresses small insular areas, the subset model may be appropriate.

At small scales within a plant community (for example, at the scale of the individual), which species are present can also be a stochastic subset of the available species. When a plant community is viewed at a larger scale, however, it may become apparent that local stochastic effects compensate for one another such that all possible species are present somewhere all the time. Hence, at small scales we have multiple states and at larger scales we have only one possible composition. Small-scale stochastic turnover has been described in vegetation science as the carousel model: stochastic change at the scale of the individual, but stability at the scale of the community (van der Maarel 1996). Study of directional succession, however, tends to be implicitly large-scale, and the phenomenon of local subsets is usually ignored. If multiple pathways of succession are reported, there are low numbers of alternative compositions compared to those reported in assembly-rules experiments (see Chapter 4).

There is a further deterministic effect driven by succession itself: the changes in environment and resources during succession are a filter that restricts the possible outcomes. For this reason, some authors have considered succession to be driven by environmental change rather than by species interactions (Lockwood 1997). However, interactions in narrow-sense assembly rules must also affect the resource base and environment, if only at the scale of the individual. The difference we do see is that resources and environments are changing more, and in a more predictable and directional manner, during succession than during the process described by narrow-sense assembly rules. In the context of a changing environment during succession, there are multiple alternative combinations of species (and hence several possible successional pathways)—it is the environment that predictably changes, with the predictability of species combinations and turnover open to question, just as it is in investigations of assembly rules.

After Young et al. (2001) (see also Chapter 7), we group filters, assembly rules, and succession under the broader term *community assembly* (Figure 17.3). Species availability is determined by environmental, dispersal, and disturbance filters. Among species available at the site, assembly rules determine niche division in the context of a given environment and resource base. Succession describes community development within the context of a chang-

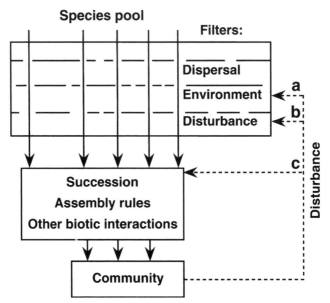

FIGURE 17.3 The threefold role of disturbance. Community assembly consists of filters and biotic interactions that include succession and assembly rules. Flows of species are represented as solid arrows and influences as dashed arrows. Disturbance influences environmental filters (a) and acts as a filter itself (b) (see Figure 17.4), but it also modifies the process of assembly in ongoing community dynamics (c).

ing overall environment (allogenic succession) or in the context of community effects on environment and resource availability (autogenic succession). Studies of succession have been largely descriptive but have sometimes generated and tested rules (for example, Noble and Slatyer 1980). Together, succession and disturbance are the fundamental processes of vegetation dynamics. We now turn to the relationship of disturbance to filters, assembly rules, and succession in more detail.

The Significance of Disturbance for Community Assembly

Disturbances act on community assembly in three ways (see Figure 17.3): they modify environmental filters, act directly as filters (see Chapter 19; also see Keddy 1992, Díaz et al. 1999), and are dynamic mechanisms that modify the process of community assembly over time. As a filter, disturbance acts on plant traits (Díaz et al. 1999).

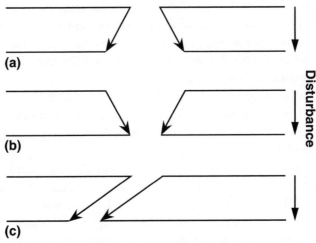

FIGURE 17.4 Disturbance modifies environmental filters in three ways. (a) Disturbance expands the number of species that can establish. (b) Disturbance contracts the number of species that can establish. (c) Disturbance shifts which species can establish, regardless of its effect on the number of species that can establish.

Disturbances Modify Environmental Filters

The environmental filter consists of the physical conditions and other measures of the abiotic environment on the restored site. If this filter has a "mesh" that determines what species are available, disturbance is a process that changes the nature of the mesh by changing the physical environment. Disturbance can increase, decrease, or shift the mesh (Figure 17.4). Many disturbances increase mesh size by increasing resource availability or space. For example, disturbance by small mammals creates patches of open mineral soil in a grassland, thereby expanding establishment to include species that are intolerant of soil litter (Vogl 1974). Sometimes this effect simply creates open space for establishment without affecting resources, as in an intertidal community disturbed by wave action (Sousa 1984) or in dry acidic grassland covered by cryptogams (Jentsch et al. 2002b). Disturbances can also contract mesh size. For example, fire volatilizes nitrogen, thereby favoring nitrogen fixers or species tolerant of low nitrogen (Vinton et al. 1993). Finally, disturbances can also change environmental conditions qualitatively—that is, they can shift the environmental position of the mesh and shift resources on the site without necessarily having any effects on mesh size itself. For example, debris avalanches remove forest soils and initiate primary succession on open rock (White et al. 1985, Dale 1991). In this case, the species pool is not expanded or contracted but rather shifted to an entirely new set of species. Regardless of whether disturbance changes which species can establish, it almost always increases the number of individuals that can establish by removing predisturbance individuals.

Disturbance changes the filter abruptly. Thus, the ability to colonize rapidly can have a large effect on species composition. This can lead to stochastic effects in which order of arrival determines which species dominate. If early colonizers determine the pattern of succession, changes are said to follow the model of "initial floristic composition" (Glenn-Lewin and van der Maarel 1992). The early dominants can become an inertial force (for example, under the successional model called inhibition, where pioneer species inhibit the establishment of later successional species; Connell and Slatyer 1977). The abrupt nature of disturbance and the location of disturbed patches can determine which species are able to pass another filter—that of dispersal.

Disturbance Acts as a Filter in Its Own Right

If filters alter species availability as selective forces on adaptations, disturbance can act as a direct filter in the sense that the effect on species availability is contingent on the nature of the disturbance and is not predictable from the postdisturbance environment. We define four ways that disturbance can act as a direct filter: the survival filter, the reproduction filter, the colonizing filter, and the disturbance-adaptation filter.

DISTURBANCE MAGNITUDE AS A SURVIVAL FILTER

Disturbance acts as a survival filter because species and individuals have different abilities to survive disturbance events such as wind, fire, grazing, and pollution. For example, allocation of tree growth to bark within several meters of the ground increases survival after fire (Harmon 1984). Recurrent disturbances would also act as a filter in this way, selecting species by their survival abilities. This filter acts through disturbance intensity.

DISTURBANCE FREQUENCY AS A FILTER FOR REPRODUCTION

Different species reach reproductive maturity at different times. Disturbance that causes mortality at a particular rate acts as a recurrent filter on community membership by determining which species reproduce (Grime 1979). Species not able to reproduce may be maintained in the community by dispersal from other populations (the mass effect; Shmida and Elner 1984), but they are transient. In the absence of dispersal from external populations, they will disappear.

DISTURBANCE ABRUPTNESS AND TIMING AS A FILTER FOR COLONIZING ABILITY

Because disturbances open space abruptly, they select for species that can colonize quickly. Disturbances that make resources more available also promote fast colonization, leading to rapid establishment and quick early growth rates. Seasonal patterns of disturbance may select for species that disperse at particular times. Spatial patterns of disturbance may select for species that disperse over particular distances.

DISTURBANCE AS A FILTER FOR DIRECT ADAPTATIONS

Some species traits are direct adaptations to disturbance (Díaz et al. 1999). Examples include serotinous cones and fruiting structures that release seeds after the passage of fire and seeds that germinate after exposure to smoke or because of high nitrate levels caused by disturbance (for example, Marks 1974, Dixon et al. 1995).

The role of disturbance as a filter on community membership recalls the contrast between two views of succession: the gradient-in-time view and the competitive-sorting view (Peet 1992). In the gradient-in-time view, the turnover of species during succession reflects the change in resources and the environment that accompanies succession (termed *reaction* by Clements 1916). In the current context, this is equivalent to the view that successional changes produce a changing environmental filter on community membership. In the competitive-sorting view, chance determines initial establishment, with competition determining which species dominate over time. Disturbance gives room to both perspectives. In keeping with the competitive-sorting view, disturbance can increase the importance of stochastic effects on initial species composition. In keeping with the gradient-in-time view, specific traits—such as fast colonizing ability—are selected by the environmental conditions created by disturbance, leading to turnover of species during succession that is the result of both stochastic and deterministic factors.

Disturbance and Succession as Filters on Plant Traits

Plant traits are what filters act upon (Díaz et al. 1999), and they have been classified in many ways that relate to disturbance (Grime 1979, Noble and Slatyer 1980, Pavlovic 1994, Walker et al. 1999). We will discuss generalized plant strategies in a historical context, starting with the r-K continuum (MacArthur and Wilson 1967) and proceeding to Grime's triangular scheme of C, S, and R strategies (Grime 1979). We will then extend Grime's basic argument by describing adaptation to the circumstances of regeneration (Grubb 1977) and disturbance itself.

The r-K continuum describes trade-offs between reproductive potential and persistence potential. Reproductive potential (r) is increased by early sexual maturity and by large and consistent reproductive output—among temperate trees, early-successional species such as *Betula lenta* and *Populus tremuloides* are examples. Persistence potential (K) is increased by greater longevity and better competitive ability—among temperate trees, late-successional species such as *Fagus sylvatica* and *F. grandifolia* are examples. r-traits are advantageous when the restored site is far from the carrying capacity, when resources are relatively abundant, and when competition is low. K-traits convey fitness in crowded and competitive environments when resources are limiting. In this perspective, further disturbance is assumed to be absent and the carrying capacity is fixed.

Restoration sites differ in resource availability and carrying capacity. Nonetheless, r-traits (typical of pioneer species) would be advantageous early after disturbance (and early in restoration) and K-traits (typical of successor species) later in community assembly. Grime improved upon the r-K scheme by making disturbance and resource availability explicit. He described three extremes in plant strategies: ruderals, competitors, and stress tolerators. R (ruderal) species are those with rapid growth, establishment, and reproduction in environments that are both productive and frequently disturbed. C (competitor) species are those with rapid growth, rapid establishment, and delayed reproduction in environments that are productive and infrequently disturbed. S (stress tolerator) species are those that are tolerant of low resource availability. We suggest here, however, that S species include two important subcases. The first occurs in low-resource environments where species are tolerant of physical limits (cold temperature, scarce water, or poor mineral and organic nutrition). The second occurs in high-resource environments where resources are low for a particular portion of the community because they are largely contained within or preempted by living biomass or are only slowly released from organic detritus (for example, in a closed forest). In low-resource environments, disturbance may have little effect on resource availability, since the total resources in living or dead components of the ecosystem are low. In high-resource environments, disturbance can both reduce competition and increase resource availability by resulting in greater sunlight and the mineralization of nutrients from organic matter. (Chapter 16 describes the important difference between internal resources such as soil nutrients that must be recycled from biomass and external resources such as sunlight and water.)

Grime's scheme is congruent with the frequent observation that secondary successions are more directional in productive habitats than in stressful ones (for example, Churchill and Hanson 1958 for tundra ecosystems). For instance, successional roles are well marked in temperate humid forests and poorly marked in dry forests, grasslands, deserts, and tundra. If we view succession as an environmental gradient in time (Peet and Christensen 1980), that gradient is longer in more-productive places than in less-productive ones in the sense that more species participate in successional turnover. In Grime's scheme, the differentiation in response to disturbance occurs only under nonstressful conditions, where disturbance frequency separates R (frequent disturbance) and C (infrequent disturbance) strategies.

We can make further improvements in the Grime scheme by discussing two ways in which plant strategies show adaptation to disturbance: niche differentiation by regeneration requirements and traits that respond to specific disturbances (Table 17.2; Díaz et al. 1999). Patchy disturbances produce heterogeneity within the community in terms of places for establishment of new individuals (Grubb 1977). Such microsites can differentially affect the success of different plant strategies (Menges and Waller 1983) and can be limiting in restoration (Connett et al. 2001). In describing disturbance as a direct filter, we have pointed out that it is not just disturbance rate but other disturbance characteristics that are responsible for differential species performance (Díaz et al. 1999). Bond and van Wilgen (1999) noted the failure of the Grime scheme to consider specific characteristics of species in fire systems.

When resources are high after disturbance, such as in forest blowdowns, colonizing ability and growth rate are important and can have a lasting impact on ecosystem composition and structure. Rapid establishment supports rapid uptake of resources and stabilization of soil. In areas with high resource availability and low competition, species with these characteristics play key roles by conserving nutrients on-site. For example, Marks (1974) showed that *Prunus pensylanica* in North American deciduous forests could rapidly take up nitrogen released during mineralization after disturbance and thus could stabilize the nutrient capital of the forest, which would otherwise be vulnerable to leaching. This species lives only about 60 years. After it has vanished, its rapid uptake of nitrogen has produced a legacy for the late-successional ecosystem—and has performed the role of nexus species (see Chapter 4). Fast-growing species of poor competitive ability such as *P. pensylvanica* are important for stabilizing resources, but after immobilization of nutrients, restoration managers can also speed succession by removing inhibition from early-successional species by felling them (but leaving nutrient-rich remains on-site) and introducing later successional species if they are not already present. Although they are not part of the mature community, early-successional

TABLE 17.2

Disturbance Can Act as a Filter for Plant Traits

Disturbance characteristics	Filter	Examples
Intensity	Survival	Bark thickness promotes survival after fire
Frequency	Reproduction	Early maturation promotes reproduction under frequent disturbance
Abruptness/ resource increase	Colonizing ability	Dormant seeds are triggered to germinate by elevated nitrate
Various	Direct adaptations	Smoke exposure stimulates germination in fire-dependent specie

species may play an important role in achieving restoration goals—for example, in the case of *P. pensylvanica*, by holding nutrients on the site.

Because some species reproduce only on recently disturbed patches, spatial scale becomes critical to understanding population stability. For example, Busing and White (1993) suggested that shade-tolerant species exhibited lower variability in population density at small scales than did shade-intolerant species. White (1996) reviewed other cases in which scale was an important factor in population structure and metapopulation dynamics. There are two consequences if disturbance is missing in the restoration process. First, some species should not be expected to reproduce successfully in the presence of adults (that is, they can establish only on disjunct, disturbed patches). Second, some species may not be expected to reproduce in any given time period (if they rely on episodic disturbance).

Díaz et al. (1999) pointed out the importance of understanding the functional traits of plants and examining correlations among these traits to define plant functional types. The traits that are most important with regard to disturbance and succession within a particular ecosystem include relative growth rates, growth form, biomass partitioning, leaf size, leaf duration, shade tolerance, nutrient foraging, moisture relationships, life span, age at sexual maturity, seed size, seed-crop size, seed dormancy, reproductive frequency, vegetative reproduction, and stress tolerance (see Keddy 1992).

Disturbances Modify the Process of Succession and Community Assembly

The role of disturbance depends on when it occurs relative to succession (Wilson and Gitay 1995). As stated earlier, disturbances can set succession back to earlier stages, accelerate succession ahead to later stages, hold succession in place, or cause change that has no relationship to directional succession. Disturbances that set back or accelerate succession do so by removing the inertia represented by the dominant individuals.

Connell and Slatyer (1977) described three models of succession: tolerance, inhibition, and facilitation. If the turnover of species simply reflects the different initial growth rates and longevities of species that are present at the start, succession follows the tolerance model. Within the tolerance model, the influence of the first established species on the pattern of succession has been called the initial floristic composition hypothesis (see Glenn-Lewin and van der Maarel 1992), although both deterministic and stochastic factors can affect the beginning point. If the turnover of species occurs because the early dominants suffer mortality, thus opening up space for invasion by other species, succession follows the inhibition model: early dominants inhibit the arrival of later dominants. Succession proceeds by facilitation when early-successional species facilitate the establishment of later successional species. In the context of plant communities, facilitation has been called "relay floristics" because each compositional stage is essential to the transition (see Glenn-Lewin and van der Maarel 1992).

It was not proposed that the three models are mutually exclusive. In fact, experiments have shown that in some cases, removal of early species increased the growth and establishment of later species, even if those species were initially present on the site (nominally, the tolerance model). In other cases, early species were required for establishment of later species (nominally, the facilitation model; see Armesto and Pickett 1986, DeSteven 1991). Generally, mutualism and facilitation are important in stressful environments, and competition and inhibition are important in environments with abundant resources and favorable physical conditions (see Chapters 10, 13, 14, and 16). Interactions between species can also change from competition and inhibition during favorable times to facilitation and mutualism in stressful times (Holmgren et al. 1997).

Synthesis: Guidelines for Incorporating Principles of Disturbance and Succession into Restoration

Although ecosystems and restoration goals vary widely, we can ask three questions about any restoration project: First, is the trajectory of change moving in the desired direction? Second, what is the rate of change? Third, what actions are needed to correct the trajectory or adjust the rate of change? How ecosystems respond to disturbance can help us answer these questions. We organize our summary under three headings: initial conditions, productivity, and ongoing dynamics. We phrase each summary point as a statement of general expectations, from which the answers to the three questions can be deduced.

In developing these expectations, we assume that the goal of the restoration has been articulated and is feasible with regard to initial conditions (Zedler and Callaway 1999) and spatial scale (White and Walker 1997). In that sense, all of our summary statements relate to the feasible goal and to ongoing dynamics that are self-sustaining to the degree possible (that is, require minimum ongoing human effort). For example, productivity or nutrient conditions should not be judged relative to some possible maximum but relative to inherent site conditions and the restoration goal. Indeed, in some cases—such as when avoiding or reversing eutrophication by ameliorating anthropogenic nutrient deposition—the goal is to reduce productivity, because artificial enrichment often decreases diversity, changes ecosystem structure, and produces conditions that exceed the physiological abilities of the target species. We also recognize that restoration projects can include multiple goals, including maintenance of a variety of successional states or age classes. Our summary statements do not, therefore, point to mutually exclusive actions, although each of the following 20 guidelines is phrased in the context of a single site and goal.

Initial Conditions

Initial conditions relate to the ecological legacy the disturbance leaves behind and the historical contingency, or similarity of the disturbance to those in the past.

LEGACY

In natural disturbances, the legacy of the predisturbance ecosystem, determined by disturbance severity, influences recovery in three ways:

1. The greater the retention of nutrients and soil from the predisturbance ecosystem and/or the closer the initial conditions are to the target conditions, the more likely it is that the trajectory and rate are acceptable and the less human effort is required (see Chapters 13 and 16 for the contrasting limits imposed by nutrition and toxicity).
2. The higher the survival rate of seeds and living plants, the higher the rate of recovery and the lower the dependence on regeneration or new dispersal.
3. The greater the range of microsite conditions, the higher the number of species that establish.

Restoration actions to correct for inadequate ecological legacy include (1) altering site physical conditions (usually by adding soil, nutrients, or water and removing toxic substances), and (2) altering biotic conditions (usually by adding mycorrhizae, seeds, seedlings, or new species).

HISTORIC CONTINGENCY

Restoration should seek to understand the relationship between current anthropogenic disturbances and historic ones and to understand the requirements of the target species.

4. The closer the disturbance is to historical precedent, the greater the set of functional traits available to cope with disturbance effects.

Productivity

Productivity (or resource availability, correlated with soil resources, water, and temperature regime) affects assembly in seven ways. We emphasize again that the ideal productivity is not judged in absolute terms but relative to each site and to restoration goals. Restoration actions can establish the goals for productivity and resource levels; add species appropriate to conditions; and use disturbance to remove inhibition, increase resources, decrease resources, hold resources constant, or create turnover to accelerate, decelerate, or halt succession if appropriate. In any case, disturbance increases the variety of patch types and supports the continued occurrence of disturbance-dependent species.

5. The greater the resource supply relative to the predisturbance ecosystem or restoration target, the higher the initial rates of establishment and growth.
6. The greater the abruptness and magnitude of increase in resources, the greater the selection for rapid colonization, the higher the initial growth rates (which leads to critical uptake of soil elements that are otherwise vulnerable to leaching), and the greater the stochastic effects on composition.
7. The greater the productivity, the greater the differentiation of successional roles and the greater the amount of successional turnover during assembly. These relationships are reflected in changes in life history traits: resource use efficiency, longevity, and age at sexual maturity increase, while relative investment in reproduction decreases (this approximates Rule 8 of Belyea and Lancaster 1999).

8. As resources become immobilized in biomass and organic detritus, further establishment is reduced. At the same time, disturbance becomes a mechanism for periodic increases in resource availability, which can benefit growth and reestablishment depending on successional role (larger patches and greater resource abundance favor early-successional traits; smaller patches and lower resource abundance favor late-successional traits).

9. The greater the resource supply, the greater the importance of inhibition (disturbance can increase turnover by removal of inhibition), whereas the greater the stress, the greater the importance of facilitation and mutualism (see Chapters 13, 14, and 16).

10. In low-resource-supply ecosystems in which the total resource content of the ecosystem (biotic and abiotic compartments) is low, disturbance does not result in dramatic increases in establishment or growth, although disturbance may be needed to promote desired composition and dynamics by allowing regeneration of desired species or by removing or preventing invasion by unwanted species (such systems are also vulnerable to artificial increases in productivity).

11. In low-resource-supply ecosystems in which the total resource content of the ecosystem is high but resources are immobilized in biomass and organic detritus, disturbance increases resource availability (and hence establishment and growth rates) and is an opportunity to modify assembly.

Ongoing Dynamics

Restoration actions must establish the disturbance specificity, frequency, and spatial pattern to promote the desired community composition and structure.

SPECIFICITY

How disturbance modifies succession can depend on specificity. Restoration actions include introducing disturbance at the right stage of successional turnover to accelerate or arrest succession or to produce episodic reproduction in target species.

12. Disturbance accelerates succession when early-successional species are disturbed, allowing later successional species to establish or dominate.

13. Disturbance sets succession back to earlier stages when late-successional species are disturbed, resources increase, and early-successional species establish.

14. Disturbance holds succession in check when invading late-successional species are disturbed.
15. Disturbance produces no effect on successional direction through small-scale random or cyclic replacements.

PATCH SIZE

Restoration should address both the size of the disturbed patches and the heterogeneity of multipatch areas to support the restoration goal.

16. Response varies with patch size: the smaller the disturbance patch, the more likely small-scale patch dynamics, without changes in composition, will dominate; the larger the disturbance patch (within the spatial context of the restoration site), the more regeneration succession will dominate.
17. The greater the spatial heterogeneity, the greater the species diversity of the restoration site.

FREQUENCY

Restoration actions include introducing disturbance at the appropriate rate for the target composition. Chapter 19 presents an example of the importance of ongoing disturbance to ecosystem function.

18. The greater the frequency of disturbance, the greater the selection for early-reproducing and fast-growing species.
19. The greater the diversity of disturbance frequencies across the site, the greater the diversity of species and successional stages present on the restoration site.

PATTERN AND PROCESS

The spatial pattern of disturbance can affect the processes of restoration. Restoration must attempt to compensate through management if the requisite relationship of pattern to process cannot be restored (Radeloff et al. 2000).

20. Early-successional species intolerant of competition and low resource supply require the appropriate temporal frequency (within reproductive life span or the persistence of dormant seeds) and spatial pattern (within dispersal distances) for persistence, whereas late-successional species are often able to establish within late-

successional communities, leading to development of an all-aged structure.

Acknowledgments

We thank the participants in the Jena restoration workshop, Vicky Temperton, Richard Hobbs, Susan Wiser, and Christoph Matthaei for helpful discussions and comments in the preparation of this chapter.

REFERENCES

Armesto, J. J.; and Pickett, S. T. A. 1986. Removal experiments to test mechanisms of plant succession in oldfields. *Vegetatio* 66:85–93.

Belyea, L. R.; and Lancaster, J. 1999. Assembly rules within a contingent ecology. *Oikos* 86:402–416.

Bond, W. J.; and van Wilgen, B. J. 1999. *Fire and Plants*. New York: Chapman and Hall.

Busing, R. T.; and White, P. S. 1993. Effects of area on old-growth forest attributes: implications for the equilibrium landscape concept. *Landscape Ecology* 8:119–126.

Churchill, E. C.; and Hanson, H. C. 1958. The concept of climax in arctic and alpine vegetation. *Botanical Review* 24:127–191.

Clements, F. C. 1916. *Plant Succession*. Publication No. 242. Washington, D.C.: Carnegie Institution.

Connell, J. H. 1978. Diversity in tropical rain forests and coral reefs. *Science* 199:1302–1310.

Connell, J. H.; and Slatyer, R. O. 1977. Mechanisms of succession in natural communities and their role in community stability and organization. *Nature* 111:1119–1144.

Connett, M. W.; Puettmann, K. J.; Frelich, L. E.; and Reich, P. B. 2001. Comparing the seedbed and canopy type in the restoration of upland *Thuja occidentalis* forests of northeastern Minnesota. *Restoration Ecology* 9:386–396.

DeSteven, D. 1991. Experiments on mechanisms of tree establishment in old-field succession: seedling emergence. *Ecology* 72:1066–1075.

Diamond, J. M. 1975. Assembly of species communities. In *Ecology and Evolution of Communities*, ed. M. I. Cody and J. M. Diamond, 342–444. Cambridge, Mass.: Harvard University Press.

Díaz, S. M.; Cabido, J.; and Casanoves, F. 1999. Functional implications of trait-environment linkages in plant communities. In *Ecological Assembly Rules: Perspectives, Advances, Retreats*, ed. E. Weiher and P. Keddy, 338–362. Cambridge: Cambridge University Press.

Dixon, K. W.; Roche, S.; and Pate, J. S. P. 1995. The promotive effect of smoke derived from burnt vegetation on seed germination of western Australian plants. *Oecologia* 101:185–192.

Glenn-Lewin, D. C.; and van der Maarel, E. 1992. Patterns and processes of vegetation dynamics. In *Plant Succession: Theory and Prediction*, ed. D. C. Glenn-Lewin, R. K. Peet, and T. T. Veblen, 11–59. London: Chapman and Hall.

Grime, J. P. 1979. *Plant Strategies and Vegetation Processes*. Chichester, England: John Wiley and Sons.

Grubb, P. J. 1977. The maintenance of species-richness in plant communities: the importance of the regeneration niche. *Biological Reviews of the Cambridge Philosophical Society* 52:107–145.

Harmon, M. E. 1984. Survival of trees after low-intensity surface fires in Great Smoky Mountains National Park. *Ecology* 65:796–802.

Holmes, P. M.; and Richardson, D. M. 1999. Protocols for restoration based on recruitment dynamics, community structure, and ecosystem function: perspectives from South African fynbos. *Restoration Ecology* 7:215–230.

Holmgren, M.; Sheffer, M.; and Huston, M. A. 1997. The interplay of facilitation and competition in plant communities. *Ecology* 78:1966–1975.

Jentsch, A. 2001. *The significance of disturbance for vegetation dynamics. a case study in dry acidic grasslands.* Ph.D. diss., Bielefeld University, Germany.

Jentsch, A.; Beierkuhnlein, C.; and White, P. S. 2002a. Scale, the dynamic stability of forest ecosystems, and the persistence of biodiversity. *Silva Fennica* 36.

Jentsch, A.; Friedrich, S.; Beyschlag, W.; and Nezadal, W. 2002b. Significance of ant and rabbit disturbances for seedling establishment in dry acidic grasslands dominated by *Corynephorus canescens. Phytocoenologia* 32:553–580.

Johnson, E. A.; and Miyanishi, K. 1995. The need for consideration of fire behavior and effects in prescribed burning. *Restoration Ecology* 3:271–278.

Keddy, P. A. 1992. A pragmatic approach to functional ecology. *Functional Ecology* 6:621–626.

Keddy, P. A.; and Weiher, E. 1999. Introduction: the scope and goals of research on assembly rules. In *Ecological Assembly Rules: Perspectives, Advances, Retreats*, ed. E. Weiher and P. Keddy, 1–22. Cambridge: Cambridge University Press.

Lockwood, J. L. 1997. An alternative to succession. *Restoration and Management Notes* 15:45–51.

MacArthur, R. H.; and Wilson, E. O. 1967. *The Theory of Island Biogeography.* Princeton: Princeton University Press.

Marks, P. L. 1974. The role of pin cherry (*Prunus pensylvanica* L.) in the maintenance of stability in northern hardwood ecosystems. *Ecological Monographs* 44:73–88.

Menges, E. S.; and Waller, D. M. 1983. Plant strategies in relation to elevation and light in floodplain herbs. *American Naturalist* 122:454–473.

Noble, J. R.; and Slatyer, R. O. 1980. The use of vital attributes to predict successional changes in plant communities subject to recurrent disturbances. *Vegetatio* 43:5–21.

Paine, R. T. 1974. Intertidal community structure: experimental studies on the relationship between a dominant competitor and its principal predator. *Oecologia* 15:93–120.

Pavlovic, N. B. 1994. Disturbance-dependent persistence of rare plants: anthropogenic impacts and restoration implications. In *Recovery and Restoration of Endangered Species*, ed. M. L. Boels and C. Whelan, 159–193. Cambridge: Cambridge University Press.

Peet, R. K. 1992. Community structure and ecosystem function. In *Plant Succession: Theory and Prediction*, ed. D. C. Glenn-Lewin, R. K. Peet, and T. T. Veblen, 103–151. London: Chapman and Hall.

Peet, R. K.; and Christensen, N. 1980. Succession, a population process. *Vegetatio* 43:131–140.

Pickett, S. T. A.; and Thompson, J. N. 1978. Patch dynamics and the design of nature reserves. *Biological Conservation* 13:27–37.

Radeloff, V. C.; Mladenoff, D. J.; and Boyce, M. S. 2000. A historical perspective and future outlook on landscape scale restoration in the northwest Wisconsin pine barrens. *Restoration Ecology* 8:119–126.

Shmida, A.; and Elner, S. 1984. Coexistence of plant species with similar niches. *Vegetatio* 58:29–55.

Sousa, W. P. 1984. The role of disturbance in natural communities. *Annual Review of Ecological Systems* 15:353–391.

Swanson, F. J.; and Franklin, J. F. 1992. New forestry principles from ecosystem analysis of Pacific Northwest forests. *Ecological Applications* 2:262–274.

Turner, M. G.; Romme, W. H.; Gardner, R. H.; O'Neill, R. V.; and Kratz, T. K. 1993. A revised concept of landscape equilibrium: disturbance and stability on scaled landscapes. *Landscape Ecology* 8:213–227.

van der Maarel, E. 1996. Vegetation dynamics and dynamic vegetation science. *Acta Botanica Neerlandica* 45:421–442.

van der Maarel, E.; and Sykes, M. T. 1993. Small-scale plant species turnover in grasslands: the carousel model and a new niche concept. *Journal of Vegetation Science* 4:179–188.

Vinton, M. A.; Hartnett, D. C.; Finck, E. J.; and Briggs, J. M. 1993. Effect of fire, bison (Bison bison) grazing and plant community composition in tallgrass prairie. *American Midland Naturalist* 129:10–18.

Vogl, R. J. 1974. Effects of fire on grasslands. In *Fire and Ecosystems*, ed. T. T. Kozlowski and C. E. Ahlgren, 139–194. New York: Academic Press.

Walker, B.; Kinzig, A.; and Langridge, J. 1999. Plant attribute diversity, resilience, and ecosystem function: the nature and significance of dominant and minor species. *Ecosystems* 2:95–113.

White, P. S. 1996. Spatial and biological scales in reintroduction. In *Restoring Diversity*, ed. D. A. Falk, C. Millar, and M. Olwell, 49–86. Washington, D.C.: Island Press.

White, P. S.; Harrod, J.; Romme, W.; and Betancourt, J. 1999. The role of disturbance and temporal dynamics. In *Ecological Stewardship*, vol. 2, ed. R. C. Szaro, N. C. Johnson, W. T. Sexton, and A. J. Malk, 281–312. Kidlington, UK: Elsevier.

White, P. S.; and Jentsch, A. 2001. The search for generality in studies of disturbance and ecosystem dynamics. *Progress in Botany* 62:399–450.

White, P. S.; MacKenzie, M. D.; and Busing, R. T. 1985. Natural disturbance and gap phase dynamics in southern Appalachian spruce-fir. *Canadian Journal of Forest Research* 15:233–240.

White, P. S.; and Pickett, S. T. A. 1985. Natural disturbance and patch dynamics: an introduction. In *The Ecology of Natural Disturbance and Patch Dynamics*, ed. S. T. A. Pickett and P. S. White, 3–13. New York: Academic Press.

White, P. S.; and Walker, J. L. 1997. Approximating nature's variation: selecting and using reference sites and reference information in restoration ecology. *Restoration Ecology* 5:338–249.

Wilson, J. B.; and Gitay, H. 1995. Limitations to species coexistence: evidence for competition from field observations using a patch model. *Journal of Vegetation Science* 6:369–376.

Young, T. P.; Chase, J. M.; and Huddleston, R. T. 2001. Community succession and assembly. *Ecological Restoration* 19:5–17.

Zedler, J. B.; and Callaway, J. C. 1999. Tracking wetland restoration: do mitigation sites follow desired trajectories? *Restoration Ecology* 7:69–73.

Chapter 18

Disturbance, Assembly Rules, and Benthic Communities in Running Waters: A Review and Some Implications for Restoration Projects

CHRISTOPH D. MATTHAEI, COLIN R. TOWNSEND,
CHRIS J. ARBUCKLE, KATHI A. PEACOCK,
CORNELIA GUGGELBERGER, CLAUDIA E. KÜSTER,
AND HARALD HUBER

Our aim is to examine the role of disturbance in shaping community assembly in running-water ecosystems and the implications of disturbance ecology for river restoration. This chapter is divided into five sections. We first give a brief overview of research on assembly rules in running waters. In doing so, we emphasize the importance of disturbance as an environmental filter in most streams and rivers. Second, we summarize the present state of knowledge about the role of high-flow events in the community ecology of streams and rivers. These events are the most common type of natural disturbance in these ecosystems. We also demonstrate how these disturbances can affect community assembly at large spatial scales. Next we focus on bed movement patterns and invertebrate refugia, important aspects of disturbance research in running waters. We then turn to the effects, usually unmeasured and unrecognized, of local disturbance history (the specific stability or instability of small bed patches during high-flow events) on the key ecosystem components of invertebrates and algae. In these two sections, we provide examples showing that disturbance can also have a strong influence on community assembly at smaller spatial scales. Finally, we demonstrate some practical applications of our disturbance history research for river restoration using the specific example of a large-scale restoration project.

Research on Assembly Rules in Running Waters

Surprisingly few publications in running-water ecology have actually mentioned the term *assembly rules* as a specific research objective. The only papers we could find concerned a study of fish invasions in streams (Moyle and Light 1996) and a study of plant assembly in riparian wetlands (Toner and Keddy 1997). Nevertheless, stream and river ecologists have been addressing issues arising from the concept of assembly rules for several decades. This situation provides good evidence that exchange of information and active debate is often lacking not only between plant and animal ecologists (see Chapter 3) but also between running-water ecologists and other ecological disciplines. That stream ecologists have long been concerned with assembly rules holds particularly true if the term is not restricted to the outcome of interactions between the living components of an ecosystem (for example, Wilson and Whittaker 1995, Belyea and Lancaster 1999) but includes the abiotic constraints acting on a given species pool (see, for example, Keddy 1992; also see Chapter 3). Biotic interactions such as competition (McAuliffe 1984, Feminella and Resh 1991, Kohler and Wiley 1997, Kuhara et al. 1999), grazing (Feminella and Resh 1991, Power 1992, Kohler and Wiley 1997), or predation (Englund and Evander 1999, Nakamo et al. 1999, Diehl et al. 2000) can be very important in determining the structure and function of stream and river communities.

Almost all existing examples of biotic interactions in running waters have been reported from streams with permanently stable flow conditions (small spring-fed streams or lake outlets), during long periods of stable flow in periodically disturbed streams, or in stable experimental channels. Townsend (1989) and Poff and Ward (1989) hypothesized that although biotic interactions are likely to be important in stable streams, their importance ought to decrease with increasing frequency of hydrologic disturbance (that is, high-flow events). In streams and rivers with high disturbance frequencies, community assembly and development ought to be determined mainly by the frequent disturbance events. Poff and Ward (1989) also conducted a large-scale analysis of discharge regimes in 78 streams and rivers across the United States, in which they classified only 27 percent of the investigated systems as "stable" or "periodically stable." These results imply that biotic interactions may be of little importance in the majority of running-water ecosystems, if the hypothesis of Townsend (1989) and Poff and Ward (1989) is correct. On the other hand, Chesson and Huntly (1997) showed that, at least theoretically, biotic interactions *can* play an important role in frequently disturbed ecosystems because even a relatively minor stress caused by competition or predation could be enough to "push over the edge" a species already weakened by abiotic disturbance.

Determining the interplay among abiotic disturbance, competition, and predation in more frequently disturbed streams and rivers represents one of the greatest challenges for future research in running-water ecosystems. By contrast, ecologists have already recognized the tremendous importance of disturbance itself in shaping running-water communities. We will focus on disturbance in the next three sections. Because of a general lack of knowledge about the importance of biotic interactions in frequently disturbed streams and rivers, we cannot yet say whether disturbance acts as a filter (determining which species out of a given species pool are part of the game called community assembly, which is decided mainly by biotic interactions; see Chapter 17) or disturbance in itself is one of the assembly rules in running-water ecosystems (in the sense of, for example, Keddy 1992, Zobel 1992; also see Chapter 3).

The Role of Disturbance in the Community Ecology of Streams and Rivers

Disturbances that open up space or free up resources occur in virtually all ecosystems and affect all levels of biological organization (White and Jentsch 2001; also see the other chapters in Part V, "Disturbance and Assembly"). Not surprisingly, therefore, disturbance is featured in important ecological theories; namely, the intermediate disturbance hypothesis (Connell 1978, Sousa 1979), the concept of patch dynamics (Pickett and White 1985), and several other disturbance concepts (see Chapter 17). It is only relatively recently that these theories have been extended to streams and rivers (Pringle et al. 1988, Frid and Townsend 1989, Townsend 1989; but see Downes 1990). The influence of disturbance on running-water ecosystems is particularly strong (Resh et al. 1988, Lake 2000) because disturbances in streams and rivers tend to occur more often and be more severe than they are in many other ecosystems. We focus on the effects of high-flow events because these events are the most common type of natural disturbance in running waters. Note, however, that various other types of disturbance can also affect stream and river communities, including droughts, lack of oxygen, increased sedimentation of fine particles, and extreme water temperatures (see reviews by Steinman and McIntire 1990, Wallace 1990, Wood and Armitage 1997, Lake 2000).

Highly variable and/or unpredictable patterns of discharge act as disturbances of various frequencies and intensities (Poff and Ward 1989). These disturbances can modify population, community, and ecosystem processes (Townsend 1989, Mackay 1992, Townsend and Hildrew 1994), as shown recently in a few large-scale comparative studies of different streams. For example, the disturbance regimes in 54 stream sites in the Taieri River catchment in New Zealand were quantified by measuring the movement of painted bed particles representative of those present at each site (Townsend et al. 1997a, 1997b, 1997c; but see Downes et al. 1998), and the influence of these disturbance regimes on invertebrates inhabiting the streambeds was assessed. These invertebrates (mostly the aquatic stages of various insect taxa, such as stoneflies and mayflies) represent an important component of the stream ecosystem. In addition to determining invertebrate abundances and richness, the studies also grouped invertebrates according to several morphological or life history traits (see Chapters 2 and 3). The more intensely disturbed sites had communities with a significantly higher percentage of insects possessing the following traits (Townsend et al. 1997a): small size, high adult mobility, habitat generalism (predicted to confer resilience in response to bed disturbance), the ability to cling, a streamlined or flattened body shape, and two or more life stages outside the stream (predicted to confer resistance in the face of disturbance). Furthermore, maximum invertebrate taxon richness coincided with intermediate levels of disturbance (Townsend et al. 1997b), producing a dome-shaped curve between richness and intensity of disturbance, as predicted by the intermediate disturbance hypothesis. Death and Winterbourn (1995) also reported a significant relationship between invertebrate species richness and disturbance regime in several New Zealand streams, but they described a monotonic decline in species richness with increasing disturbance. Finally, the structure of food webs in a subset of 10 of the Taieri streams was related to the disturbance regime, with a smaller number of dietary links per species in disturbed sites (Townsend et al. 1998).

Disturbance, Bed Movement Patterns, and Invertebrate Refugia

In addition to having effects on benthic communities in running waters at the catchment or regional scale, disturbance also influences the assembly of stream organisms at much smaller spatial scales, including their distributions within the streambed at the patch scale (≤ 0.1 m^2). The extent of this influence became obvious as a result of recent research on flood refugia for benthic invertebrates. Recovery of stream communities after spates (smaller high-flow events with peak flows below bank-full discharge; that is, the river does not overflow its banks) and even after floods (larger events with peak flows above bank-full discharge that inundate the floodplain) is often rapid (Fisher et al. 1982, Mackay 1992, Matthaei et al. 1997). Therefore, refugia that lessen the impacts of hydrologic disturbances on stream organisms have been predicted to occur at several spatial scales (Sedell et al. 1990, Townsend and Hildrew 1994, Lancaster and Belyea 1997). Explicit testing of these different refugium hypotheses has begun only in recent years. Proposed local refugia include lateral stream margins (Bishop 1973), large surface particles (Townsend 1989, Biggs et al. 1997), dead zones where shear stresses on the bed are always low (Lancaster and Hildrew 1993a), inundated floodplain sediments (Badri et al. 1987), and the hyporheic zone (the saturated interstitial spaces beneath the streambed that receive water from the stream and from the groundwater body below; Williams and Hynes 1974).

Increases in stream discharge may directly affect communities of fish that occupy the water column (Poff and Allan 1995). For the benthic community, the influence of increased flow is not quite as direct. For most benthic organisms, high-discharge events represent a disturbance only if they actually move the bed sediments (Poff 1992, Townsend et al. 1997c). Consequently, the search for local flood refugia is closely linked to quantifying bed movement patterns in streams and rivers. Many researchers have studied community recovery after severe floods (for example, Fisher et al. 1982, Smock et al. 1994, Matthaei et al. 1997) that had caused obvious, large-scale bed movements. By contrast, the impact of smaller high-discharge events is much more difficult to assess. In the first attempts to do so, bed movement patterns were estimated from areas where algae had been removed from surface stones (Doeg et al. 1989, Matthaei et al. 1996). These observations led ecologists to consider the possibility of patchy bed movements during smaller high-flow events. Earlier still, Townsend and Hildrew (1976) had speculated that invertebrate recolonization after smaller spates might represent a redistribution rather than true colonization from distant sources, if the event did not disturb the whole streambed. However, they did not test their hypothesis in the field. Smaller spates are likely to be more important than large floods for the spatiotemporal dynamics of the benthic community simply because they occur more frequently (see Matthaei et al. 1996, 1997).

Against this background, we now turn to the most recent advances in the search for flood refugia for benthic invertebrates, concentrating on the results of our own research in the Kye Burn, a midsized gravel-bed stream (width at mean flow 5–10 m) on the South Island of New Zealand. Our first task was to determine whether stable bed patches, potential refugia for invertebrates, persisted during high-flow events in the Kye Burn. To investigate this question, we recorded the spatial patterns of bed movements during several high-flow events (Matthaei et al. 1999a). In each of three 20-m-long stream stretches, we marked 400 surface stones in situ (Downes et al. 1998; distance between stones about 50 cm). We then monitored the stability of these stones during two high-flow events. A second set of stones was used to investigate the effects of two more high-flow events. Peak flows of three of the four events were below bank-full discharge, and the average return periods were 3 to 5 months. The fourth event was a fairly large flood, with a peak discharge that exceeded bank-full flow and a return period of about 3 years. Simultaneously, 100 metal-link scour chains (distance between chains about 1 m) were installed in each of three additional 20-m stretches of streambed that were contiguous with the three stone-monitoring stretches. The chains were used to determine the three-dimensional patterns of sediment scour and fill to assess the stability of the sediments near the surface and in the deeper hyporheic zone (another potential invertebrate refugium). Here the four abovementioned events and two additional events with peak flows below bank-full discharge were investigated (Matthaei et al. 1999b).

Both marked stones and scour chains showed that bed movement patterns were highly heterogeneous in space (Figure 18.1 shows examples). Bed patches that experienced scour were often located very close (≤ 1 m) to patches that had experienced sediment deposition (fill patches) or that had remained undisturbed (stable patches), and there were similar patterns for stable and unstable stones. During the five below-bank-full events, 18–70 percent of the chain locations or marked stones remained stable. All of these bed patches or stones were potential refugia for benthic invertebrates. Only the single event above bank-full left few marked patches or stones undisturbed (generally 15 percent or less), indicating that potential surface refugia were scarce during this large flood. (We recently found a similar lack of surface refugia during a large flood in a German stream; Matthaei and Huber 2002). Further, scour in the Kye Burn affected only the uppermost 15 cm of the bed at most of the chain locations, even during the largest event. Consequently, invertebrates could also have used most of the deeper sediment layers in the hyporheic zone as refugia.

At one of our three sites, we then tested whether stable surface stones served as refugia for benthic invertebrates during one of the smaller events. This objective was achieved by sampling 8–12 stable and 12 unstable surface stones on each of three occasions: shortly before the spate, during the declining limb of the spate, and 2 weeks after flow had receded to normal (Matthaei et al. 2000). The results obtained clearly supported our hypothesis. Before the spate, total invertebrate densities were similar on stable and unstable stones (here the terms *stable* and *unstable* refer to the predicted stability of

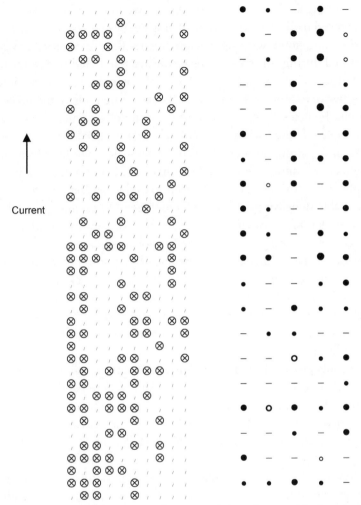

FIGURE 18.1 Spatial patterns of presence (⊗) or absence (/) of marked surface stones and scour (o), fill (●), or no change (–) at one of the sites in the Kye Burn (site 1 in Matthaei et al. 1999a, 1999b) after a bed-moving spate in December 1996. "No change" is equivalent to 1 cm or less scour or fill. The first row of stones or chains was at the downstream end of each site, and left and right in the figure represent true left and right in the stream. The size of the symbols for scour and fill is related to the degree of change: scour o > 1–5 cm, **o** > 5–10 cm; fill ● > 1–5 cm, ● > 5–10 cm, ● > 10–20 cm. (After Matthaei et al. 1999a, 1999b.)

these stones during the last previous high-flow event, about 4 months earlier). During the receding limb of the investigated spate, by contrast, invertebrate densities on stable stones were significantly higher than on stones that had been disturbed during this spate. This result was paralleled by the mean number of invertebrate taxa per sampled stone (taxon richness) and the densities of all five of the most common taxa (larvae of the midges, the mayfly *Deleatidium,* the black fly *Austrosimulium,* the stonefly *Zelandoperla,* and the Oligochaeta). These data provide striking evidence that a patchy disturbance can have a strong influence on the small-scale assembly of stream invertebrate communities. One could say that disturbance acted as a filter (see Chapter 17) in this case, creating open space by causing invertebrates to leave unstable stones or by removing them directly. On stable stones, disturbance even led to *increased* community richness, because invertebrate density and taxon richness on these stones were significantly higher than predisturbance levels, implying that a significant proportion of the invertebrate community actively sought out stable stones. Two weeks later, invertebrate densities on stable and unstable stones were similar again, indicating that the effects of local disturbance history (here, the stability or instability of individual surface stones) on invertebrate community assembly were relatively short-lived.

These results provided the first evidence that surface refugia can exist during high-flow events in gravel-bed streams. All earlier research either was conducted in other stream types or was aimed at other refugium types. Thus, inundated floodplain vegetation (Badri et al. 1987) and areas of permanently low shear stress (Lancaster and Hildrew 1993b, Robertson et al. 1995, Winterbottom et al. 1997) served as invertebrate refugia in two small, sluggish streams. Prévot and Prévot (1986) obtained the same results on inundated floodplain gravels in the lateral river margins and in a calm, hyporheically fed side arm of a French river. In a sandy-bottomed Virginia river, invertebrates associated with woody debris dams were more resistant to floods than the fauna in the sandy midchannel (Palmer et al. 1996). The evidence that certain invertebrates (for example, the midges) actively sought out stable stones in the Kye Burn contrasts with all other studies that identified refugia in the main channel of a stream (Lancaster and Hildrew 1993b, Robertson et al. 1995, Palmer et al. 1996, Winterbottom et al. 1997). The microhabitats investigated in these latter studies experienced relatively low hydraulic stress during high-discharge events, and it was possible for the fauna to be passively relocated into flow refugia or to remain there during spates. In view of these contrasting results from different streams, it seems likely that stream invertebrates use both passive and active strategies to reach flood refugia.

In contrast to the refugium types just discussed, which are fairly well documented, convincing evidence that stable sediments in the hyporheic zone can act as refugia is still lacking (see Palmer et al. 1992, Schmid-Araya 1994, Dole-Olivier et al. 1997). The only rigorous field test of this hypothesis, conducted on the meiofauna in a sandy-bottomed river in Virginia, yielded negative results (Palmer et al. 1992).

In summary, benthic invertebrates in the Kye Burn appear to use the abundant stable surface stones as refugia during smaller high-flow events, while invertebrate density and richness on unstable stones are reduced at the same time. Because such events occur quite often (about every 3–5 months), these disturbances are likely to contribute considerably to the spatiotemporal dynamics of the benthic invertebrate assembly in this stream (also see Chapter 3). By contrast, surface refugia were scarce during a large flood with a return period of about 3 years. However, the flood disturbed mainly the uppermost 15 cm of the streambed. Therefore, the deeper stable layers of the hyporheic zone may be the most important local invertebrate refugium during large, but relatively rare, disturbances. Because of the relatively small size of the interstitial spaces in the hyporheic zones of most rivers, small-bodied invertebrate taxa or early instars of larger taxa are more likely to benefit from this refugium type than relatively large taxa or individuals (also see Chapter 19).

The Influence of Local Disturbance History on Invertebrate and Algal Communities

In the next step of our research in the Kye Burn, we went beyond the search for stable bed patches acting as invertebrate refugia during hydrologic disturbance and investigated the general effects of local disturbance history (defined as the small-scale mosaic of bed patches that have experienced scouring or sediment deposition or have remained stable during high-flow events) on the benthic fauna. We found that disturbance history can have both short-term (see above) and long-term effects on stream invertebrates (Matthaei and Townsend 2000). More than 2 months after the spate studied by Matthaei et al. (2000), we collected quantitative samples from the uppermost 10 cm of each of five randomly selected bed patches that had experienced scour, fill, or no change. Density of the dominant invertebrate taxon, the highly mobile mayfly *Deleatidium* (Figure 18.2), and densities of another three of the seven most common taxa differed significantly between patch stability categories. Larvae of *Deleatidium*, the blackfly *Austrosimulium*, and the dipteran Eriopterini were most abundant in fill patches, whereas Isopoda were most abundant in scour patches. These results provided the first evidence for such long-term effects of disturbance history on stream invertebrates. The study was also among the first to show long-term effects of a patchy disturbance on an animal community dominated by organisms that are highly mobile relative to the spatial scale of the disturbance (see Chapter 17 for some examples of long-term effects of patchy disturbances on sessile plant communities). Scoured, depositional, and stable bed patches were separated by just a few meters. *Deleatidium* nymphs are rapid crawlers and good swimmers, and blackfly larvae can cover long distances by drifting (see the review by Mackay 1992). Both taxa are well known as abundant and early colonizers of disturbed reaches (Mackay 1992) and could easily have dispersed among the different patch types within a few days. The two taxa comprised 75 percent of the entire invertebrate community in the samples. The lasting effect of disturbance history on these mobile taxa contrasted with the expectations of Downes (1990), who warned that species with high dispersal ability might be unaffected by small-scale patchiness of resources (or disturbance) because they can move easily between such patches. Instead, our results supported the contention of Townsend (1989) and Frid and Townsend (1989) that patch dynamics models are applicable to streams and rivers.

In this case, disturbance is more likely to have acted as a filter by changing the physical environment of the invertebrates than by simply creating open space for the establishment of new individuals (see Chapter 17). The observed long-term patterns were unlikely to have been caused by differential dislodgment or mortality of invertebrates during the spate, because high mobility should have permitted rapid recolonization of disturbed patches. Moreover, if density patterns were caused by losses during the spate, invertebrate densities should have been highest in stable patches if these served as refugia during the event (as did stable surface stones). In fact, none of the common taxa was most abundant in stable patches 2 months after the spate. Therefore, we believe that the observed density differences were an indirect effect of the spate; scoured and (especially) depositional patches seemed to provide a more suitable habitat for particular invertebrate taxa than did stable patches. Substratum composition (Ulfstrand 1968, Holomuzki and Messier 1993), availability of food resources (Hearnden and Pearson 1991, Richardson 1992), local current velocity or shear stress (Ulfstrand 1967, Peckarsky et al. 1990), and water depth (Barmuta 1989, Ruse 1994) are known to influence the microdistribution of stream invertebrates. We do not know yet if these microhabitat parameters are themselves influenced by local disturbance history, because we did not measure them in our pilot study. However, if these parameters were influenced by disturbance history, then

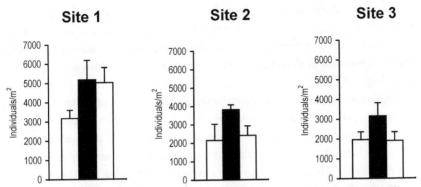

FIGURE 18.2 Mean densities (± SE) of *Deleatidium* spp. in scour (gray), fill (black), and stable (white) patches at the three Kye Burn sites more than 2 months after a bed-moving spate in August 1997. (After Matthaei and Townsend 2000.)

local disturbance history could turn out to be one of the most important factors influencing the small-scale community assembly of stream and river invertebrates.

Having found that disturbance history can influence benthic invertebrates, we turned our attention to benthic algae, which are regarded as the most important primary producers in many midsized streams and rivers (Minshall 1978, Vannote et al. 1980). The response of benthic algae to hydrologic disturbance has been investigated in many studies (see reviews by Fisher 1990, Steinman and McIntire 1990, Peterson 1996), and substrata unlikely to be moved by increased flow (for example, boulders and bedrock patches) have been reported to act as algal refugia (Douglas 1958, Power and Stewart 1987, Uehlinger 1991, Peterson et al. 1994, Francoeur et al. 1998). Nevertheless, previous research has not addressed how the small-scale patchiness of bed movements may affect benthic algae, especially the distinction between scoured and depositional bed patches.

We studied this question using buried scour chains in a relatively natural reach of the River Isar, a gravel-bed river (width about 90–120 m) in southern Germany (Matthaei et al. 2003). A total of 236 chains were installed in seven randomly chosen transects that covered the entire length of an 800-m river reach. Two floods (with return periods of ≤ 1 year) caused a mosaic of bed patches with different disturbance histories. This mosaic included both small-scale patterns (≤ 1 m distance between different patch types) and larger-scale patterns (distance of several meters). Once after the first flood and twice after the second, we sampled algal biomass and cell densities (mainly diatoms) on single surface stones in replicate bed patches that had experienced scour or fill or remained stable during the event. Three months after the first flood, biomass and total diatom densities (Figure 18.3), taxon richness, and densities of six of the nine most common diatom taxa were highest in fill patches. Six days after the second flood, biomass was highest in stable patches, indicating a refugium function of these patches. In contrast to previous studies where refugia were always large and/or immobile substrata (see references in the previous paragraph), the algae in the River Isar found flood refugia in stable bed patches consisting of average-sized stones. The stability of these patches may have been related to their position in the riverbed (mainly occupying side channels and near gravel islands). Four weeks after the second flood, diatoms tended to be most abundant in scour patches. With a single exception, the observed differences between patch types could not be attributed to differences in local near-bed current velocity or in water depth, despite the fact that these habitat parameters can influence algal microdistributions (Stevenson et al. 1996).

These results showed for the first time that local disturbance history can have short- and long-term effects not only on benthic invertebrates in a mid-sized stream but also on the community assembly of benthic algae in a much larger river. The results also provided further evidence that the patch dynamics perspective (Pickett and White 1985) can be applied to river ecosystems. The effects of disturbance history were more complex than a simple refugium function of stable patches, because algal patterns changed with time since the last disturbance, possibly depending on the successional state of the algal mats (see Peterson 1996). Differences in grazing pressure (see, for example, Power 1992) might be an explanation for the long-term patterns, but we found no correlation between algal biomass and grazer densities shortly after the second flood, the one sampling occasion when invertebrate data were available (H. Huber, unpublished data).

Further research is needed before we can fully understand the complex effects of local disturbance history on stream and river organisms. In a recent study using small steel pins to measure bed stability during base-flow conditions in a Japanese stream, Miyake and Nakano (2002) found that even the subtle erosion and deposition processes at base flow can influence invertebrate microdistribution and assembly diversity. There is a need to measure additional microhabitat parameters for invertebrates and algae, including additional components of the ecosystem (for example, primary production, benthic bacteria, and bacterial production), and also to perform manipulative experiments. All the same, our results (and those of Miyake and Nakano 2002) indicate that local disturbance history is likely to play a significant,

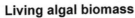

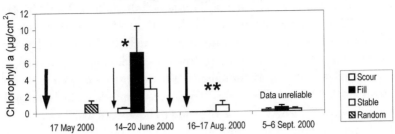

FIGURE 18.3 Living algal biomass and total diatom density (± SE) in random samples (May 2000) and in scour, fill, and stable patches (June to September 2000) in the River Isar. The arrows indicate timing and magnitudes of the high-flow events. (After Matthaei et al. 2003.)

and previously neglected, role in shaping community assembly in streams and rivers. This conclusion has important implications for restoration projects, some of which we address next.

Disturbance History and Restoration: A Case Study in a Gravel-Bed River

In this final section, we discuss some practical applications of the results of our disturbance history research to river restoration using a case study, the River Isar in Germany. In their review of advances in restoration ecology, Halle and Fattorini (see Chapter 2) analyze weaknesses and limitations of past and present restoration projects and make recommendations to improve future projects. In addition to describing our study and presenting some preliminary results, we will also assess how the Isar restoration project can be evaluated using the criteria developed by Halle and Fattorini.

The Isar-Plan, one of the largest river restoration projects in Europe, was financed by the Federal State of Bavaria and the City of Munich (Zinsser 1999). Estimated total costs are €25–30 million. From 2000 to 2005, 8 km of the River Isar within the city limits of Munich will be restored by removing all man-made channelization structures, widening the riverbed, creating several gravel islands, and allowing the river to meander freely in the 300-m- to 400-m-wide floodplain. Thus, as recommended by Halle and Fattorini, connectivity between the river and its floodplain will be reestablished. Halle and Fattorini also recommend that the natural sediment load and the river-specific flood regime should be restored. In accordance with this recommendation, the sediment load of the River Isar has been increased in recent years by reopening weirs in upstream reaches and by regularly shifting large quantities of gravel from a reservoir near the headwaters in the German Alps into the river below the dam (at a cost of more than €100,000 each year).

The discharge regime of this prealpine river is still quite dynamic, with several bed-moving floods per year (Table 18.1), despite the reservoir near its headwaters. Although part of the river water in the restoration reach is diverted for hydropower generation, the minimum flow remaining in the riverbed will be increased threefold by 2005. Nutrient levels in the river are low to moderate due to efficient sewage treatment in the catchment, although swimming is generally not recommended because of periodically high concentrations of fecal bacteria. The goals of the Isar project are clearly defined (another recommendation of Halle and Fattorini) and include (1) restoring flow regime, disturbance regime, and benthic invertebrate and fish faunas in the river; (2) restoring the semiaquatic and terrestrial fauna and flora in the floodplain; and (3) improving water quality for swimming purposes. Thus, the goals include most of those suggested for river restoration by the National Research Council in 1992 (see Chapter 2).

Several ongoing studies are testing whether these goals will be met, each using a Before-After-Control-Impact design (Stewart-Oaten and Murdoch 1986) to determine the effects of restoration. The effects of the restoration on discharge regime, disturbance regime, and benthic invertebrate fauna are the subject of a study conducted at the University of Munich (H. Huber, unpublished data). During the initial investigation period of 3 years (2000–2002; 1 year of data collecting before restoration, 2 years after restoration), first quantitative effects of the restoration on disturbance regime and invertebrates are expected (another recommendation of Halle and Fattorini). The consequences of restoration of the discharge regime for the benthic invertebrate and fish fauna will be assessed after 2005, using all previously

TABLE 18.1

Summary of Bed Movement Patterns (Scour, Fill, and Stable Bed Patches)[1]
in the Three Reaches of the Isar River Caused by the High-Flow Events in
2000[2] and in 2001[3]

Date	March 2000	August 2000	September 2000	June 2001	September 2001	December 2001
Natural reach (positive control)						
Scour (%)	24	25	33	27	41	26
Mean scour (m)	−0.21	−0.11	−0.16	−0.28	−0.15	−0.07
SD (% of mean)	84	77	54	45	73	30
Fill (%)	40	25	21	49	26	13
Mean fill (m)	0.14	0.10	0.11	0.14	0.17	0.06
SD (% of mean)	62	63	62	63	67	39
Stable (%)	36	51	46	25	33	61
Impact reach (before/after)						
Scour (%)	56	62	39	No data	20	32
Mean scour (m)	−0.20	−0.11	−0.09		−0.11	−0.10
SD (% of mean)	48	55	33		68	79
Fill (%)	29	9	29		57	22
Mean fill (m)	0.13	0.04	0.15		0.19	0.08
SD (% of mean)	47	37	61		68	45
Stable (%)	15	30	32		23	46
Channelized reach (negative control)						
Scour (%)	97	83	2	59	59	19
Mean scour (m)	−0.22	−0.16	−0.02	−0.28	−0.15	−0.04
SD (% of mean)	30	39	28	42	34	56
Fill (%)	0	7	91	22	29	64
Mean fill (m)	—	0.02	0.30	0.06	0.15	0.13
SD (% of mean)	—	23	35	35	45	52
Stable (%)	3	11	7	19	13	17

[1]Standard deviations (SD; expressed as percentage of the mean) describe the degree of variation in scouring or filling depths.
[2]Before restoration of the impact reach; $n = 267$–390).
[3]After restoration; $n = 285$–642).

collected data as "before" data. Finally, a long-term monitoring program is planned for the future (as recommended by Halle and Fattorini).

Three river reaches of similar length are under investigation in the initial project: the relatively natural 800-m reach where we studied disturbance history effects on benthic algae (Matthaei et al. 2003), one channelized reach of similar length 8 km downstream, and a second channelized reach 12 km downstream. The natural reach is used as a positive control, the second reach was restored in January–March 2001 (impact reach), and the third reach is the negative control (channelized during the entire study period). The study focuses on bed movement patterns and the invertebrate fauna in the upper-most 10 cm of riverbed.

Before restoration, differences among the three reaches in bed movement patterns or benthic fauna (for example, species richness) were determined. After restoration of the impact reach, previously existing differences between the impact reach and the positive control should diminish if restoration was successful. At the same time, differences between the impact reach and the negative control should increase. Bed movement patterns are identified using scour chains and additional measurements of net changes using a steel wire pulled tightly across each transect (H. Huber, unpublished method), and invertebrates are sampled quantitatively. The functional composition of the invertebrate community will also be assessed using the genus-level mea-sure developed by Statzner et al. (2001) (also see Chapter 2). Further, cumu-lative species lists will be created to determine whether rare species profit from the restoration.

A pilot study has indicated that invertebrates in the impact reach recov-ered within 3 months after the disturbance caused by the restoration work (Küster 2001). Invertebrate sampling after restoration started in January 2002 and continued until November 2002. Restoration effects on the invertebrates cannot be evaluated yet, but data collected during the year before restoration revealed that the natural reach is well suited to act as a positive control because species richness was highest there (Küster 2001). Further, inverte-brate microdistributions and taxon richness of the benthic invertebrate assembly were strongly influenced by local disturbance history (H. Huber, unpublished data).

The bed movement data also indicate that the natural reach serves well as a positive control. Bed movement in the natural reach was highly het-erogeneous during three floods in 2000, with small- and large-scale mosaics of differing bed stabilities and a high degree of variation in scouring and filling depths (see Table 18.1). Scoured, depositional, and stable bed patches occurred in almost equal proportions. By contrast, bed movement patterns in the two channelized reaches were more homogeneous, with lower variation in scouring and filling depths. Typically, just one type of bed change (that is, either scour or sediment deposition) dominated the entire reach, and stable patches were less common than in the natural reach. In the two control reaches, these general patterns were also found during three more floods in 2001. In the restored reach, by contrast, patterns became more diverse and started to resemble those in the natural reach. This process still continues because the restoration work (done by bulldozers) created a somewhat artificial river channel, and additional floods will probably be necessary before natural channel form and sediment dynamics are restored. Therefore, monitoring of bed movement patterns in all three reaches continued in 2002 and will also be repeated several years later as part of the planned long-term monitoring program. We expect bed movement patterns in the restored reach to become even more similar to those in the natural reach over time.

By monitoring and analyzing the effects of several individual floods before and after the restoration work, we are using a process-oriented approach to determine if restoring the natural disturbance regime of the River Isar will be successful (see Chapter 4). Because of the profound influence of local disturbance history on benthic invertebrates and algae documented in the earlier parts of this chapter and in the data collected the year before restoration, these changes in bed movement dynamics are likely to cause related changes in the benthic community. A more natural disturbance regime with spatially heterogeneous bed movement patterns should lead to a more natural, and more diverse, invertebrate fauna.

In conclusion, we believe that the River Isar restoration project meets most of the criteria and recommendations developed by Halle and Fattorini (Chapter 2). Consequently, it could become a model for future restoration projects. Furthermore, even the preliminary results show that the disturbance history adds an important, and easily quantifiable, criterion for evaluating the success of restoration projects in streams and rivers.

Acknowledgments

We thank the many people who have helped with our different research projects in various ways. We also thank Eva-Barbara Meidl, Sam Lake, and the editors of this volume for their helpful comments on earlier versions of this chapter.

REFERENCES

Badri, A.; Giudicelli, J.; and Prévôt, G. 1987. Effects of a flood on the benthic macroinvertebrate community in a Mediterranean river, the Rdat (Morocco). *Acta Oecologia Generalis* 8:481–500.
Barmuta, L. A. 1989. Habitat patchiness and macroinvertebrate community structure in an upland stream in temperate Victoria, Australia. *Freshwater Biology* 21:223–236.
Belyea, L. R.; and Lancaster, J. 1999. Assembly rules within a contingent ecology. *Oikos* 86:402–416.
Biggs, B. J. F.; Duncan, M. J.; Francoeur, S. N.; and Meyer, W. D. 1997. Physical characterisation of microform bed cluster refugia in 12 headwater streams, New Zealand. *New Zealand Journal of Marine and Freshwater Research* 31:413–422.

Bishop, J. E. 1973. Observations on the vertical distribution of the benthos in a Malaysian stream. *Freshwater Biology* 3:147–156.

Chesson, P.; and Huntly, N. 1997. The roles of harsh and fluctuating conditions in the dynamics of ecological communities. *American Naturalist* 150:519–553.

Connell, J. H. 1978. Diversity in tropical rain forests and coral reefs. *Science* 199:1302–1310.

Death, R. G.; and Winterbourn, M. J. 1995. Diversity patterns in stream benthic invertebrate communities: the influence of habitat stability. *Ecology* 76:1446–1460.

Diehl, S.; Cooper, S. D.; Kratz, K. W.; Nisbet, R. M.; Roll, S. K.; Wiseman, S. M.; and Jenkins, T. M. Jr. 2000. Effects of multiple, predator-induced behaviors on short-term population dynamics in open systems. *American Naturalist* 156:293–313.

Doeg, T. J.; Lake, P. S.; and Marchant, R. 1989. Colonization of experimentally disturbed patches by stream macroinvertebrates in the Acheron River, Victoria. *Australian Journal of Ecology* 14:207–220.

Dole-Olivier, M.-J.; Marmonier, P.; and Beffy, J.-L. 1997. Response of invertebrates to lotic disturbance: Is the hyporheic zone a patchy refugium? *Freshwater Biology* 37:257–276.

Douglas, B. 1958. The ecology of the attached diatoms and other algae in a small stony stream. *Journal of Ecology* 46:295–322.

Downes, B. J. 1990. Patch dynamics and mobility of fauna in streams and other habitats. *Oikos* 59:411–413.

Downes, B. J.; Lake, P. S.; Glaister, A.; and Webb, A. 1998. Scales and frequencies of disturbances: rock size, bed packing and variation among upland streams. *Freshwater Biology* 40:625–639.

Englund, G.; and Evander, D. 1999. Interactions between sculpins, net-spinning caddis larvae and midge larvae. *Oikos* 85:117–126.

Feminella, J. W.; and Resh, V. H. 1991. Herbivorous caddis flies, macroalgae, and epilithic microalgae: dynamic interactions in a stream grazing system. *Oecologia* 87:247–256.

Fisher, S. G. 1990. Recovery processes in lotic ecosystems: limits of successional theory. *Environmental Management* 14:725–736.

Fisher, S. G.; Gray, L. J.; and Grimm, N. B. 1982. Temporal succession in a desert stream following flash flooding. *Ecological Monographs* 52:93–110.

Francoeur, S. N.; Biggs, B. J. F.; and Lowe, R. L. 1998. Microform bed clusters as refugia for periphyton in a flood-prone headwater stream. *New Zealand Journal of Marine and Freshwater Research* 32:363–374.

Frid, C. L. J.; and Townsend, C. R. 1989. An appraisal of the patch dynamics concept in stream and marine benthic communities whose members are highly mobile. *Oikos* 56:137–141.

Hearnden, M. N.; and Pearson, R. G. 1991. Habitat partitioning among the mayfly species (Ephemeroptera) of Yuccabine Creek, a tropical Australian stream. *Oecologia* 87:91–101.

Holomuzki, J. R.; and Messier, S. H. 1993. Habitat selection by the stream mayfly *Paraleptophlebia guttata. Journal of the North American Benthological Society* 12:126–135.

Keddy, P. A. 1992. Assembly and response rules: two goals for predictive community ecology. *Journal of Vegetation Science* 3:157–164.

Kohler, S. L.; and Wiley, M. J. 1997. Pathogen outbreaks reveal large-scale effects of competition in stream communities. *Ecology* 78:2164–2176.

Kuhara, N.; Nakamo, S.; and Miyasaka, H. 1999. Interspecific competition between two stream insect grazers mediated by non-feeding predatory fish. *Oikos* 87:27–35.

Küster, C. E. 2001. Renaturierung der Isar bei München: Effekte der Renaturierungs-Baumaßnahmen auf benthische Invertebraten. Master's thesis, University of Munich.

Lake, P. S. 2000. Disturbance, patchiness, and diversity in streams. *Journal of the North American Benthological Society* 19:573–592.

Lancaster, J.; and Belyea, L. R. 1997. Nested hierarchies and scale-dependence of mechanisms of flow refugium use. *Journal of the North American Benthological Society* 16:221–238.

Lancaster, J.; and Hildrew, A. G. 1993a. Characterizing in-stream flow refugia. *Canadian Journal of Fisheries and Aquatic Sciences* 50:1663–1675.

Lancaster, J.; and Hildrew, A. G. 1993b. Flow refugia and microdistribution of lotic macroinvertebrates. *Journal of the North American Benthological Society* 12:385–393.

Mackay, R. J. 1992. Colonization by lotic macroinvertebrates: a review of processes and patterns. *Canadian Journal of Fisheries and Aquatic Sciences* 49:617–628.

Matthaei, C. D.; Arbuckle, C. J.; and Townsend, C. R. 2000. Stable surface stones as refugia for invertebrates during disturbance in a New Zealand stream. *Journal of the North American Benthological Society* 19:82–93.

Matthaei, C. D.; Guggelberger, C. and Huber, H. 2003. Effects of local disturbance history on benthic river algae. *Freshwater Biology* 48:1514–1526.

Matthaei, C. D.; and Huber, H. 2002. Microform bed clusters: are they preferred habitats for invertebrates in a flood-prone stream? *Freshwater Biology* 47:2174–2190.

Matthaei, C. D.; Peacock, K. A.; and Townsend, C. R. 1999a. Patchy surface stone movement during disturbance in a New Zealand stream and its potential significance for the fauna. *Limnology and Oceanography* 44:1091–1102.

Matthaei, C. D.; Peacock, K. A.; and Townsend, C. R. 1999b. Scour and fill patterns in a New Zealand stream and potential implications for invertebrate refugia. *Freshwater Biology* 42:41–57.

Matthaei, C. D.; and Townsend, C. R. 2000. Long-term effects of local disturbance history on mobile stream invertebrates. *Oecologia* 125:119–126.

Matthaei, C. D.; Uehlinger, U.; and Frutiger, A. 1997. Response of benthic invertebrates to natural versus experimental disturbance in a Swiss prealpine river. *Freshwater Biology* 37:61–77.

Matthaei, C. D.; Uehlinger, U.; Meyer, E. I.; and Frutiger, A. 1996. Recolonization of benthic invertebrates after experimental disturbance in a Swiss prealpine river. *Freshwater Biology* 35:233–248.

McAuliffe, J. R. 1984. Competition for space, disturbance, and the structure of a benthic stream community. *Ecology* 65:894–908.

Minshall, G. W. 1978. Autotrophy in stream ecosystems. *BioScience* 28:767–771.

Miyake, Y.; and Nakano, S. 2002. Effects of substratum stability on diversity of stream invertebrates during baseflow at two spatial scales. *Freshwater Biology* 47:219–230.

Moyle, P. B.; and Light, T. 1996. Biological invasions of fresh water: empirical rules and assembly theory. *Biological Conservation* 78:149–161.

Nakamo, S.; Miyasaka, H.; and Kuhara, N. 1999. Terrestrial-aquatic linkages: riparian arthropod inputs alter trophic cascades in a stream food web. *Ecology* 80:2435–2441.

Palmer, M. A.; Arensburger, P.; Martin, A. P.; and Denman, D. W. 1996. Disturbance and patch-specific responses: the interactive effects of woody debris and floods on lotic invertebrates. *Oecologia* 105:247–257.

Palmer, M. A.; Bely, A. E.; and Berg, K. E. 1992. Response of invertebrates to lotic disturbance: a test of the hyporheic refuge hypothesis. *Oecologia* 89:182–194.

Peckarsky, B. L.; Horn, S. C.; and Statzner, B. 1990. Stonefly predation along a hydraulic gradient: a field test of the harsh-benign hypothesis. *Freshwater Biology* 24:181–191.

Peterson, C. G. 1996. Response of benthic algal communities to natural physical disturbance. In *Algal Ecology: Freshwater Benthic Ecosystems*, ed. R. J. Stevenson, M. L. Bothwell, and R. L. Lowe, 375–402. San Diego: Academic Press.

Peterson, C. G.; Weibel, A. C.; Grimm, N. B.; and Fisher, S. G. 1994. Mechanisms of benthic algal recovery following spates: comparison of simulated and natural events. *Oecologia* 98:280–290.

Pickett, S. T. A.; and White, P. S. 1985. *The Ecology of Natural Disturbance and Patch Dynamics*. New York: Academic Press.

Poff, N. L. 1992. Why disturbances can be predictable: a perspective on the definition of disturbance in streams. *Journal of the North American Benthological Society* 11:86–92.

Poff, N. L.; and Allan, J. D. 1995. Functional organization of stream fish assemblages in relation to hydrological variability. *Ecology* 76:606–627.

Poff, N. L.; and Ward, J. V. 1989. Implications of streamflow variability and predictability for lotic community structure: a regional analysis of streamflow patterns. *Canadian Journal of Fisheries and Aquatic Sciences* 46:1805–1818.

Power, M. E. 1992. Hydrologic and trophic controls of seasonal algal blooms in northern California rivers. *Archiv für Hydrobiologie* 125:385–410.

Power, M. E.; and Stewart, A. J. 1987. Disturbance and recovery of an algal assemblage following flooding in an Oklahoma stream. *American Midland Naturalist* 117:333–345.

Prévot, G.; and Prévot, R. 1986. Impact d'une crue sur la communauté d'invertébrés de la Moyenne Durance: role de la dérive dans la reconstitution du peuplement du chenal principal. *Annales de Limnologie* 22:89–98.

Pringle, C. M.; Naiman, R. J.; Bretschko, G.; Karr, J. R.; Oswood, M. W.; Webster, J. R.; Welcomme, R. L.; and Winterbourn, M. J. 1988. Patch dynamics in lotic systems: the stream as a mosaic. *Journal of the North American Benthological Society* 7:503–524.

Resh, V. H.; Brown, A. V.; Covich, A. P.; Gurtz, M. E.; Li, H. W.; Minshall, G. W.; Reice, S. R.; Sheldon, A. L.; Wallace, J. B.; and Wissmar, R. C. 1988. The role of disturbance in stream ecology. *Journal of the North American Benthological Society* 7:433–455.

Richardson, J. S. 1992. Food, microhabitat, or both? Macroinvertebrate use of leaf accumulations in a montane stream. *Freshwater Biology* 27:169–176.

Robertson, A. L.; Lancaster, J.; and Hildrew, A. G. 1995. Stream hydraulics and the distribution of microcrustacea: a role for refugia? *Freshwater Biology* 33:469–484.

Ruse, L. R. 1994. Chironomid microdistribution in gravel of an English chalk river. *Freshwater Biology* 32:533–551.

Schmid-Araya, J. M. 1994. Temporal and spatial distribution of benthic meiofauna in sediments of a gravel streambed. *Limnology and Oceanography* 39:1813–1821.

Sedell, J. R.; Reeves, G. H.; Hauer, F. R.; Stanford, J. A.; and Hawkins, C. P. 1990. Role of refugia in recovery from disturbances: modern fragmented and disconnected river systems. *Environmental Management* 14:711–724.

Smock, L. A.; Smith, L. C.; Jones, J. B.; and Hooper, S. M. 1994. Effects of drought and a hurricane on a coastal headwater stream. *Archiv für Hydrobiologie* 131:25–38.

Sousa, W. P. 1979. Disturbance in intertidal boulder fields: the non-equilibrium maintenance of species diversity. *Ecology* 60,1225–1239.

Statzner, B.; Bis, B.; Dolédec, S.; and Usseglio-Polatera, P. 2001. Perspectives for biomonitoring at large spatial scales: a unified measure for the functional composition of invertebrate communities in European running waters. *Basic and Applied Ecology* 2:73–85.

Steinman, A. D.; and McIntire, C. D. 1990. Recovery of lotic periphyton communities after disturbance. *Environmental Management* 14:589–604.

Stevenson, R. J.; Bothwell, M. L.; and Lowe, R. L. 1996. *Algal Ecology: Freshwater Benthic Ecosystems*. San Diego: Academic Press.

Stewart-Oaten, A.; and Murdoch, W. W. 1986. Environmental impact assessment: "pseudoreplication" in time? *Ecology* 67:929–940.

Toner, M.; and Keddy, P. 1997. River hydrology and riparian wetlands: a predictive model for ecological assembly. *Ecological Applications* 7:236–246.

Townsend, C. R. 1989. The patch dynamics concept of stream community ecology. *Journal of the North American Benthological Society* 8:36–50.

Townsend, C. R.; and Hildrew, A. G. 1976. Field experiments on the drifting, colonization and continuous redistribution of stream benthos. *Journal of Animal Ecology* 45:759–772.

Townsend, C. R.; and Hildrew, A. G. 1994. Species traits in relation to a habitat template for river systems. *Freshwater Biology* 31:265–275.

Townsend, C. R.; Dolédec, S.; and Scarsbrook, M. R. 1997a. Species traits in relation to temporal and spatial heterogeneity in streams: a test of habitat templet theory. *Freshwater Biology* 37:367–387.

Townsend, C. R.; Scarsbrook, M. R.; and Dolédec, S. 1997b. The intermediate disturbance hypothesis, refugia, and biodiversity in streams. *Limnology and Oceanography* 42:938–949.

Townsend, C. R.; Scarsbrook, M. R.; and Dolédec, S. 1997c. Quantifying disturbance in streams: alternative measures of disturbance in relation to macroinvertebrate species traits and species richness. *Journal of the North American Benthological Society* 16:531–544.

Townsend, C. R.; Thompson, R. M.; McIntosh, A. R.; Kilroy, C.; Edwards, E. D.; and Scarsbrook, M. R. 1998. Disturbance, resource supply and food-web architecture in streams. *Ecology Letters* 1:200–209.

Uehlinger, U. 1991. Spatial and temporal variability of the periphyton biomass in a pre-alpine river (Necker, Switzerland). *Archiv für Hydrobiologie* 123:219–237.

Ulfstrand, S. 1967. Microdistribution of benthic species (Ephemeroptera, Plecoptera, Trichoptera, Diptera: Simuliidae) in Lapland streams. *Oikos* 18:293–310.

Ulfstrand, S. 1968. Benthic animal communities in Lapland streams. *Oikos* 10 (suppl.): 1–120.

Vannote, R. L.; Minshall, G. W.; Cummins, K. W.; Sedell, J. R.; and Cushing, C. E. 1980. The river continuum concept. *Canadian Journal of Fisheries and Aquatic Sciences* 37:130–137.

Wallace, J. B. 1990. Recovery of lotic macroinvertebrate communities from disturbance. *Environmental Management* 14:605–620.

White, P. S.; and Jentsch, A. 2001. The search for generality in studies of disturbance and ecosystem dynamics. *Progress in Botany* 62:399–450.

Williams, D. D.; and Hynes, H. B. N. 1974. The occurrence of benthos deep in the substratum of a stream. *Freshwater Biology* 4:233–256.

Wilson, J. B.; and Whittaker, R. J. 1995. Assembly rules demonstrated in a saltmarsh community. *Journal of Ecology* 83:801–807.

Winterbottom, J. H.; Orton, S. E.; Hildrew, A. G.; and Lancaster, J. 1997. Field experiments on flow refugia in streams. *Freshwater Biology* 37:569–580.

Wood, P. J.; and Armitage, P. D. 1997. Biological effects of fine sediment in the lotic environment. *Environmental Management* 21:203–217.

Zinsser, T. 1999. Der Isar-Plan München: auf zu neuen Ufern. In *Die Isar. Wildfluss in der Kulturlandschaft,* ed. C. Margerl and D. Rabe, 74–77. Vilsbiburg, Germany: Verlag Kiebitz Buch.

Zobel, M. 1992. Plant species coexistence: the role of historic, evolutionary and ecological factors. *Oikos* 65:314–320.

Chapter 19

How Structure Controls Assembly in the Hyporheic Zone of Rivers and Streams: Colmation as a Disturbance

EVA-BARBARA MEIDL AND WILFRIED SCHÖNBORN

The organisms that live in running waters play an important role in the cycling of nutrients, because they turn over organic matter and pollutants that enter the water. Most species live their whole lives in the water, and many others—especially insects—spend a part of their lives there. All of them are adapted to the natural disturbance regime of rivers. Hence, the regeneration of rivers is intricately linked with a natural disturbance regime. Human constructions such as weirs, barriers, reservoirs, and other artificial alterations of river channels and floodplains often affect the natural disturbance regime in negative ways. Many ecological restoration projects have tried to reinstate natural disturbance dynamics (see Chapter 18). Such human disturbances usually cause flooding regimes: more frequent weak floods and less frequent severe floods that destroy river habitats and communities of organisms. Disturbance through colmation, in contrast, has not generally been recognized as an important disturbance. Colmation is the physical clogging of sediment pores by fine-grained material and perhaps biofilm. Although it is not a discrete event on a short time scale like flooding, it is a disturbance. Both flooding and colmation destroy available living space for organisms and reduce biomass in the hyporheic zone. Therefore, we consider colmation to be a disturbance that is as important as flooding in terms of its effect on the ecosystem.

One of the most important tasks in ecology is to find out which organisms live where and why. Illies and Botosaneanu (1963) showed that streams and rivers have a biocoenotic longitudinal zonation, which means that they are structured in successive zones along different environmental gradients. Each zone has its own identifiable community of organisms (biocoenoses) and forms a functional unit. Illies (1961) attributed this zonation to temperature differences resulting from changes in altitude as a stream flows from its source to its mouth. Other authors considered current velocity (Zimmermann 1961, 1962), oxygen content (Ruttner 1926, Ambühl 1959), or simply elevation (Ward 1986) to be the most important factor. Vannote et al. (1980) emphasized in their "river continuum concept" the functional aspects (feeding types) of communities and related them to the production-respiration ratio in different river sections.

Because river organisms experience frequent disturbance, they must constantly reassemble into functioning communities. Thus, river communities can be seen as assemblages. All of the research just outlined includes the concept of community assembly, even though the authors may never have used the word *assembly* explicitly. One approach to community assembly is based on the view that it is ruled by abiotic as well as biotic factors (see Figure 3.1 in Chapter 3 for more details). Because the influence of abiotic conditions plays a crucial role in running-water systems, especially during disturbance events, we also adopt this view.

Many rivers must compensate for disturbances such as floods; these floods may be strong or weak and occur yearly or more often. Flooding results in an immense outwash of animals from the sediment surface (Hynes 1968, Hoopes 1974, Dolédec 1987). It is astounding that the recovery time— between 1 and 6 months—always seems to be very short (Sheldon 1984, van Treek and Rennerich 1990, Mackay 1992). One of the mechanisms for recovery is recolonization of the benthic habitat by animals of the hyporheic zone (Schwoerbel 1964, 1967), which is analogous to recolonization from the seed bank in terrestrial systems (see Chapter 6). The hyporheic zone is the living space inside the river sediment under and beneath the channel (Schwoerbel 1964; Hynes 1983; Stanford and Ward 1988, 1993) (Figure 19.1). Brunke and Gonser (1997) gave a very good overview of the living conditions in this zone. The chemistry of the zone and the invertebrate species that inhabit it change slowly with depth from the sediment surface. Thus, the community consists mostly of surface-water animals, although groundwater animals can also be found. Water velocity is very slow in the hyporheic zone. If the sediment is porous enough, the zone provides several good conditions that serve certain functions. First, it acts as a nursery for early instars of invertebrates. Second, it acts as a refuge that provides stable temperature, spatial structure, and wetness (Dole-Oliver et al. 1997). Third, it works like a "biological reactor" because of the turnover and degradation of organic matter and pollutants. Finally, and perhaps most important, the hyporheic zone represents an ecotone, because it forms a link between the surface water and the groundwater (see Figure 19.1). The ecotone is a basis

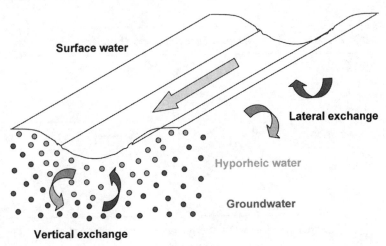

FIGURE 19.1 Water exchange between the surface water and groundwater through the hyporheic zone.

for the functions just described and is also a contact zone for groundwater and surface-water animals, thus enriching species diversity. In addition, it has a buffering effect when nutrients are transferred from the river to the groundwater. This buffering effect causes, among other things, groundwater to remain a valuable habitat.

The quality and distribution of sediments form the structural matrix for the animals living in the hyporheic zone (Brunke and Gonser, 1998), and the quality and quantity of life depend explicitly on the porosity of these sediments (Figure 19.2). Fine-grained and very-fine-grained fractions play a very important role in the hyporheic zone of rivers. These fine particles offer a large surface area for the growth of biofilm, and they are food for sediment-feeding invertebrates. The amount of these fine-grained fractions is an important and sensitive factor in the hyporheic zone. If the fine-grained fractions are too high, the sediment will become colmated, making it too dense to offer organisms enough room to live in or to support the necessary water exchange between the channel and the hyporheic zone and between the groundwater and the hyporheic zone (Findlay 1995) (see Figures 19.1 and 19.2). The denser the pores, the more fine-grained material will be retained. Thus, there is a positive feedback between colmation and the amount of fine-grained fractions. Most of the organisms are sensitive to these changes in their environment; as a result, species may die off or at least be reduced in abundance. Therefore, we can suppose that structure rules assembly of organisms in the hyporheic zone and that the composition of grain sizes, the packing of sedi-

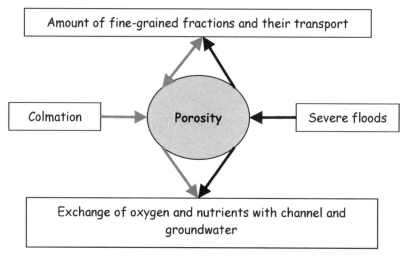

FIGURE 19.2 Hypothesis for the regulation of sediment porosity by colmation and severe floods. Porosity is important for determining macroinvertebrate habitat quality. The black arrows indicate the influence of severe floods; and gray arrows indicate the influence of colmation. Single arrows indicate the direction of an effect; arrows in two directions indicate a feedback effect.

ment, the content of total organic carbon (TOC) as a nutrition factor, and the presence of oxygen are very important for community assembly.

The question arises, how is a stream able to compensate for colmation effects? Do any restoring mechanisms exist and, if so, what are they? We hypothesize that severe floods, which are strong enough to turn the sediment around and to move even very-coarse-grained material, can wash out fine material and so increase pore size (Hoehn 2001). Water can then flow faster even in deep zones, promoting the transport of fine grains and the exchange of water, including dissolved nutrition and oxygen (see Figure 19.2). The animal community, finding good living conditions, will reassemble and the abundance of the various species will increase. To test our hypotheses, we investigated the frequency and intensity of severe floods, the distribution of grain sizes, the TOC content, and the abundance of animals from the sediment surface to the hyporheic zone. Using these data, we examined whether severe floods are able to compensate colmation effects, as proposed in Figure 19.2. If so, we need to ascertain how long this process takes and how the dynamics of the two opposite forces (severe flooding and colmation) work to preserve the river's functions. In addition, we were interested in identifying any factors that rule the assembly of these communities.

Study Site

The Schwarza brook is a low mountain stream flowing through a V-shaped valley in the *Thüringer Schiefergebirge* (Thuringian Slate Mountains), Germany. It has a total length of 52.65 km and a mean slope of 9.78‰ (Bauer 1961). Average discharge is 5 m³/s. The catchment area is characterized by slate and its derivatives.

In this slightly regulated river with a permanent water supply from a small reservoir upstream (to prevent drying out in summertime), occasional large boulders and bedrock indicate severe floods with high forces. The sediment layer in the channel is mostly between 20 and 30 cm thick, but there are many gravel banks along the brook, which are usually dry and are flooded only once or twice a year. Here, the sediment layer can be up to 60 or 80 cm thick.

We selected two sampling sites, S6 and S8, in a stretch about 15 km long. The sites appear superficially similar: both are located at riffles in the rhithral (upper region of river) where the sediment layer is at least 60 cm and coarse gravel dominates. S8 is peculiar because of the many crossbars of slate found in the bedrock of the riverbed, which are responsible for frequent changes in habitat structure caused even by weaker floods over the years. The crossbars form a resistance to the current, causing increased turbulence that can move coarse gravel from deeper sediment layers and transport it downstream.

Methods

The hyporheic zone of the Schwarza brook was studied quantitatively to a depth of 60 cm at two sampling sites (S6 and S8) using the freeze-core technique with in situ electropositioning at 600 V (Klemens 1983, Bretschko and Klemens 1986). All samples were taken in October of 1997, 1999, 2000, and 2001. The analysis is based on five biological and five sedimentological freeze-cores at each sampling site per year (Figure 19.3). The cores were taken at the slip of slope banks at riffles, where sediment accumulates. Still frozen, they were cut into slices 10 cm long and fixed in 4 percent formalin for biological samples or dried for sedimentological samples. Animals were sorted from biological samples, counted, identified under 50X magnification, and fixed in 70 percent ethanol. The sediment was wet-sieved to the grain-size fractions of gravel (> 2 mm), sand (2 mm–63 μm), and silt/clay (< 63 μm). The TOC of silt/clay was analyzed following Bretschko and Leichtfried (1987).

FIGURE 19.3 Freeze-core sample of the Schwarza brook, Thuringia, Germany, showing sediment packing. Core diameter is about 30–35 cm, length about 60 cm.

We related particle size, TOC, and animal abundance to the mean values of all samples per sampling time from riverbed to gravel banks that are flooded only once or twice a year. The presence of oxygen was confirmed by qualitative analysis using rust patches on steel rods.

Discharge data to detect the peaks and the frequency of high floods and dry periods were measured by government offices in Thuringia (SUA Gera) at a water depth gauge in Schwarzburg (located between the two sampling sites) and were made available by emc-GmbH, an engineering consultancy.

Results

Two severe floods occurred, one in early November 1998 (discharge of 57.3 m^3/s) and the other in early March 1999 (discharge of 72.3 m^3/s) (Figure 19.4a). The 1999 flood was strong enough to turn the sediment around even in deep layers, and so a habitat structure was established at the two sampling sites that was different from the structure that existed before the floods. We observed that the surface of sediment at S6 reacted more conservatively than did the sediment surface at S8, which seemed to be more dynamic. This difference was due mainly to the large crossbars of slate in the bedrock at S8.

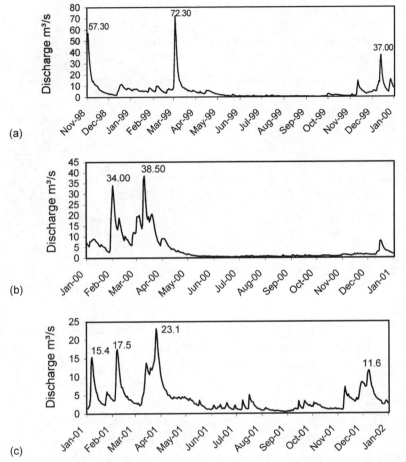

FIGURE 19.4 Discharge for the years (a) 1998–1999, (b) 2000, and (c) 2001 at the water depth gauge in Schwarzburg, Thuringia, Germany.

Sediment packing changed at this site even with moderate floods, as occurred in early 2000 (Figure 19.4b), at least in the upper layers up to 20-cm depth (data on morphological measurements from emc-GmbH). However, the weak floods in early 2001 (Figure 19.4c) did not influence the sediment structure remarkably.

After the two big floods in 1998 and 1999, the amount of coarse gravel and silt/clay at S8 increased slightly, whereas the amount of sand decreased. At S6, there was no change in the amount of coarse gravel, only a slight decrease of sand, and a strong increase of silt/clay. Remarkably, the largest amounts of silt/clay at S6 and S8 were observed in 1999 after the two big

TABLE 19.1

Percent Silt and Clay[1] at Sites S6 and S8 on the Schwarza Brook,
Thuringia, Germany, from Sediment Surface to the Hyporheic Zone[2]

	Depth (cm)	1997	1999	2000	2001
Site S6	10	0.42 (0.33)	3.75 (2.90)	1.65 (0.90)	1.82 (1.13)
	20	0.84 (0.18)	1.79 (0.50)	1.70 (0.46)	2.01 (1.72)
	30	0.32 (0.08)	3.88 (3.30)	3.16 (1.80)	1.52 (0.38)
	40	0.81 (0.47)	4.92 (2.63)	2.55 (1.11)	0.99 (0.56)
	50	0.81 (0.38)	2.93 (1.05)	3.03 (1.32)	3.51 (4.10)
	60	1.09 (0.50)	3.44 (0.88)	2.04 (0.67)	2.73 (0.66)
Site S8	10	0.23 (0.26)	0.75 (0.64)	1.61 (0.88)	0.89 (0.73)
	20	0.30 (0.20)	1.76 (1.19)	1.93 (1.47)	1.53 (0.84)
	30	0.73 (0.34)	3.21 (1.61)	3.14 (0.92)	3.42 (1.75)
	40	0.74 (0.29)	4.34 (2.00)	2.89 (1.06)	2.97 (1.27)
	50	0.78 (0.35)	3.40 (2.16)	2.97 (0.93)	2.94 (2.30)
	60	1.05 (0.73)	3.97 (1.70)	2.14 (0.92)	3.19 (0.70)

[1]Average of grain fraction <63 μm; SD in parentheses.
[2]Depth classes are, for example, 0–10 cm, 10–20 cm, and so forth.

floods (Table 19.1 and Figure 19.5). In all years of this study, we observed more coarse gravel and less sand in the upper layers at S8 than in the upper layers at S6. At depths of 30 to 60 cm, the situation changed at S8: coarse gravel was reduced and the other grain fractions—especially the sand fraction—increased up to the level of S6.

TOC was always negatively correlated with the silt/clay fraction and decreased strongly with depth (see Figure 19.5). Conspicuously, the increasing or decreasing spatial trends of all grain-size fractions and TOC were very strong in the layers between depths of 20 and 40 cm. TOC was lowest at S6 after the big floods, when it was highest at S8. In 2000, however, TOC was at the lowest level at S8, but it increased at S6. Thus, TOC and silt/clay were strongly negatively correlated spatially, but they were not always strongly correlated temporally. There must be other important factors that control retention and resuspension of organic carbon in the hyporheic zone, such as temperature, pH, conductivity, oxygen, and biofilm (which is probably a large part of the TOC). Generally however, there was no great difference of TOC between the two sampling sites.

Oxygen was detected everywhere in the hyporheic zones of both sites.

The abundance of most animals strongly correlated with sediment structure and TOC. With decreasing TOC and porosity (and increasing sand and silt/clay) with depth, the abundance of animals and especially of insect larvae decreased along the vertical scale, especially at depths of 20 to 60 cm

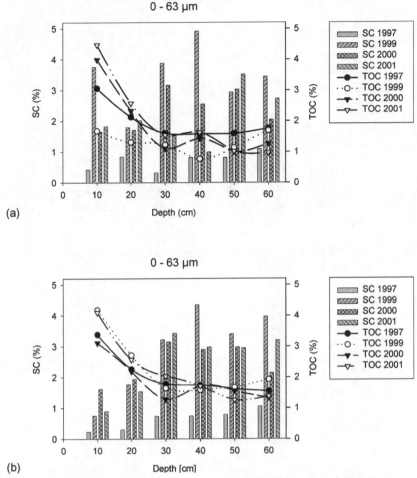

FIGURE 19.5 Distribution of the silt/clay fraction (SC) and total organic carbon (TOC) in 1997–2001 from sediment surface to the hyporheic zone at sites (a) S6 and (b) S8 at the Schwarza brook, Thuringia, Germany.

(Figures 19.6, 19.7, and 19.8). Nevertheless, there were animals present throughout the sediment from its surface to 60-cm depth. The animal colonization at site S6 was often at least 2 times higher than it was at S8 for the duration of this study (see Figure 19.6). Nearly every animal taxonomic group increased in abundance in the upper 30 cm of the sediment layer. At a depth of 40 cm, some taxonomic groups increased in abundance whereas others decreased. In the lower parts of the sediment, however, the animal community structure barely changed over the study period. The vermiform

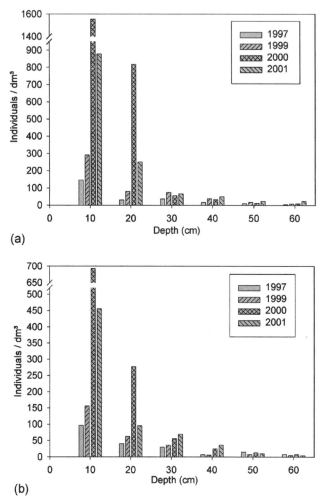

FIGURE 19.6 Total abundance and distribution of all macroinvertebrates from surface to the hyporheic zone at sites (a) S6 and (b) S8 at the Schwarza brook, Thuringia, Germany.

taxa—such as Oligochaeta (see Figures 19.7a and 19.8a), Nematomorpha (see Figures 19.7c and 19.8c), and Chironomidae (see Figures 19.7b and 19.8b)—as well as the most important predators in the hyporheic zone, the Acari (see Figures 19.7d and 19.8d), were abundant even in the deeper layers. In contrast, most merolimnic insects—for example, Plecoptera (see Figures 19.7f and 19.8f)—were very sensitive to colmation and so were less abundant in areas with higher density of sediment packing in the hyporheic

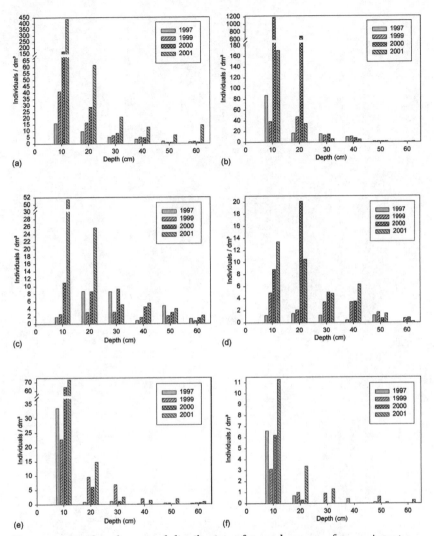

FIGURE 19.7 Abundance and distribution of several groups of macroinvertebrates from surface to the hyporheic zone at site S6 at the Schwarza brook, Thuringia, Germany. (a) Oligochaeta, (b) Chironomidae, (c) Nematomorpha, (d) Acari, (e) Ephemeroptera, (f) Plecoptera.

zone. Hence, they were restricted to the upper layers (to 20-cm depth). In contrast, very young and small larvae of the Ephemeroptera families Baetidae and Leptophlebiidae (see Figures 19.7e and 19.8e) were able to live in deeper layers for a short part of their life.

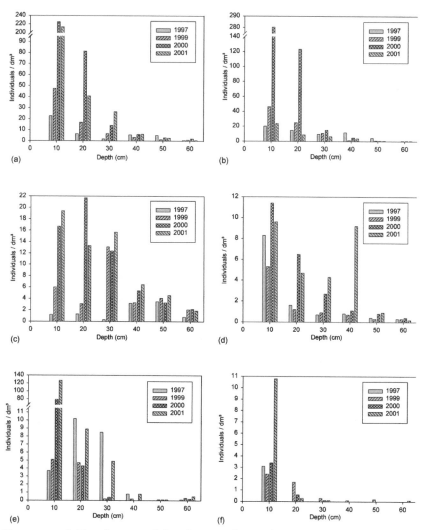

FIGURE 19.8 Abundance and distribution of several groups of macroinvertebrates from surface to the hyporheic zone at site S8 at the Schwarza brook, Thuringia, Germany. (a) Oligochaeta, (b) Chironomidae, (c) Nematomorpha, (d) Acari, (e) Ephemeroptera, (f) Plecoptera.

Discussion and Conclusions

The effect of colmation was obvious: animals were low in abundance in 1997 at both sites compared to the situation after the two big floods in November 1998 and March 1999. There was a greater tendency to colmate at S6 than at S8 during the preflood period of 1997 (Figure 19.9, modified from Meidl

1998 and from the poster of the North American Benthological Society 2000 meeting). Differences in the sediment clogging behavior at the two sites are just as clear in the postflood period from 1999 to 2001 (see Figure 19.5). It seems that S6 tended to experience more sediment clogging because the amount of the fine-grained material was greater in the upper layers at S6 than it was at S8 after the severe floods. Although this clogging is not yet colmation, silt/clay-particles can be a substrate for biofilm, which can help to bind together fine particles and therefore increases sediment clogging. However, S8 was probably more influenced by the movement of sediment at the surface, and so colmation effects there occurred less often than they did at site S6. Nevertheless, the animal colonization at site S6 was often at least 2 times higher than at S8 (see Figure 19.6). At depths of 50 and 60 cm, animal abundances were low at both sites. Because environmental factors change with depth, they act differently at the sediment surface than they do in the hyporheic zone. In deeper layers, the sediment is less porous because fine-grained material accumulates; on the surface, the situation is reversed. Spatially, this abiotic factor clearly acts on the vertical distribution of the animals (see also Schwoerbel 1967) and makes it evident that this abiotic factor controls assembly in the hyporheic zone.

The predatory Acari can cope with almost any sediment grain size. Some of their prey taxa—such as Oligochaeta, some species of Nematomorpha, and Chironomidae—have the least problems with fine-grained material of all the inhabitants of the hyporheic zone, because of their vermiform body (Schwoerbel, 1967; Meidl 1999, 2000). However, we did not find a clear correlation between these predators and their prey at this stage of analysis.

Besides Chironomidae, Ephemeroptera is the only other group of the merolimnic insects that can advance at least partly into the deeper layers of the hyporheic zone. These are the smallest larvae of the families Baetidae and Leptophlebiidae, which were already observed close to the bank and in subsurface space by Orghidan (1959) and Schwoerbel (1967). The other insects tend to avoid clogged sediment. These results lead us to suggest that biotic factors do not rule community assembly in the hyporheic zone.

As noted above, very dynamic sites such as S8—at which the sediment changes more often due to weak floods, especially at the surface—tend to experience less colmation than do more static sites. If the flood is strong enough, it can turn around the sediment and enlarge the pores even at sectors of a stream that tend to colmate, such as S6. There are grounds for supposing that sediment does not get clogged because of an accumulation of fine material, especially silt and clay, as previously assumed (for example, Bretschko and Leichtfried 1987). Instead, clogging occurs through an

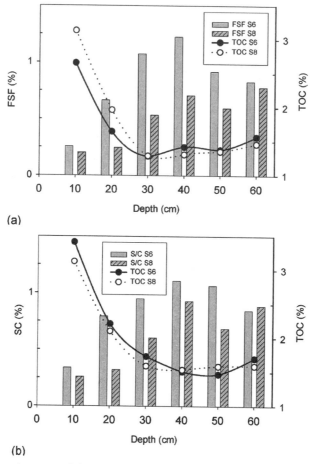

(a)

(b)

FIGURE 19.9 Amount of the finest-grain-size fractions and their TOC content (mean values) at the S6 and S8 sites in the pilot study from August 1997 to February 1998 at the Schwarza brook, Thuringia, Germany. (a) Fine sandy fraction (63–125 μm, FSF) and (b) silt and clay fraction (0–63 μm, SC). After Meidl 1998, and the poster from the North American Benthological Society meeting of 2000.

enrichment of sand in the upper layers of sediment, especially up to 20-cm depth (Meidl and Daut, unpublished data). Immediately after the floods in 1998 and 1999, the amount of silt and clay was at the highest level, without any presence of colmation effects. In fact, the percentage of silt/clay was at the lowest level in times of colmation. The wet-sieving procedure was always performed on a high-quality standard, but it is possible that it was not high enough during the pilot study in 1997. Therefore, it could be that the amount of silt and clay detected was slightly underestimated. Nevertheless, it can be taken as correct that the lowest level of silt and clay was found in 1997, irrespective of any methodological error, because the magnitude of the difference with other years was so large (see Figure 19.5). In time, sandy fractions arrive and act together with silt and clay fractions to reduce porosity, as was shown at S6 during the pilot study for the finest sandy fraction of 63–125 μm in Figure 19.9. Reduced porosity hinders the exchange of oxygen, particulate and dissolved organic matter, and animals between the surface, the hyporheic zone, and the groundwater (see Figure 19.2). Without flood disturbances, colmation effects would appear at both sites in the Schwarza brook, which would lead to a qualitative and quantitative impoverishment of the fauna, as detected in the 1997–1998 sampling period. How long it takes for the sediment to become clogged is influenced by a multitude of mainly abiotic factors and is just as unpredictable in this study as the fluctuation of amplitudes of abundance.

One of the most important results of the present study is that there is a steady increase of total animal abundance after the big floods, with a decline of flood peaks from 1999 up to the present (see Figure 19.6). Only in 2001 did the trend in abundance reverse, because Chironomidae decreased after their enormous increase in 2000 (see Figure 19.7b). We do not know exactly why these animals increased and decreased so fast. It could be that temperature, which was about 2°C above the long-term mean value for the region in 2000, played an important role in affecting the production of Chironomidae. Hence, in all likelihood, the decrease of total animal abundance in 2001 was not caused by the onset of colmation. To clarify such relationships, we need to analyze the data further with respect to seasonal factors. We can conclude, however, that total regeneration after colmation and flooding requires many years and not only a few weeks or months (Mackay 1992). In considering the opportunity for recolonization out of the hyporheic zone (Schwoerbel 1964, 1967) and the present data of 1997–1998 before the floods, one can assume that most past investigations of flood events or sediment disturbances have used reference data sampled during times of colmation before flood disturbance. The sites sampled were probably not seen as disturbed themselves, because nobody thought of colmation as a disturbance. Hence, the very low animal abundances during colmation would have been taken as reference situations, explaining why such short regeneration times in rivers have been reported.

Finally, the fast compensation of animal abundance during 1999, after the two severe flood events, shows the importance of abiotic factors. The abundances are higher even a short time after the floods than they were in

1997–1998. We conclude that colmation has a large negative influence on animals because of its negative effects on sediment porosity (see Figure 19.2). Biotic factors such as competition or predation, which seem to play a more important role in other ecosystems (see Chapter 7), are less important in the hyporheic zone.

From our results, we conclude that it is necessary to develop a new concept to understand animal dynamics in streams in relation to colmation and severe floods as disturbances of the system. Colmation and severe floods have strong and opposite effects on porosity (see Figure 19.2). Both control sediment dynamics, and the sediment structure and its dynamics have strong effects on the hyporheic animal community and their assembly or reassembly. Figure 19.10 is a conceptual representation of how we suggest this kind of control might affect the abundance of the animals at the surface of the riverbed and in the hyporheic zone. With increasing colmation, animal abundance decreases. Presumably, this process starts slowly and then speeds up because of the accumulation of sticky biofilms, which need time to grow. If a severe flood occurs, sediment is turned around and fine particles and animals are washed out, so animal abundance decreases still further. After the flood, sediment with greater porosity and increased hydraulic conductivity offers the animals very good living conditions. Water, oxygen, and nutrients are transported again, and animals have enough space to live in. Return of the animal abundance to levels from before the flood takes only a few months, as is often described in the literature (for example, Mackay 1992). Complete regeneration, however, takes place over a period of years. As a result, we need to investigate a stream over several years to get an idea about the possible amplitudes of animal abundance and about the characteristics of the fully regenerated state.

The system oscillates between both extreme disturbances: severe flooding and colmation. If there is only one kind of disturbance, we predict that animal abundance would decline to a very low level. But we cannot show this with our data, because both kinds of disturbance were present during the study. Laboratory experiments would be necessary to validate our prediction, because natural situations without flooding or colmation seldom occur.

In general, the dynamics and distribution of animals in the hyporheic zone seem to confirm the relationships shown in Figures 19.2 and 19.10. This makes our conceptual model (Figure 19.10) plausible, and we think it is worthy of more thought in the future. To get further insights into disturbance relationships in rivers, we must analyze the data in detail with multivariate statistical methods including seasonal factors.

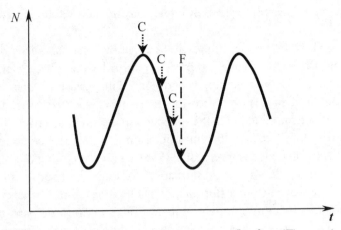

FIGURE 19.10 Effect of colmation (C) and severe flooding (F) on animal abundance (N) over time (t).

In our study, we tried to find patterns of organisms' distribution and abundance and their changes over time. These patterns depend on a mechanism or filter (see Part II, "Ecological Filters as a Form of Assembly Rule"), which here seems to be the structure and dynamics of sediment. In the hyporheic zone, the forces that induce the mechanism are the abiotic factors, especially the presence or absence of such disturbances as severe floods and colmation, because they are able to change the "mesh" of the filter. Because environmental gradients are consistent in longitudinal, vertical, and lateral directions, the mechanisms of both flooding and colmation appear to work in one direction at every point in time. For example, after a flood, sediment is always more porous. On the other hand, colmation always results in a clogged sediment structure. Thus, the two counteracting disturbances modify the porosity filter but are not filters in themselves. Our view is thus slightly different from that of Matthaei et al. (Chapter 18), where disturbance itself is seen as the filter. The two views are not necessarily inconsistent, because pattern, mechanism, and inducing forces are not the same in different parts of a river ecosystem.

As discussed by Temperton and Hobbs (see Chapter 3), assembly rules can be biotic or abiotic. In the special case of the hyporheic zone, the abiotic rules dominate. It is questionable whether the distinction made by many authors between categorical and topological approaches (whether trophic or related to whole systems; see Chapter 3) is necessary or correct. Even if the ruling factors for assembly were only biotic, making such a distinction would not be obligatory.

We conclude that severe floods and colmation, as antagonistic disturbances, are the factors that act as principles or rules that, first, determine how a community is put together; second, control the reassembly of the community; and third, make possible the functions of the hyporheic zone that are described in the introduction to this chapter. It is essential for a river's health not to disturb this dynamic. If it must be disturbed, it should be done very carefully. If the dynamic is interrupted (for example, by cutting the highest flow peaks during floods), one disturbance (for example, floods) will have less influence than the other (for example, colmation). But streams cannot avoid colmation without the help of floods. Both are intrinsic parts of the system and form two extreme poles between which the whole system moves.

The hyporheic zone, like the floodplain, is an important part of the river ecosystem, but this fact has not been considered in restoration projects until now. With respect to our results and to the ecological functions described in the introduction, it has to be taken into account. Such consideration would be an important step forward in understanding the regeneration of rivers and streams and in restoration ecology. Our findings support and broaden existing work (Gore 1985, Friedrich 1986) that calls for preserving or restoring running-water systems in terms of their own dynamics, including the hyporheic zone. To restore hyporheic zones in rivers and brooks, it is necessary not only to rebuild river channels by removing obstructions in and beneath the riverbed (for example, weirs, barriers, artificial changes of river channels and floodplains) but also to give the stream room to move laterally and to offer enough sediment deposits to be transported. Especially, the last two points are not considered in most restoration projects (except the Isar-Plan; see Chapter 18).

Acknowledgments

This study was financed primarily by the Thuringian Reservoir Management Office (*Thüringer Talsperrenverwaltung*). Other funding was provided by the *Deutsche Forschungsgemeinschaft* (DFG, German Research Council) as part of the Graduate Research Group (*Graduiertenkolleg*) on Functioning and Regeneration of Degraded Ecosystems (Project GRK 266). We offer many thanks for discussion to everyone who attended the workshop in Dornburg, Germany, and especially to Vicky Temperton for her great help in correcting and discussing the manuscript. We also thank Georg Becker and Carmen Rojo for their helpful comments on earlier versions of this chapter and Tim Nuttle for correcting the manuscript as a native English speaker.

REFERENCES

Ambühl, H. 1959. Die Bedeutung der Strömung als ökologischer Faktor. *Zeitschrift für Hydrologie* 21:133–264.

Bauer, L. 1961. Zur Hydrogeographie des Schwarza und Rodagebietes. Ein Beitrag zu Gewässerkunde und Gewässerpflege in Thüringen. *Archiv für Naturschutz und Landschaftsforschung* 1(2): 99–141.

Bretschko, G.; and Klemens, W. E. 1986. Quantitative methods and aspects in the study of the interstitial fauna of running waters. *Stygologia* 2:297–316.

Bretschko, G.; and Leichtfried, M. 1987. The determination of organic matter in river sediments. *Archiv für Hydrobiologie* 68 (suppl.):403–417.

Brunke, M.; and Gonser, T. 1997. The ecological significance of exchange processes between rivers and groundwater. *Freshwater Biology* 37:1–33.

Brunke, M.; and Gonser, T. 1998. Biodiversität in Sedimenten von Fließgewässern. *Deutsche Gesellschaft für Limnologie* 1:139–143. Tagungsbericht 1997, Krefeld 1998.

Dolédec, S. 1987. Etude des peuplements de macroinvertebrets benthiques de l'Ardeche dans son cours inferieur. *Archiv für Hydrobiologie* 109:541–565.

Dole-Oliver, M. J.; Marmonier, P.; and Beefy, J.-L. 1997. Response of invertebrates to lotic disturbance: is the hyporheic zone a patchy refugium? *Freshwater Biology* 37:257–276.

Findlay, S. 1995. Importance of surface-subsurface exchange in stream ecosystems: the hyporheic zone. *Limnology and Oceanography* 40:159–164.

Friedrich, G. 1986. Was bedeutet Renaturierung von Fließgewässern. In *Aktuelle Fragen der Unterhaltung von Fließgewässern*, 23–35, Landesamt für Wasser und Abfall NRW, 104 p, Düsseldorf.

Gore, J. A. ed. 1985. *The Restoration of Rivers and Streams: Theories and Experience.* Boston: Butterworth.

Hoehn, E. 2001. Exchange processes between rivers and ground waters: the hydrological and geochemical approach. In *Groundwater Ecology: A Tool for Management of Water Resources*, ed. C. Griebler, D. L. Danielopol, J. Gibert, H. P. Nachtnebel, and J. Notenboom, 55–68, Luxembourg: Office for Official Publications of the European Communities.

Hoopes, R. L. 1974. Flooding, as a result of Hurricane Agnes, and its effect on a macrobenthic community in an infertile headwater stream in central Pennsylvania. *Limnology and Oceanography* 19:853–857.

Hynes, H. B. N. 1968. Further studies on the invertebrate fauna of a Welsh mountain stream. *Archiv für Hydrobiologie* 57:344–388.

Hynes, H. B. N. 1983. Groundwater and stream ecology. *Hydrobiologia* 100:93–99.

Illies, J. 1961. Versuch einer allgemeinen biozönotischen Gliederung der Fließgewässer. *Internationale Revue der gesamten Hydrobiologie* 46(2): 205–213.

Illies, J.; and Botosaneanu, L. 1963. Problèmes et méthods de la classification et de la zonation écologique des eaux courantes, considerées surtout du point de vue faunistique. *Internationale Vereinigung für theoretische und angewandte Limnologie, Mitteilungen* 12:1–59.

Klemens, W. E. 1983. Zur Problematik quantitativer Probennahmen in Bettsedimenten von Schotterbächen unter besonderer Berücksichtigung des Zoobenthos. *Jahresbericht der Biologischen Station Lunz* 6:25–47.

Mackay, R. J. 1992. Colonization by lotic macroinvertebrates: a review of processes and patterns. *Canadian Journal of Fisheries and Aquatic Sciences* 49:617–628.

Meidl, E.-B. 1998. Das hyporheische Interstitial eines schotterreichen Mittelgebirgs- baches (Schwarza, Thüringen): Sedimentologische Eigenschaften. *Deutsche Gesellschaft für Limnologie* 2:848–852. Tagungsbericht 1998, Tutzing 1999.

Meidl, E.-B. 1999. Verteilung aquatischer Invertebraten im hyporheischen Interstitial eines Mittelgebirgsbaches (Schwarza, Thüringen, Deutschland). Distribution of aquatic invertebrates in the hyporheic zone of a mountain brook (Schwarza, Thuringia, Germany). *Zoology* 102 (suppl. II): 86.

Meidl, E.-B. 2000. Distribution of aquatic insects in the hyporheic zone of a mountain brook (Schwarza, Thuringia, Germany). *Mitteilungen der Deutschen Gesellschaft für Allgemeine und Angwandte Entomologie* 12:241–244. DGaaE Basel (1999), Tagungs- bericht (Gießen 2000).

Orghidan, T. 1959. Ein neuer Lebensraum des unterirdischen Wassers: das hyporheische Biotop. *Archiv für Hydrobiologie* 55:392–414.

Ruttner, F. 1926. Bemerkungen über den Sauerstoffgehalt der Gewässer und dessen res- piratorischen Wert. *Naturwissenschaften* 14:1237–1239.

Schwoerbel, J. 1964. Die Bedeutung des Hyporheals für die benthische Lebensgemein- schaft der Fließgewässer. *Internationale Vereinigung für theoretische und angewandte Limnologie, Verhandlungen* 15:215–226.

Schwoerbel, J. 1967. Das hyporheische Interstitial als Grenzbiotop zwischen oberirdis- chem und subtranem Ökosystem und seine Bedeutung für die Primärevolution von Kleinsthöhlenbewohnern. *Archiv für Hydrobiologie* 33 (suppl.):1–62.

Sheldon, A. L. 1984. Colonization dynamics of aquatic insects. In *The Ecology of Aquatic Insects*, ed. V. H. Resh and D. M. Rosenberg, 401–429. New York: Praeger.

Stanford, J. A.; and Ward, J. V. 1988. The hyporheic habitat of river ecosystems. *Nature* 335:64–66.

Stanford, J. A.; and Ward, J. V. 1993. An ecosystem perspective of alluvial rivers: con- nectivity and the hyporheic corridor. *Journal of the North American Benthological Society* 12:48–60.

Vannote, R. L.; Minshall, G. W.; Cummins, K. W.; Sedell, J. R.; and Cushing, C. E. 1980. The river continuum concept. *Canadian Journal of Fisheries and Aquatic Sci- ences* 37, 130–177.

van Treek, P.; and Rennerich, J. 1990. Vergleichende Untersuchung zum Einfluß von Strömung und Substratstruktur auf die Wiederbesiedlung der Bachsohle durch Makroinvertebraten. *Deutsche Gesellschaft für Limnologie* 468–472. Erweiterte Zusammenfassung der Jahrestagung 1990, Essen, Germany.

Ward, J. V. 1986. Altitudinal zonation in a Rocky Mountain stream. *Archiv für Hydrobi- ologie* 74 (suppl.): 133–199.

Zimmermann, P. 1961. Experimentelle Untersuchungen über die ökologische Wirkung der Strömungsgeschwindigkeit auf die Lebensgemeinschaften des fliessenden Wassers. *Zeitschrift für Hydrologie* 23:1–81.

Zimmermann, P. 1962. Der Einfluß der Strömung auf die Zusammensetzung der Lebensgemeinschaften im Experiment. *Zeitschrift für Hydrologie* 24:408–411.

Synthesis

Chapter 20

Assembly Rules and Ecosystem Restoration: Where to from Here?

Tim Nuttle, Richard J. Hobbs,
Vicky M. Temperton, and Stefan Halle

In this final chapter, we critically assess the likely value for ecological restoration of ecosystem assembly theory in light of some major conclusions that can be drawn from the work presented in this book. Many chapters explored the substantial gap between basic ecology theory and restoration practice. One aim of the book was to overcome this gap by searching for a theoretically based conceptual framework. Ecological restoration is the attempt to reach a desired ecosystem state by speeding up and directing natural succession and by helping the system to move over otherwise insurmountable thresholds. Clearly, these tasks are in the realm of community ecology, with obvious relationships to disturbance and succession theory in particular. Disturbance is a tricky issue here because it is such a driving influence: the difficulty stems from the fact that disturbance can be both a primary degrading factor and a useful tool to overcome the degradation (see Chapter 17).

Another strong and interesting relationship exists between assembly rules and succession. Whereas succession describes the dynamics of changes in species composition, assembly rules deal with the interactions between organisms that determine the trajectory of those changes. It is here that we meet the basic distinction between pattern (that is, succession) and process (that is, the assembly rules). The dynamic filter model (Chapter 6) points out that the rules change with time and the state of the system, so there is a feedback from the pattern to the process.

The fact that there are many ways of viewing assembly rules is apparent in the review of Chapter 3 and the different approaches of the various chapters. Nevertheless, certain aspects of assembly rules theory are particularly helpful in guiding restoration practice, even if ecologists still disagree about the validity of the different approaches; the alternative stable state concept and the dynamic response approach are examples. From the four chapters of Part I, "Assembly Rules and the Search for a Conceptual Framework for Restoration Ecology," we can construct the following scenario. In a hypothetical starting point with no species present, there are many possible trajectories and final stable states. Which species can enter the system at this stage is a complex question in itself and is generally driven by abiotic conditions and species availability, especially in severely degraded areas (see also Part IV, "Assembly Rules in Severely Disturbed Environments"). As the assembly process continues, new opportunities for some species are created whereas other species are constrained or eliminated. Regardless, the set of remaining possible trajectories is reduced. Because species differ in their importance (that is, their effects on other species), the decrease in options will be a stepwise process rather than a continuous one, especially when keystone or nexus species are involved (see Chapter 4).

Basically, ecological restoration means that we guide a system along a desired trajectory to reach a specific, desired stable state. We may either attempt to mimic the original trajectory that we believe led to the system that we want to restore or can follow an alternative trajectory—one, perhaps, with no natural and historical equivalent—toward the same endpoint. In any case, the task is enormous, and we have to admit that our present understanding is far from complete enough to let us manipulate these processes with full control. Nonetheless, promising first steps have been taken, especially during the last 10 years, and some of them have been outlined in this book.

Assembly Rules Described in This Book

Part II, "Ecological Filters as a Form of Assembly Rule," addressed the concept of ecological filters that constrain the ability of species to arrive, survive, and persist in any given ecosystem. Both abiotic and biotic filters were discussed. There continues to be disagreement over whether the concept of a biotic filter is useful. Biotic filters encompass the total effect of many biological interactions—not just competition but also symbiosis, mutualism, and parasitism—on a species trying to enter a system. In the dynamic filter model (Chapter 6), all these individual interactions are summarized as a binary or probabilistic filter controlling the establishment of a species at a site. Hence, all the specific interactions are treated as a black box. A related issue is whether we need to consider the community interactions that occur after establishment. Are these important in a restoration context, or can we consider them as a black box, too? The use of a black-box model would definitely make things easier in terms of applying assembly ideas to restoration within specific systems, because the exact identities of species would be considered less important. Many of the case studies presented in the book, however, indicate that it might be necessary to shine some light into the black box and consider what happens in there.

Chapter 5 explored the array of filters possible in ecosystems in relation to how these filters can form a barrier to restoration success. Given the plethora of possible filtering effects on species trying to establish themselves in an ecosystem (particularly in very disturbed ecosystems), the filter concept may be useful only in identifying factors that prevent the establishment of particular species. Fattorini and Halle (Chapter 6) reached the same conclusion, but also stressed that a *dynamic* filter model (one that changes over time) is necessary for ecological restoration, since the importance of filtering effects on species will change as the system recovers from the original disturbance. Belyea (Chapter 7) warned against relying exclusively on the filter approach to community development in restoration projects because it ignores potentially important interactions within the system and looks only at the overall effect on species membership in the system. She advocated supplementing the filter approach with a systems approach, which looks in detail at species requirements, characteristics, and interactions within a system. In other words, one can potentially ignore the black box (of all the many interactions between organisms within a system) at the beginning of a project, but more detailed knowledge of species interactions will be required for successful restoration of ecosystem structure—not just function—over time (see also Lockwood and Pimm 1999). Belyea proposed a general model of system feedback to explain the dynamics of very different systems or of one system undergoing major environmental changes (as in restoration projects): Positive feedbacks act at small scales and amplify fluctuations that are initially random; at larger scales, negative feedbacks eventually serve to keep these fluctuations in check.

Part III, "Assembly Rules and Community Structure," explored how investigating the structure of ecological communities could help restoration practitioners by providing biomarkers of the extent of ecosystem health and recovery. The approaches taken here assessed structure at higher levels of abstraction than the species level. The presence or absence of certain taxonomic groups (see Chapters 9 and 10) could thus be used as indicators of ecosystem health (see also Rapport et al. 1998). Rojo (Chapter 8) showed that in many freshwater aquatic systems, phytoplankton communities lacked structure at the species level, but patterns emerged at higher levels of abstraction (functional groups or biomass size spectra) when assessing degradation in eutrophic wetlands.

Voigt and Perner (Chapter 9) investigated different functional groups across many trophic levels in a degraded and a seminatural grassland and concluded that generalist primary producers will dominate regenerating terrestrial systems and produce bottom-up control of species at all consumer levels. The successful establishment of diverse perennial plant species may speed development of a more diverse community by allowing establishment of more specialist herbivores and carnivores. In keeping with recommendations from Belyea (Chapter 7), Chapters 8, 9, and 10 advocated the use of functional groups as a means of reducing ecological complexity without missing out on key developments in the regeneration of a system. Renker et al. (Chapter 10) also discussed the importance of priority effects in regenerating ecosystems: arbuscular mycorrhizal fungi often positively affect plant establishment on degraded sites, thereby affecting community structure and development. The presence of a vascular plant species in a community might depend on the presence of a particular arbuscular mycorrhizal species. Effectively, the presence or absence of such a fungal species acts as a biotic filter for the establishment of certain plant species.

In Chapter 11, Heil abstracted from the species level of assembly to the community level of assembly to describe the influence of spatially restricted human impacts on disturbance and therefore on transitions between community types. He formulated rules based on likelihood of human access that modified the transition rates from one community type to another.

Rothe and Gleixner (Chapter 12) proposed the use of stable isotopes to describe food-web development, since the isotopic signature of soil in terrestrial systems acts as a kind of ecosystem memory related to a system's functioning. Ecosystem recovery was correlated over time with stabilization of $\delta^{15}N$ differences between trophic levels at around 3.4‰. This difference does not vary from one trophic level to the next, so it may provide a general tool for assessing trophic stability in a variety of ecosystems.

Part IV, "Assembly Rules in Severely Degraded Environments," is perhaps a good application of the adage that the exception proves the rule. Here, the exception is the status of the ecosystems themselves—that is, the fact that they are severely degraded. It seems that any assembly rules or guidelines, such as the filtering out of species that cannot establish themselves at a site, are applicable in both relatively healthy and severely degraded ecosystems. What changes in different systems is the relative effect of abiotic and biotic conditions on the development of the community. Wagner's main conclusion in this regard (see Chapter 13) was that assembly rules that focus on species interactions alone are useful only for stable, nonstressful habitats. In severely degraded habitats, the nature of the biotic stress and species availability determine community composition. Similarly, Temperton and Zirr (Chapter 14) formulated a rule that early colonizers ameliorate the environment and hence facilitate establishment of subsequent colonists. It is interesting that this effect seemed to be independent of species identity, a finding in direct contradiction to the usual species-specific focus of the biotic-interactions-only school of thought (see Chapter 3). Sänger and Jetschke (Chapter 15) proposed rules that interacted: species composition depended first on dispersal to the site, then on substrate, and finally on priority effects (which species established first influenced which successional pathway would develop). Thus, biotic factors interacted with abiotic factors in determining assembly trajectories. Bradshaw (Chapter 16) emphasized this point further with respect to the importance of understanding nutrient conditions in restoration sites. Nutrients can be considered as both biotic and abiotic issues, and they influence not only the type of community that will develop but also whether proposed restoration activities will succeed or fail completely.

Thus, perhaps in contrast to less degraded systems, it seems that most of the assembly rules proposed in Part IV can be generalized to other, similarly degraded habitats, because these rules focus on how functional traits influence development of the system rather than on interactions between specific species. The main difference between less degraded and severely degraded ecosystems seems to be that the abiotic and biotic effects are inextricably linked in the initial stages of development after a severe disturbance. This conclusion is in agreement with CSR plant strategy theory (Grime 1979): as disturbance or stress in a system decreases, competition becomes the driving influence.

Part V, "Disturbance and Assembly," began with White and Jentsch's review (Chapter 17) of how understanding natural disturbance regimes can aid not only in devising effective restoration techniques but also in understanding where the system is in terms of restoration of function. The idea that the disturbance regime constitutes an assembly rule is further elaborated for stream ecosystems in Chapters 18 and 19. Both Matthaei et al. (Chapter 18) and Meidl and Schönborn (Chapter 19) see disturbance itself as a filter that drives which species can establish at different time points in a river ecosystem. Other approaches to assembly see disturbance more as an influence on other filters (see Chapter 6 and the review in Chapter 17).

Thus, it seems that most ecologists, at least the ones who contributed to this book, would agree that patterns are what we can observe, whereas assembly rules explain how the patterns arise (although one school of thought would argue that the pattern *is* the rule; see Chapter 3). Yet the search for such universal rules has not borne much fruit over the past decades. Moreover, how assembly rules work depends on the organisms and communities of interest. Therefore, we might ask whether the search for rules is particularly useful at all. Perhaps the main value of the search is that in considering rules, we are forced to develop a deeper understanding of the dynamics of how species spread, interact, and are affected by various components of the environment. The results presented in this book lead us to conclude that the most constructive approach to employing assembly theory in restoration practice involves using the patterns we find in different ecosystems as guidelines for ecosystem development and as indicators of the current state of an ecosystem.

Consensus or Confusion?

A key aim of this book, reflected in its subtitle, was to explore the theoretical aspects of ecosystem assembly and examine how these aspects might guide or influence restoration practice. As Halle and Fattorini (Chapter 2) pointed out, restoration ecology has lacked a firm conceptual basis, which has hindered its ability to provide general guidelines that can apply to a variety of different systems or situations. Although this situation has undoubtedly been improving, as evidenced by the recent publication of a book specifically addressing principles for restoration (Perrow and Davy 2002), there is still a recognized need to continue work on restoration ecology's conceptual framework.

To what extent have we succeeded in contributing to a sound conceptual basis for restoration ecology and in linking theory and practice? The first thing that needs to be said is that this book contains relatively few contributions about projects in which active restoration efforts have been undertaken. Several chapters examine recovery in systems subject to intense degradation, but the recovery examined has been undirected. Nevertheless, manipulative experiments have been used to investigate the mechanisms behind the assembly of ecosystems (for example, see Chapter 14). More general approaches in other chapters supplement these experiments and provide interesting insights into whether assembly theory has anything useful to say about how ecosystems assemble following disturbance. As Hobbs and Norton (Chapter 5) point out, however, restoration efforts need not depend passively on the results of the assembly process. Instead, such efforts can try to allow desired species in and keep undesired species out; they can also try to speed up or direct the natural recolonization process where it is too slow or is impeded in some way. We need to consider whether the material presented here can aid in this process.

One of the main conclusions of the theoretical chapters in Parts I and II was that the most useful aspects of assembly theory as a background to the practice of restoration include the concept of alternative stable states in community development (Chapters 3, 4, and 5). There is also increasing evidence that community dynamics might be influenced more by stochastic processes than by deterministic ones that depend on traits of individuals and species (for instance, Hubbell 2001; also see Chapter 15 for an example). Hence, the old ecological paradigm of equilibrium ecosystems with predictable endpoints to succession needs to be cast away and replaced with the idea that a system can develop along alternative paths and that the restoration practitioner can potentially manipulate the system to guide it along a desired one.

Often, new disturbances in a system are necessary to alter filters, both to allow species to establish in communities and to push systems across thresholds to force a transition between states. Dynamic assembly models focusing on development over time (as opposed to a categorical focus on species) would be most constructively applied to restoration projects. Evidence from the studies in this book, however, suggest that a categorical approach to assembly that looks at functional groups as well as species is a promising means of assessing how far along the regeneration trajectory a system may be. The use of such indicators would allow restoration practitioners to assess how much manipulation of a system is required, depending on how degraded it is.

Clearly, an understanding of how biotic communities are put together is essential if we are to achieve sensible management of the reassembly process. Active restoration is necessary when it is not possible to wait for natural assembly of communities in degraded ecosystems when they exist in an undesirable state. Restoration must choose options and define goals and a time frame. Recognizing what combinations of species are possible within any given set of abiotic and biotic limitations helps to constrain and guide the development of restoration goals. Recognizing abiotic and biotic filters might also allow their purposeful modification to achieve a particular composition (by modifying soil structure or chemical composition, for instance, or by removing competition from dominant, weedy species). It is apparent that, beyond relatively simple generalities about species assembly, much of what happens will still depend on the specific system and situation. Hence, we still require information about the responses of individual species, or at least types of species, and about the functioning of undamaged ecosystems similar to those we wish to restore.

Linking Theory with Practice

Has this book produced a common thread of ideas and approaches, or has it resulted in more confusion? As stated in the introduction, we did not intend to develop a consensus view that held sway throughout the book, and so we have presented a variety of ideas and interpretations, not all of which agree. Some readers may find this approach annoying and unsatisfying—particularly, perhaps, readers from the practical sphere of restoration ecology. Practitioners generally want answers, preferably straightforward ones. Is this, then, another in the long line of scientific publications that fails to deliver, sliding instead into obfuscation and weasel words?

Before answering, it is perhaps useful to consider again the current status of the science this book discusses. As noted in the introductory chapters, the questions of why particular groups of species are found where they are and how these assemblages develop have been discussed for some considerable time and have been at the heart of many ecological debates. Temperton and Hobbs (Chapter 3) indicated that these debates rarely reached satisfactory conclusions and that the current emphasis on ecosystem assembly was a further attempt to explore the issues. Although the questions are old, our understanding of how ecosystems and communities work has changed dramatically in the past few decades—marking a shift from a deterministic, equilibrium view of natural systems to a recognition of their complexity, stochasticity, and nonlinearity (Botkin 1990, Pahl-Wostl 1995, Bak 1996, Levin 1999, Hubbell 2001). The science relating to this changed view of how ecosystems work is still evolving rapidly, but it emphasizes the difficulty of reaching definitive conclusions about what might happen where, because of the complex interactions and contingencies involved. The science is thus certainly not textbook science yet, but it can be regarded as frontier science (in the sense of Pickett et al. 1994). In other words, the relevant connections are not all in place, many bits of the picture do not make sense, and everything does not add up to a coherent whole.

That is the state in which we find ourselves with the subject matter of this book. We are reporting on a topic that deals with complex systems for which the scientific approach and detailed methodologies and conceptual frameworks are far from fully worked out. This is the reality of the situation, and it is important that practical people looking for definitive answers are not deceived by spurious agreement and simplicity. For decades, managers have done their work hoping that the ecosystems they watch over are simple and deterministic, despite the increasingly common knowledge that they aren't. The lack of a unifying conceptual framework, however, has left them with no alternative but to adopt such a view and alter single factors in postulated isolation from everything else in the system. The continuing degradation of many ecosystems provides clear evidence that this approach does not work. To be effective, we must recognize the inherent complexity of the systems we are trying to manage. The approaches discussed in this book start to provide a means toward this end.

Nevertheless, it is important that science and theory do not renege on their responsibility to deliver useful concepts and approaches. Scientists cannot ignore the complexity we have been discussing in an attempt to deliver the kind of certainty managers and policymakers crave. Neither can they hide behind the complexity and say that nothing can be done. In this book, we have tried to ensure that the more theoretical discussions have been balanced by the practical realities of ecosystem restoration.

Getting back to our question: What can this book deliver to restoration ecologists? Most apparent is the almost universal agreement among the various authors that the alternative stable states concept is decidedly more relevant than either the deterministic or individualistic concepts of assembly. Furthermore, the dynamic response approach is more useful for restoration than the categorical approach because of the former's focus on changing function over time. In this light, the dynamic filter model (Chapter 6) might be useful in designing restoration projects, to focus attention on important factors that need to be considered for planning. More detailed system-specific models are more appropriate for understanding processes at later stages, particularly if specific target species are part of restoration goals. Finally, in severely degraded ecosystems, we should not underestimate the importance of restoring some level of ecosystem function by ameliorating abiotic conditions (for example, pH, hydrology), because without such functions, projects aimed at restoring structure (that is, a certain species composition) will fail.

Future Directions

Although it is often called for, collaboration between restoration practitioners and ecological theorists remains rare. This phenomenon results in an inability to escape the pattern we see in this book: all experimental studies remain relegated to small scales, whereas large-scale studies remain purely descriptive. To cross this threshold, restoration projects must be carried out in conjunction with large-scale experiments, so that the anecdotally and observationally based patterns can be tested and formulated into rigorous principles. One way forward would be for ecologists to repeat experiments in different ecosystems to validate proposed assembly rules based on patterns found in regenerating ecosystems. Even more important, restoration practitioners working on degraded ecosystems should actively test some of the assembly theories by implementing them in restoration actions. Such testing would provide experimental (rather than merely observational or anecdotal) evidence for the validity and applicability of assembly concepts in ecological restoration and thus contribute to progress in ecological theory (Bradshaw's 1987 acid test for community ecology).

In many ways, we are living in a brave new world, where old rules and assumptions must be reconsidered and changed. This is particularly true in the area of ecosystem assembly because human activities have greatly altered the movement patterns of all elements of the biota (Bright 1998, Mooney and Hobbs 2000, Baskin 2002). World trade and the transport of species for agriculture, horticulture, and the pet industry, together with the inadvertent spread of species from one place to another, have resulted in a situation where biogeographic boundaries mean little in terms of species dispersal. Deliberate and inadvertent species movement has resulted in the spread of many species that are able to invade native and managed ecosystems and become pests or problems. These new species have become part of the species pool available for ecosystem assembly, and they must be factored into management and restoration decisions. For instance, can they play a useful role in the restoration of an ecosystem? Are they "desirable," or should they be prevented from entering? The dynamics of invasion and the importance of understanding the potential impacts of invasive species have been discussed often (see, for example, recent accounts in Cox 1999, Ewel et al. 1999, Lonsdale 1999, Meinesz 1999, Stohlgren et al. 1999, Williamson 1999). Invasive species have also been considered in a restoration context (for example, Berger 1993, Hobbs and Mooney 1993, Holmes and Richardson 1999).

The fact that nonnative species are now part of so many ecosystems has led some ecologists to believe that we have entered a new phase in the dynamics of ecosystems, where novel combinations of species are appearing and occupying altered environments, giving rise to a new set of "emerging ecosystems." The concept of emerging ecosystems has important ramifications, both for the theoretical considerations of ecosystem assembly and for the practical side of restoration. For instance: What should restoration goals be in light of the potential for the emergence and persistence of alternative ecosystems? What are the implications of novel sets of species for the development of assembly rules? Do introduced species follow the same rules or make up their own?

Some ecologists have gone further to suggest that we are actually entering an era in which humans are deliberately or inadvertently creating new ecosystems; these have been dubbed "designer ecosystems" (MacMahon 1998). Perhaps designer ecosystems are where assembly rules and restoration ecology will increasingly intersect in the future. The purposeful creation and management of ecosystems that are assembled (or reassembled) to fulfill particular goals will require both the theoretical understanding of ecosystem assembly and the practical skills of restoration. We hope that this book has provided a stepping stone toward achieving at least some of this understanding and some of these skills.

REFERENCES

Bak, P. 1996. *How Nature Works: The Science of Self-Organized Criticality*. New York: Copernicus.

Baskin, Y. 2002. *A Plague of Rats and Rubber Vine*. Washington, D.C.: Island Press.

Berger, J. J. 1993. Ecological restoration and non-indigenous plant species: a review. *Restoration Ecology* 1:74–82.

Botkin, D. B. 1990. *Discordant Harmonies. A New Ecology for the Twenty-First Century*. Oxford: Oxford University Press.

Bradshaw, A. D. 1987. Restoration: an acid test for ecology. In *Restoration Ecology: A Synthetic Approach to Ecological Research*, ed. W. R. I. Jordan, M. E. Gilpin, and J. D. Aber, 23–30. Cambridge: Cambridge University Press.

Bright, C. 1998. *Life Out of Bounds: Bioinvasion in a Borderless World*. New York: W. W. Norton.

Cox, G. W. 1999. *Alien Species in North America and Hawaii: Impacts on Natural Ecosystems*. Washington, D.C.: Island Press.

Ewel, J. J.; O'Dowd, D. J.; Bergelson, J.; Daehler, C. C.; D'Antonio, C. M.; Gomez, L. D.; Gordon, D. R.; Hobbs, R. J.; Holt, A.; Hopper, K. R.; Hughes, C. E.; LaHart, M.; Leakey, R. R. B.; Lee, W. G.; Loope, L. L.; Lorence, D. H.; Louda, S. M.; Lugo, A. E.; McEvoy, P. B.; Richardson, D. M.; and Vitousek, P. M. 1999. Deliberate introductions of species: research needs. *BioScience* 49:619–630.

Grime, J. P. 1979. *Plant Strategies and Vegetation Processes*. Chichester, England: John Wiley and Sons.

Hobbs, R. J.; and Mooney, H. A. 1993. Restoration ecology and invasions. In *Reconstruction of Fragmented Ecosystems: Global and Regional Perspectives*. Vol. 3 of *Nature Conservation*, ed. D. A. Saunders, R. J. Hobbs, and P. R. Ehrlich, 127–133. Chipping Norton, NSW, Australia: Surrey Beatty and Sons.

Holmes, P. M.; and Richardson, D. M. 1999. Protocols for restoration based on recruitment dynamics, community structure, and ecosystem function: perspectives from South African fynbos. *Restoration Ecology* 7:215–230.

Jones, C. G.; Lawton, J. H.; and Shachak, M. 1994. Organisms as ecosystem engineers. *Oikos* 69:373–386.

Jones, C. G.; Lawton, J. H.; and Shachak, M. 1997. Positive and negative effects of organisms as physical ecosystem engineers. *Ecology* 78:1946–1957.

Levin, S. A. 1999. *Fragile Dominion: Complexity and the Commons*. Cambridge, Mass.: Perseus Publishing.

Lockwood, J. L.; and Pimm, S. L. 1999. When does restoration succeed? In *Ecological Assembly: Advances, Perspectives, Retreats*, ed. E. Weiher and P. Keddy, 363–392. Cambridge: Cambridge University Press.

Lonsdale, W. M. 1999. Global patterns of plant invasions and the concept of invasibility. *Ecology* 80:1522–1536.

MacMahon, J. A. 1998. Empirical and theoretical ecology as a basis for restoration: an ecological success story. In *Successes, Limitations, and Frontiers in Ecosystem Science*, ed. M. L. Pace and P. M. Groffman, 220–246. New York: Springer-Verlag.

Meinesz, A. 1999. *Killer Algae: The True Tale of a Biological Invasion*, trans. D. Simberloff. Chicago: University of Chicago Press.

Mooney, H. A.; and Hobbs, R. J. eds. 2000. *Invasive Species in a Changing World.* Washington, D.C.: Island Press.

Pahl-Wostl, C. 1995. *The Dynamic Nature of Ecosystems: Chaos and Order Entwined.* Chichester, England: John Wiley and Sons.

Perrow, M. R.; and Davy, A. J. eds. 2002. *Handbook of Ecological Restoration.* Vol. 1, *Principles of Restoration.* Cambridge: Cambridge University Press.

Pickett, S. T. A.; Kolasa, J.; and Jones, C. G. 1994. *Ecological Understanding: The Nature of Theory and the Theory of Nature.* New York: Academic Press.

Rapport, D. J.; Constanza, R.; and McMichael, A. J. 1998. Assessing ecosystem health. *Trends in Ecology and Evolution* 13(10): 397–401.

Stohlgren, T. J.; Binkley, D.; Chong, G. W.; Kalkahn, M. A.; Schell, L. D.; Bull, K. A.; Otsuki, Y.; Newman, G.; Bashkin, M.; and Son, Y. 1999. Exotic plant species invade hot spots of native plant diversity. *Ecological Monographs* 69:25–46.

Williamson, M. 1999. Invasions. *Ecography* 22:5–12.

About the Contributors

Edith B. Allen is professor at the Department of Botany and Plant Sciences at the University of California–Riverside. Her research interests concentrate on restoration of plant communities with particular emphasis on invasive species and mycorrhizae.

Michael F. Allen is professor for plant pathology and biology and chair of the Center for Conservation Biology at the University of California–Riverside. His research interests focus on the dynamics and role of mycorrhizal fungi in ecosystems and the use of mycorrhizae in the conservation and restoration of disturbed wildlands.

Chris J. Arbuckle is an ecologist at the Otago Regional Council, Dunedin, New Zealand. His research interests lie in water resource management, the effects of irrigation on aquatic ecosystems and water quality, native fish, and the use of GIS in environmental research.

Lisa R. Belyea is an ecologist working at the Institute of Geography, School of GeoSciences, University of Edinburgh, United Kingdom. Her research interests are in the links between community structure and ecosystem processes, particularly in peatland systems.

François Buscot is professor and chair of terrestrial ecology at the Institute of Botany, University of Leipzig, Germany. His main research topic is soil microbial ecology with special emphasis on fungi and mycorrhizal symbioses.

Tony Bradshaw has worked on the mineral nutrition of higher plants since the 1960s, and for thirty years lead a major research group looking at the theoretical and practical aspects of the restoration of derelict land in the United Kingdom and worldwide. He has been president of the British Ecological Society.

Marzio Fattorini is a plant ecologist whose research interests are principally in restoration of degraded ecosystems. He is currently living in South America (Perú and Venezuela), where he is learning more about the tropical flora.

Gerd Gleixner from the Max Planck Institute for Biogeochemistry in Jena, Germany, is a biochemist investigating biogeochemical processes in the pedosphere, hydrosphere, and biosphere of natural systems. His current research interests comprise isotope fractionation, the importance of inter- and intramolecular isotope distributions, molecular turnover of soil organic matter, and climate reconstruction.

Cornelia Guggelberger studied biology at the University of Munich, Germany. She teaches biology and chemistry at a secondary school in Landshut, Germany.

Stefan Halle is a behavioral ecologist, head of the Institute of Ecology, and speaker of the graduate research group "Functioning and Regeneration of Degraded Ecosystems" at the University of Jena, Germany.

Gerrit W. Heil is a plant ecologist in the Faculty of Biology, Utrecht University, The Netherlands. His principle interest is in understanding how biodiversity is caused at the level of plant communities and ecosystems. His research focuses on the process of growth of plant species, the spatiotemporal dynamics of sites, and the dispersal of seeds.

Richard Hobbs is professor of Environmental Science at Murdoch University in Perth, Western Australia, and is currently a member of the executive board of the Society for Ecological Restoration International. His research interests focus on the management and restoration of altered ecosystems and landscapes.

Harald Huber is a freshwater ecologist working as resource manager at Munich International Airport. He is also a Ph.D. student at the University of Munich. His research interests lie in river restoration, groundwater flow modeling, and other aspects of applied aquatic ecology.

Anke Jentsch is a vegetation ecologist working at the Centre for Environmental Research in Leipzig, Germany. Her main research interests are in disturbance ecology, system dynamics, plant biodiversity, and ecosystem functioning.

Claudia E. Küster works for the Regional Authority for Nature Conservation and Landscape Management in Tübingen, Germany. Her research interests include river restoration projects and their influence on benthic invertebrates.

Julie Lockwood is an ecologist in the Environmental Studies Department, University of California–Santa Cruz. Her research interests are community assembly, conservation of rare birds, invasion ecology, and biogeography.

Christoph D. Matthaei is an aquatic ecologist at the University of Otago, Dunedin, New Zealand. His research interests include the role of disturbance in streams and rivers, river restoration, and more recently the effects of human land uses on running water ecoystems.

Barbara E. Meidl is a freshwater ecologist working at the Department of Limnology at the Friedrich-Schiller-University of Jena, Germany. Her research interests are in ecology of running waters with emphasis on the hyporheic zone and the relationships between disturbance and recovery therein.

David Norton is a conservation biologist at the University of Canterbury. New Zealand, with a particular interest in issues associated with restoration and fragmentation ecology and the integration of biodiversity conservation in agricultural and forestry landscapes.

Tim Nuttle is a research fellow at the University of Jena, Institute for Ecology. His research interests include ecological modeling, restoration ecology, seed dispersal, and ecological education.

Maarja Öpik is a Ph.D. student at the Institute of Botany and Ecology at the University of Tartu, Estonia. Her research interests are focused on the molecular ecology of mycorrhizal fungi.

Kathi A. Peacock is a fluvial geomorphologist working on sediment accumulation and mobilization changes due to hydro-regulation for Devine Tarbell & Associates in Bellingham, Washington. Her research interests include effects of land use on stream morphology and linkages between aquatic communities and physical processes in streams.

Jörg Perner is an animal ecologist working at the Institute of Ecology, at the University of Jena, Germany. His favorite topics are analyzing functional groups of invertebrates as ecological indicators of succession and restoration processes, statistical data analysis, and modelling.

Carsten Renker works as a scientist at the Department for Terrestrial Ecology at the Institute of Botany at the University of Leipzig, Germany. His research interests focus on the molecular ecology of mycorrhizal fungi.

Carmen Rojo is a plankton ecologist at the Cavanilles Institute of Biodiversity and Evolutionary Biology in Valencia, Spain. Her research interests are plankton community dynamics, assembly rules concerning plankton community, and self-organization.

Jan Rothe is a vegetation ecologist currently working at the Max Planck Institute for Biogeochemistry in Jena, Germany. His research interests are in analysis of ecosystem development, especially in natural succession and regeneration following disturbance, soil ecology, and food web reconstruction using stable isotope techniques.

Jana Rydlová is a scientist at the Institute of Botany, Academy of Sciences of the Czech Republic. Her research interests concern arbuscular mycorrhizal symbiosis in plant succession in anthropogenic habitats as well as interactions of mycorrhizal fungi with heavy metals, xenobiotics, and other stress factors in the soil.

Hartmut Sänger is an ecologist and managing director of BIOS-Office, an environmental consulting firm in Crimmitschau, Germany. His research interests are in plant community assembly and its relation to restoration ecology and disturbance, plant biodiversity, and ecosystem functioning, particularly at former mining sites.

Wilfried Schönborn is a freshwater ecologist who was head of the limnology section of the Insitute of Ecology at the University of Jena, Germany, until 1999. His main research interests are protozoan ecology and ecology and restoration of running waters.

Colin R. Townsend is an ecologist at the University of Otago, Dunedin, New Zealand. He is coauthor of one of the standard textbooks in ecology. His research interests include ecology and behavior of invertebrates and fish in streams, ecology of invaders, influence of land use, and impact of flow disturbances on river ecosystems.

Vicky M. Temperton is a plant ecologist at the Max Planck Institute for Biogeochemistry in Jena, Germany. Her research interests are in plant community assembly and its relation to restoration ecology and disturbance, and plant biodiversity and ecosystem functioning.

Winfried Voigt is a community ecologist at the Institute of Ecology at the University of Jena, Germany. His main research interests are grassland communities, climate effects on trophic structures, global change, interaction webs, biodiversity, ecological succession, and statistical modelling.

Miroslav Vosátka is the head of a research group at the Institute of Botany, Academy of Sciences of the Czech Republic. His research topics concern the ecophysiology of mycorrhizal fungi in degraded ecosystems and the practical aspects of application of mycorrhizal fungi inocula.

Markus Wagner is a plant ecologist at the Institute of Ecology, University of Jena, Germany. His research focuses on plant population ecology and on the role of soil conditions in plant community assembly, especially in the context of grassland restoration.

Peter S. White is professor of biology at the University of North Carolina and Director of the North Carolina Botanical Garden in Chapel Hill, North Carolina. His research interests are disturbance ecology, conservation biology, ecology of invasions, the conservation garden, species richness, and scale.

Martin Zobel is professor of plant ecology and head of the Institute of Botany and Ecology at the University of Tartu, Estonia. His research is focused on biodiversity patterns and the processes affecting it.

Kerstin Zirr is a nutrition scientist at the Institute of Nutrition Science, University of Jena, Germany. She researches on plant response to stress and interaction situations.